Water Resources Engineering

Water Resources Engineering

Second Edition

Larry W. Mays

Professor
Civil, Environmental, and Sustainable Engineering Group
School of Sustainable Engineering and the Built Environment
Arizona State University
Tempe, Arizona

WILEY

John Wiley & Sons, Inc.

VP and Publisher	Don Fowley
Acquisition Editor	Jenny Welter
Editorial Assistant	Alexandra Spicehandler
Production Manager	Janis Soo
Assistant Production Editor	Elaine S. Chew
Senior Marketing Manager	Christopher Ruel
Marketing Assistant	Diana Smith
Media Editor	Lauren Sapira
Designer	RDC Publishing Group Sdn. Bhd.
Cover Image	© Larry W. Mays

This book was set in 9.5/12 Times Roman by Thomson Digital and printed and bound by Courier Kendallville. The cover was printed by Courier Kendallville.

This book is printed on acid free paper.

Library of Congress Cataloging in Publication Data

Mays, Larry W.
  Water resources engineering / Larry W. Mays.—2nd ed.
    p. cm.
  Includes index.
   ISBN 978-0-470-46064-1 (cloth : alk. paper)
   1. Hydraulic engineering. 2. Hydrology. I. Title.

TC145.M383 2010
627—dc22   2010005952

Printed in the United States of America
11

About the Author

Larry W. Mays is Professor in the Civil, Environmental, and Sustainable Engineering Group in the School of Sustainable Engineering and the Built Environment at Arizona State University (ASU), and former chair of the Department of Civil and Environmental Engineering. Prior to ASU he was Director of the Center for Research in Water Resources at the University of Texas at Austin, where he held an Engineering Foundation–endowed professorship. A registered professional engineer in several states, and a registered professional hydrologist, he has served as a consultant to many national and international organizations.

Professor Mays has published extensively in refereed journal publications and in the proceedings of national and international conferences. He was the author of the first edition of this book and *Optimal Control of Hydrosystems* (published by Marcel Dekker), and co-author of *Applied Hydrology* and *Hydrosystems Engineering and Management* (both from McGraw-Hill) and *Groundwater Hydrology* (published by John Wiley & Sons, Inc). He was editor-in-chief of *Water Resources Handbook, Water Distribution Systems Handbook, Urban Water Supply Management Tools, Stormwater Collection Systems Design Handbook, Urban Water Supply Handbook, Urban Stormwater Management Tools, Hydraulic Design Handbook, Water Supply Systems Security*, and *Water Resources Sustainability*, all published by McGraw-Hill. In addition, he was editor-in-chief of *Reliability Analysis of Water Distribution Systems* and co-editor of *Computer Methods of Free Surface and Pressurized Flow* published by Kluwer Academic Publishers.

Professor Mays developed the book, *Integrated Urban Water Management: Arid and Semi-arid Regions*, published by Taylor and Francis. This book was the result of volunteer work for the United Nations UNESCO-IHP in Paris. He recently was editor of the fourth edition of *Water Transmission and Distribution*, published by the American Water Works Association.

One of his major efforts is the study of ancient water systems and the relation that these systems could have on solving our problems of water resources sustainability using the concepts of traditional knowledge, not only for the present, but the future. His most recent book is *Ancient Water Technology*, published by Springer Science and Business Media, The Netherlands.

Among his honors is a distinguished alumnus award from the Department of Civil and Engineering at the University of Illinois at Champaign-Urbana and he is a Diplomate, Water Resources Engineering of the American Academy of Water Resources Engineering. He is also a Fellow of the American Society of Civil Engineers and the International Water Resources Association. He loves the mountains where he enjoys alpine skiing, hiking, and fly-fishing. In addition he loves photographing ancient water systems around the world and gardening. Professor Mays lives in Mesa, Arizona and Pagosa Springs, Colorado.

Acknowledgments

Water Resources Engineering is the result of teaching classes over the past 34 years at the University of Texas at Austin and Arizona State University. So first and foremost, I would like to thank the many students that I have taught over the years. Several of my past Ph.D. students have helped me in many ways through their review of the material and help in development of the solutions manual. These former students include Drs. Aihua Tang, Guihua Li, John Nicklow, Burcu Sakarya, Kaan Tuncok, Carlos Carriaga, Bing Zhao, El Said Ahmed, and Messele Ejeta. I would like to give special thanks to Professor Y.K. Tung of the Hong Kong University of Science and Technology. He has been a long time friend and was my very first Ph.D. student at the University of Texas at Austin. Y. K. was very gracious in providing me with some of the end of chapter problems for the hydrology chapters. I would like to acknowledge Arizona State University, especially the time afforded me to pursue this book.

I would like to thank Wayne Anderson for originally having faith in me through his willingness to first publish the book and now Jenny Welter who has worked to get this edition published.

During my academic career as a professor I have received help and encouragement from so many people that it is not possible to name them all. These people represent a wide range of universities, research institutions, government agencies, and professions. To all of you I express my deepest thanks.

Water Resources Engineering has been a part of a personal journey that began years ago when I was a young boy with a love of water. This love of water resources has continued throughout my life, even in my spare time, being an avid snow skier, fly-fisherman and hiker. Books are companions along the journey of learning and I hope that you will be able to use this book in your own exploration of the field of water resources. Have a wonderful journey.

Larry W. Mays
Mesa, Arizona
Pagosa Springs, Colorado

I would like to dedicate this book to humanity and human welfare.

Preface

AUDIENCE

Water Resources Engineering can be used for the first undergraduate courses in hydraulics, hydrology, or water resources engineering and for upper level undergraduate and graduate courses in water resources engineering design. This book is also intended as a reference for practicing hydraulic engineers, civil engineers, mechanical engineers, environmental engineers, and hydrologists.

TOPICAL COVERAGE

Water resources engineering, as defined for the purposes of this book, includes both water use and water excess management. The fundamental water resources engineering processes are the hydrologic processes and the hydraulic processes. The common threads that relate to the explanation of these processes are the fundamentals of fluid mechanics using the control volume approach. The hydraulic processes include pressurized pipe flow, open-channel flow, and groundwater flow. Each of these in turn can be subdivided into various processes and types of flow. The hydrologic processes include rainfall, evaporation, infiltration, rainfall-runoff, and routing, all of which can be further subdivided into other processes. Knowledge of the hydrologic and hydraulic processes is extended to the design and analysis aspects. This book, however, does not cover the water quality management aspects of water resources engineering.

HISTORY OF WATER RESOURCES DEVELOPMENT

Water resources development has had a long history, basically beginning when humans changed from being hunters and food gatherers to developing of agriculture and settlements. This change resulted in humans harnessing water for irrigation. As humans developed, they began to invent and develop technologies, and to transport and manage water for irrigation. The first successful efforts to control the flow of water were in Egypt and Mesopotamia. Since that time humans have continuously built on the knowledge of water resources engineering. This book builds on that knowledge to present state-of-the-art concepts and practices in water resources engineering.

NEW TO THIS EDITION

The *Second Edition* provides the most up-to-date information along with a remarkable range and depth of coverage. In addition to other changes, two new chapters have been added that explore water resources sustainability and water resources management for sustainability:

Chapter 2: Water Resources Sustainability, defines water resources sustainability, discusses challenges and specific examples of water resources systems, as well as examples of water resources unsustainability.

Chapter 19: Water Resources Management for Sustainability, introduces the idea of integrated water resources management, law related to water resources, methodologies for both arid and semi-arid regions, economics, systems analysis techniques, and uncertainty and risk-reliability analysis for sustainable design.

Principles of Flow in Hydrosystems, which was previously Chapter 2 in the *First Edition*, has now been integrated with Chapter 3 in the *Second Edition*.

Homework Problems: There are over 300 new problems in the *Second Edition*, resulting in a total of over 670 end-of-chapter problems, expanding the applications to which students are exposed.

New and updated graphics and photos: Over 50 new diagrams, maps and photographs have been integrated throughout the chapters to reinforce important concepts, and support student visualization and appreciation of water resources systems and engineering.

HALLMARK FEATURES

Breadth and Depth: The text includes a breadth and depth of topics appropriate for under-graduate courses in hydraulics, hydrology, or water resources engineering, or as a comprehensive reference for practicing engineers.

Control Volume Approach: Hydrologic and hydraulic processes are explained through their relationship to the control volume approach in fluid mechanics.

Visual program: Hundreds of diagrams, maps, and photographs illustrate concepts, and reinforce the importance and applied nature of water resources engineering.

CHAPTER ORGANIZATION

Water Resources Engineering is divided into five subject areas: Water Resources Sustainability, Hydraulics, Hydrology, Engineering Analysis and Design for Water Use, and Engineering Analysis and Design for Water Excess Management.

Water resources sustainability includes: Chapter 1 which is an introduction to water resources sustainability; Chapter 2 addresses water resources sustainability; and Chapter 19 water resources management for sustainability. Chapter 11 on water withdrawals and uses, Chapter 13 on water for hydroelectric generation, and Chapter 14 on water excess management also contain material related to water resources sustainability.

Hydraulics consists of five chapters that introduce the basic processes of hydraulics: Chapter 3 presents a basic fluid mechanics review and the control volume approach for continuity, energy, and momentum; and Chapters 4, 5, and 6 cover pressurized flow, open-channel flow, and groundwater flow, respectively. Chapter 18 covers the basics of sedimentation and erosion hydraulics.

Hydrology is covered in four chapters: Chapter 7 on hydrologic processes; Chapter 8 on rainfall-runoff analysis; Chapter 9 on routing; and Chapter 10 on probability and frequency analysis.

Engineering analysis and design for water use consists of three chapters: Chapter 11 on water withdrawals and uses; Chapter 12 on water distribution systems; and Chapter 13 on water for hydroelectric generation.

Engineering analysis and design for water excess management includes four chapters: Chapter 14 on water excess management; Chapter 15 on stormwater control using storm sewers and detention; Chapter 16 on stormwater control using street and highway drainage and culverts; and Chapter 17 on the design of hydraulic structures for flood control storage systems.

COURSE SUGGESTIONS

Several first courses could be taught from this book: a first course on hydraulics, a first course on hydrology, a first course on water resources engineering analysis and design, and a first course on hydraulic design. The flowcharts on the following pages illustrate the topics and chapters that could be covered in these courses.

This is a comprehensive book covering a large number of topics that would be impossible to cover in any single course. This was done purposely because of the wide variation in the manner in which faculty teach these courses or variations of these courses. Also, to make this book more valuable to the practicing engineer or hydrologist, the selection of these topics and the extent of coverage in each chapter were considered carefully. I have attempted to include enough example problems to make the theory more applicable, more understandable, and most of all more enjoyable to the student and engineer.

Students using this book will most likely have had an introductory fluid mechanics course based on the control volume approach. Chapter 3 should serve as a review of basic fluid concepts and the control volume approach. Control volume concepts are then used in the succeeding chapters to introduce the hydrologic and hydraulic processes. Even if the student or engineer has not had an introductory course in fluid mechanics, this book can still be used, because the concepts of fluid mechanics and the control volume approach are covered.

MOTIVATION

I sincerely hope that this book will be a contribution toward the goal of better engineering in the field of water resources. I constantly remind myself of the following quote from Baba Diodum: "In the end we will conserve only what we love, we will love only what we understand, and we will understand only what we are taught."

This book has been another part of a personal journey of mine that began as a young boy with an inquisitive interest and love of water, in the streams, creeks, ponds, lakes, rivers, and oceans, and water as rain and snow. Coming from a small Illinois town situated between the Mississippi and Illinois Rivers near Mark Twain's country, I began to see and appreciate at an early age the beauty, the useful power, and the extreme destructiveness that rivers can create. I hope that this book will be of value in your journey of learning about water resources.

WEB SITE

The Web site for this book is located at www.wiley.com/college/mays and includes the following resources:

- *Errata listing*: a list of any corrections that may be found in this book.
- *Figures from text*: non-copyrightable figures are available for making lecture slides or transparencies.
- *Solutions Manual for Instructors*: Includes solutions to all problems in the book. This resource is password-protected, and available only to instructors who have adopted this book for their course. Visit the Instructor Companion site portion of the Web site at www.wiley.com/college/mays to register for a password.

First Undergraduate Hydraulics Course

Outcome

Introduction to book

Review flow processes using control volume concepts

Introduction to pipe flow

Introduction to open-channel flow

Introduction to groundwater flow

Introduction to hydrologic processes

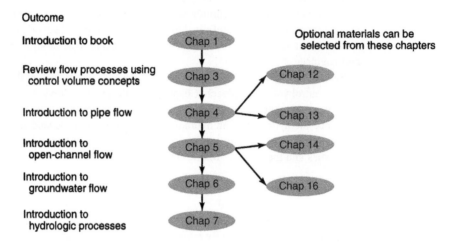

Optional materials can be selected from these chapters

First Undergraduate Hydrology Course

Learning outcomes

Course introduction

Introduction to ground water flow processes

Hydrologic process

Rainfall-runoff analysis based upon unit hydrograph

Reservoir and river routing

Probability and frequency analysis

Floodplain analysis

Hydrologic design: storm sewer design and storm water detention

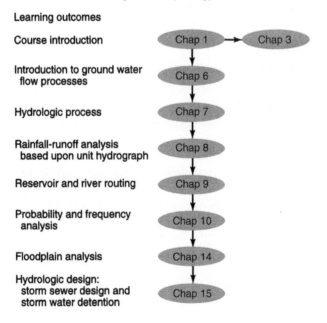

Undergraduate Hydraulic Design Course

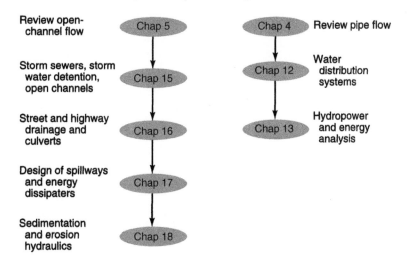

Review open-channel flow — Chap 5

Storm sewers, storm water detention, open channels — Chap 15

Street and highway drainage and culverts — Chap 16

Design of spillways and energy dissipaters — Chap 17

Sedimentation and erosion hydraulics — Chap 18

Chap 4 — Review pipe flow

Chap 12 — Water distribution systems

Chap 13 — Hydropower and energy analysis

Water Resources Engineering and Sustainability

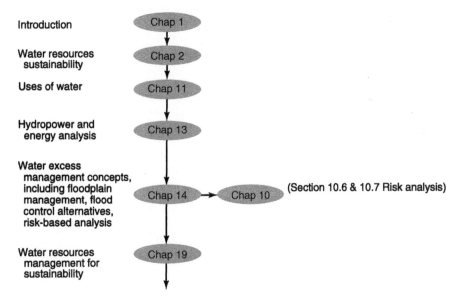

Introduction — Chap 1

Water resources sustainability — Chap 2

Uses of water — Chap 11

Hydropower and energy analysis — Chap 13

Water excess management concepts, including floodplain management, flood control alternatives, risk-based analysis — Chap 14 → Chap 10 (Section 10.6 & 10.7 Risk analysis)

Water resources management for sustainability — Chap 19

Brief Contents

Contents

the ozone layer; destruction of rain forests; threats to wetland, farmland, and other renewable resources; and many others.

These problems are very different from those that humans have faced before. The fact that there are so many things undiscovered by the human race leads me to the statement by Sir Isaac Newton, shortly before his death in 1727:

> *I do not know what I may appear to the world, but to myself I seem to have been only like a boy playing on the sea shore, and diverting myself in now and then finding a smoother pebble or a prettier shell than ordinary, while the great ocean of truth lay all undiscovered before me.*

REFERENCES

Adams, R. M., *Heartland of Cities, Surveys of Ancient Settlement and Land Use in the Central Floodplain of the Euphrates*, University of Chicago Press, Chicago, 1965.

Butzer, K. W., *Early Hydraulic Civilization in Egypt*, University of Chicago Press, Chicago, 1976.

Dziegielewski, B., E. M. Opitz, and D. R. Maidment, "Water Demand Analysis," Chapter 23 in *Water Resources Handbook* edited by L. W. Mays, McGraw-Hill, New York, 1996.

Gleick, P. H., *Water in Crisis*, Oxford University Press, Oxford, 1993.

Mays, L. W., "Water Resources: An Introduction," in *Water Resources Handbook* edited by L. W. Mays McGraw-Hill, New York, 1996.

Mays, L. W., "Introduction," in *Hydraulic Design Handbook* edited by L. W. Mays, McGraw-Hill, New York, 1999.

Mays, L. W., "Introduction," in *Water Distribution Systems Handbook* edited by L. W. Mays, McGraw-Hill, New York, 2000.

Mays, L. W., D. Koutsoyiannis, and A. N. Angelakis, "A Brief History of Urban Water Supply in Antiquity," *Water and Science Technology: Water Supply*, vol. 7, no. 1, pp. 1–12, IWA Publishing, 2007.

Mays, L. W., "Water Sustainability of Ancient Civilizations in Mesoamerica and the American. Southwest," *Water and Science Technology: Water Supply*, vol. 7, no. 1, pp. 229–236, IWA Publishing, 2007.

Mays, L. W., "A Very Brief History of Hydraulic Technology during Antiquity," *Environmental Fluid Mechanics*, vol. 8, no. 5, pp. 471–484, 2008.

Mays, L. W., (Editor-in-Chief), *Ancient Water Technology*, Springer, New York, 2010.

Plate, E. J., "Sustainable Development of Water Resources: A Challenge to Science and Engineering," *Water International*, vol. 18, no. 2, pp. 84–94, International Water Resources Association, 1993.

Shiklomanov, I., "World Fresh Water Resources," in *Water in Crisis* edited by P. H. Gleick, Oxford University Press, New York, 1993.

Solley, W. B., R. R. Pierce, and H. A. Perlman, *"Estimated Use of Water in the United States in 1990,"* U.S. Geological Survey Circular 1081, Washington, DC, 1993.

Turney, O. S., Map of Prehistoric Irrigation Canals, Map No. 002004, Archaeological Site Records Office, Arizona State Museum, University of Arizona, Tucson, 1922.

Chapter 2

Water Resources Sustainability

2.1 WHAT IS WATER RESOURCES SUSTAINABILITY?

Traditionally, sustainability explores the relationships among economics, the environment, and social equity, using the three-legged stool analogy that includes not only the technical, but also the economic and social issues.

The term "sustainable development" was defined in 1987 by the World Commission on Environment and Development as "development that can meet the needs of the present generation without compromising the ability of future generations to meet their own needs. Some of the questions related to sustainable systems and sustainable design are:

- What are the characteristics of sustainable systems?
- How does the design process encourage sustainability?
- What is sustainable water resources development?
- What are the components of sustainable development?

2.1.1 Definition of Water Resources Sustainability

We live in a world where approximately 1.1 billion people lack safe drinking water, approximately 2.6 billion people lack adequate sanitation, and between 2 and 5 million people die annually from water-related diseases (Gleick, 2004). The United Nations Children's Fund's (UNICEF) report, "The State of the World's Children 2005: Childhood under Threat," concluded that more than half the children in the developing world are severely deprived of various necessities essential to childhood. For example, 500 million children have no access to sanitation and 400 million children have no access to safe water. One might ask how sustainable is this? The key to sustainability is the attention to the survival of future generations. Also important is the global context within which we must think and solve problems. The future of water resources thinking must be within the context of water resources sustainability.

The overall goal of water resources management for the future must be water resources sustainability. Mays (2007) defined water resources sustainability as follows:

"Water resources sustainability is the ability to use water in sufficient quantities and quality from the local to the global scale to meet the needs of humans and ecosystems for the present and the future to sustain life, and to protect humans from the damages brought about by natural and human-caused disasters that affect sustaining life."

The Brundtland Commissions's report, "Our Common Future" (World Commission on Environment and Development, WCED), defined sustainability as focusing on the needs of both current and future generations. A development is sustainable if "it meets the needs of the present without compromising the ability of future generations to meet their own needs."

Because water impacts so many aspects of our existence, there are many facets that must be considered in water resources sustainability including:

- Water resources sustainability includes the *availability of freshwater supplies* throughout periods of climatic change, extended droughts, population growth, and to leave the needed supplies for the future generations.
- Water resources sustainability includes having the *infrastructure*, to provide water supply for human consumption and food security, and to provide protection from water excess such as floods and other natural disasters.
- Water resources sustainability includes having the *infrastructure* for clean water and for treating water after it has been used by humans before being returned to water bodies.
- Water sustainability must have adequate *institutions* to provide the management for both the water supply management and water excess management.
- Water sustainability must be considered on a *local, regional, national, and international basis.*
- To achieve water resources sustainability, the principles of *integrated water resources management (IWRM)* must be implemented.

Sustainable water use has been defined by Gleick et al. (1995) as "the use of water that supports the ability of human society to endure and flourish into the indefinite future without undermining the integrity of the hydrological cycle or the ecological systems that depend on it." Seven sustainability requirements are presented in Section 11.1.

2.1.2 The Dublin Principles

The following four simple, but yet powerful messages, were provided in 1992 in Dublin and were the basis for the Rio Agenda 21 and for the millennium Vision-to-Action:

1. *Freshwater is a finite and vulnerable resource, essential to sustain life, development and the environment*, i.e. one resource, to be holistically managed.
2. *Water development and management should be based on a participatory approach, involving users, planners, and policy-makers at all levels*, i.e. manage water with people—and close to people.
3. *Women play a central role in the provision, management and safeguarding of water*, i.e. involve women all the way!
4. *Water has an economic value in all its competing uses and should be recognized as an economic good*, i.e. having ensured basic human needs, allocate water to its highest value, and move towards full cost pricing, rational use, and recover costs.

Poor water management hurts the poor most! The Dublin principles aim at wise management with focus on poverty.

2.1.3 Millennium Development Goals (MDGs)

The *Millennium Development Goals* (MDGs), adopted in September 2000 during the Millennium Summit of the United Nations General Assembly, is comprised of eight goals (see Table 2.1.1). All of the goals can be translated directly or indirectly into water-related terms (Gleick, 2004). For example, Goal No. 1—"Eradicate extreme poverty and hunger"—and No. 7—"Ensure

Table 2.1.1 UN Millennium Development Goals and Targets for Goal 7

Goal 1 Eradicate Extreme Hunger and Poverty
Goal 2 Achieve Universal Primary Education
Goal 3 Promote Gender Equality and Empower Women
Goal 4 Reduce Child Mortality
Goal 5 Improve Maternal Health
Goal 6 Combat HIV/AIDS, Malaria, and Other Diseases
Goal 7 Ensure Environmental Sustainability
　　Target 9 Integrate the principles of sustainable development into country policies and programs and reverse the loss of environmental resources.
　　Target 10 Halve, by 2015, the proportion of people without sustainable access to safe drinking water and basic sanitation.
　　Target 11 Achieve by 2020 a significant improvement in the lives of at least 100 million slum dwellers.
Goal 8 Develop a Global Partnership for Development

Source: http://www.mdgmonitor.org/browse_goal.cfm

environmental sustainability" have direct relevance to water; whereas Goal No. 2—"Achieve universal primary education" and No. 3—"Promote gender equality and empower women" are water-related as millions of women and young girls spend many hours every day to fetch water. The health related Goals 4, 5, and 6 also have strong relevance to water, or the lack of it.

The MDG Goal 7, target 10 of halving, by the year 2015, the proportion of people without sustainable access (to reach or to afford) to safe drinking water seems unlikely to be met. The international community has made little progress to meet the similar part of target 10—to halve, also by 2015, the proportion of people without access to basic sanitation—adopted at the *World Summit on Sustainable Development* (WSSD), in Johannesburg in 2002 (United Nations,). An interesting fact is that this goal did not specifically emphasize wastewater treatment and disposal, because in many parts of the world wastewater treatment does not exist even though sanitation services exist and the sewage is used to irrigate agricultural crops. It is estimated that in Latin America 1.3 million hectares of agricultural land is irrigated with raw wastewater and has related health and disease issues. In countries with water shortages, the reuse of untreated wastewater will likely increase in the future.

2.1.4 Urbanization – A Reality of Our Changing World

Urban populations demand high quantities of energy and raw material, water supply, removal of wastes, transportation, etc. Urbanization creates many challenges for the development and management of water supply systems and the management of water excess from storms and floodwaters. Many urban areas of the world have been experiencing water shortages, which are expected to explode this century unless serious measures are taken to reduce the scale of this problem (Mortada, 2005). Most developing countries have not acknowledged the extent of their water problems, as evidenced by the absence of any long-term strategies for water management.

Changes Caused by Urbanization

Urbanization is a reality of our changing world. From a water resources perspective, urbanization causes many changes to the hydrological cycle including radiation flux, amount of precipitation, amount of evaporation, amount of infiltration, increased runoff, etc. Changes brought about by urbanization can be summarized briefly as follows (Marsalek et al., 2006):

- Transformation of undeveloped land into urban land (including transportation corridors);
- Increased energy release (i.e., greenhouse gases, waste heat, heated surface runoff); and
- Increased demand on water supply (municipal and industrial).

2.2   CHALLENGES TO WATER RESOURCES SUSTAINABILITY

Urban populations are growing rapidly around the world with the addition of many mega-cities (populations of 10 million or more inhabitants). In 1975 there were only four mega-cities in the world and by 2015 there may be over 22 mega-cities in the world (Marshall, 2005). Other cities that will not become mega-cities are also growing very rapidly around the world. By 2010, more than 50% of the world's population is expected to live in urban areas (World Water Assessment Program, 2006).

Mega-cities mean mega problems of which urban water supply management and water excess management are among the largest. Mega-cities and other large cities will be a drain on the Earth's dwindling resources, while at the same time significantly contributing to the environmental degradation. Many of the large cities around the world are prone to water supply shortages, others are prone to flooding, and many are prone to both. A large number of the cities of the world do not have adequate wastewater facilites and most of the waste is improperly disposed or used as irrigation of agricultural lands. As the Earth's population continues to grow, so will the growth of cities continue across the globe, stretching resources and the ability to cope with disasters such as floods and droughts. These factors, coupled with the consequences of global warming, create many challenges for future generations.

There are many factors that affect water resources sustainability including: urbanization, droughts, climate change, flooding, and human-induced factors. Developed areas of the world such as the United States are not exempt from the need for water resources sustainability. Figure 2.2.1 shows areas in the western United States with potential water supply crisis by 2025.

2.2.1   Urbanization

The Urban Water Cycle

The overall urban water cycle is illustrated in Figure 2.2.2 showing the main components and pathways. How does the urbanization process change the water budget from predevelopment to developed conditions of the urban water cycle in arid and semi-arid regions? This change is a very complex process and very difficult to explain.

Urban Water Systems

Urban water system implies that there is a single urban water system and the reality of this is that it is an integrated whole. The concept of a single "urban water system" is not fully accepted because of the lack of integration of the various components that make up the total urban water system. For example, in municipalities it is common to plan, manage, and operate urban water into separate entities such as by service, i.e. water supply, wastewater, flood control, and stormwater. Typically there are separate water organizations and management practices within a municipality, or local or regional government because that is the way they have been historically. Grigg (1986) points out that integration could be achieved by functional integration

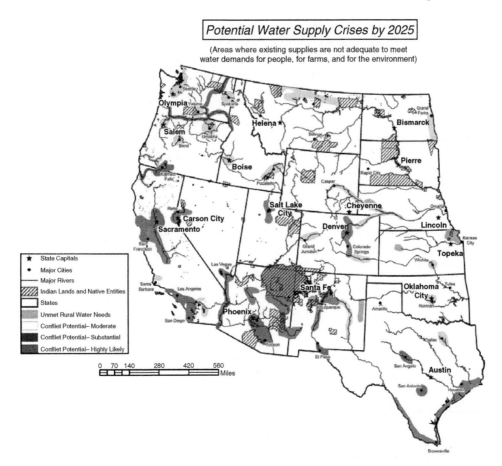

Figure 2.2.1 Areas in the western United States with potential water supply crisis by 2025.
Source: U.S. Department of the Interior (2003).

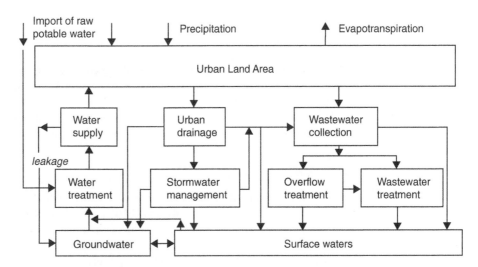

Figure 2.2.2 Urban water cycle: Main components and pathways (from Marsalek et al. (2006)).

and area-wide integration. There are many linkages of the various components of the urban water system with the hydrologic cycle being what connects the urban water system together. There are many reasons for considering the urban water system in an integrated manner. Two of the principal reasons are (a) the natural connectivity of the system through the hydrologic cycle and (b) the real benefits that are realized through integrated management rather than by independent action.

The urban water management system is considered herein as two integrated major entities, water supply management and water excess management. The various interacting components of water excess and water supply management in conventional urban water infrastructure are:

Water Supply Management

— Sources (groundwater, surface water, reuse)
— Transmission
— Water treatment (WT)
— Distribution system
— Wastewater collection
— Wastewater treatment (WWT)
— Reuse

Water Excess Management

— Collection/drainage systems
— Storage/treatment
— Flood control components (levees, dams, diversions, channels)

Sustainable Urban Water Systems

Sustainable urban water systems are being advocated because of the depletion and degradation of urban water resources coupled with the rapid increases in urban populations around the world. Marsalek et al. (2006) defined the following basic goals for sustainable urban water systems:

• Supply of safe and good-tasting drinking water to the inhabitants at all times.
• Collection and treatment of wastewater in order to protect the inhabitants from diseases and the environment from harmful impacts.
• Control, collection, transport, and quality enhancement of stormwater in order to protect the environment and urban areas from flooding and pollution.
• Reclamation, reuse, and recycling of water and nutrients for use in agriculture or households in case of water scarcity.

In North America and Europe many of the above goals have been achieved or are within reach. In many developing parts of the world these goals are far from being achieved. Climate change will be a major factor in both the developed and undeveloped parts of the world that has not been addressed for the future of water resources sustainability. The Millennium Development Goals put a strong emphasis on poverty reduction and reduced child mortality.

Urban Stormwater Runoff

Urban stormwater runoff includes all flows discharged from urban land uses into stormwater conveyance systems and receiving waters. Urban runoff includes both dry-weather, non-storm-water sources (e.g., runoff from landscape irrigation, dewatering, and water line and hydrant flushing) and wet-weather stormwater runoff. Water quality of urban stormwater runoff can be affected by the transport of sediment and other pollutants into streams, wetlands, lakes, estuarine and marine waters, or groundwater. The costs and impacts of water pollution from urban runoff are significant and can include fish kills, health concerns of human and/or terrestrial animals, degraded drinking water, diminished water-based recreation and tourism opportunities, economic losses to commercial fishing and aquaculture industries, lowered real estate values, damage to habitat of fish and other aquatic organisms, inevitable costs of clean-up and pollution reduction, reduced aesthetic values of lakes, streams, and coastal areas, and other impacts (Leeds et al., 1993).

Increased stormwater flows from urbanization have the following major impacts (FLOW, 2003):

- acceleration of stream velocities and degradation of stream channels,
- declining water quality due to washing away of accumulated pollutants from impervious surfaces to local waterways, and an increase in siltation and erosion of soils from pervious areas subject to increased runoff,
- increase in volume of runoff with higher pollutant concentrations that reduces receiving water dilution effects,
- diminished groundwater recharge, resulting in decreased dry-weather flows; poorer water quality of streams during low flows; increased stream temperatures; and greater annual pollutant load delivery,
- increased flooding,
- combined and sanitary sewer overflows due to stormwater infiltration and inflow,
- damage to stream and aquatic life resulting from suspended solids accumulation, and increased health risks to humans from trash and debris which can also endanger, and
- destroying food sources or habitats of aquatic life (FLOW, 2003).

Groundwater Changes

Urbanization often causes changes in groundwater levels as a result of decreased recharge and increased withdrawal. In rural areas, water supplies are usually obtained from shallow wells, while most of the domestic wastewater is returned to the ground through cesspools or septic tanks. Thus the quantitative balance in the hydrologic system remains. As urbanization occurs many individual wells are abandoned for deeper public wells. With the introduction of sewer systems, stormwater, and (treated or untreated) wastewater are discharged to nearby surface water bodies. Three conditions disrupt the subsurface hydrologic balance and produce declines in groundwater levels.

1. Reduced groundwater recharge due to paved surface areas and storm sewers
2. Increased groundwater discharge by pumping wells
3. Decreased groundwater recharge due to export of wastewater collected by sanitary sewers

Groundwater quality is certainly another challenge to water resources sustainability resulting in many cases from urbanization. Groundwater quality can be affected by residential and commercial development as illustrated in Figure 2.2.3. The U.S. Geological Survey's National Water Quality Assessment (NAWQA) program (http://water.usgs.gov/nawqa) seeks to determine how shallow groundwater quality is affected by development (Squillace and Price, 1996). Residential developments have taken up very large tracts of land, and as a consequence, have widespread influence on the quality of water that recharges aquifers, streams, lakes, and wetlands. Liquids discharged onto the ground surface in an uncontrolled manner can migrate downward to degrade groundwater. Septic tanks and cesspools are another source of groundwater pollution. Polluted surface water bodies that contribute to groundwater recharge are sources of groundwater pollution.

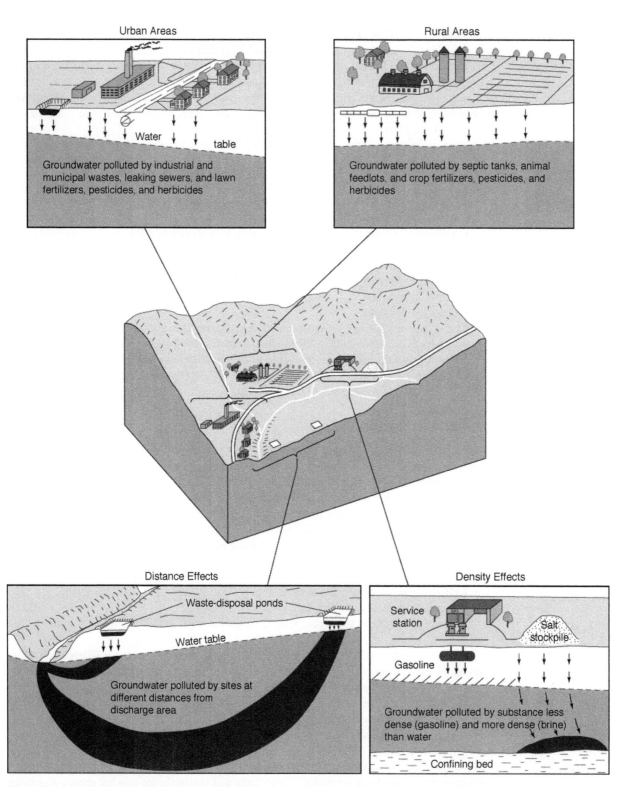

Figure 2.2.3 Groundwater pollution affected by differences in chemical composition, biological, and chemical reactions, and distance from discharge areas (from Heath, 1998).

many hydrologic implications related to evaporation, infiltration, snow melt, surface water runoff, aquifer recharge, and streamflows. One important result of these impacts is drought. The drought of the early 2000s has taken place in warmer conditions, comparing the temperature departures for the 6-year period (2000–2005) as compared to the 1895–2000 averages. Both in terms of absolute degrees and in terms of annual standard deviation, the Colorado River basin has warmed more than any region of the United States (Water Science and Technology Board, 2007).

Climate Change Affects the Basin

Climate change models have been used to project future precipitation and temperature changes for the Colorado River Basin. Long-term projections of precipitation are a greater modeling challenge than temperature projections. Precipitation projections for the Colorado River Basin using climate models suggest a wide range of potential changes in annual precipitation. Models have forecasted a slight decrease (less than 10 percent below current values) in annual precipitation in southwestern United States, with relatively little change in annual precipitation amounts forecast for headwaters regions of the Colorado River (Water Science and Technology Board, 2007). The changes in seasonality of precipitation or changes in the type of precipitation (rain or snow) can be just as important as changes in annual amounts of precipitation.

Future projections and past trends indicate a strong likelihood of a warmer future climate across the Colorado River Basin. Projected temperature increases across the Colorado River Basin have important direct and indirect implications for hydrology and streamflow, irrespective of the amount of the precipitation increases or decreases. The effects of warmer temperatures across the Colorado River Basin for hydrology include the following (Water Science and Technology Board, 2007):

- freezing levels at higher elevations, which means more winter precipitation will fall as rain rather than snow;
- shorter seasons of snow accumulation at a given elevation;
- less snowpack accumulation as compared to the present;
- earlier melting of snowpack;
- decreased baseflows from groundwater during late summer, and lowered water availability during the important late-summer growing season;
- more runoff and flood peaks during the winter months;
- longer growing seasons;
- reductions in soil moisture availability in the summer and increases in the spring and winter;
- increased water demand by plants; and
- greater losses of water to evapotranspiration.

A Short History of Water Development of the Basin

Due to the doctrine of prior appropriation (first in time, first in right, see Section 19.1), the states in the upper Colorado River Basin became worried that the rapidly developing California would obtain a large portion of the appropriated water, leaving them with a shortage in the future. As an attempt to settle the issues, the upper basin states agreed to support California on the Hoover Dam proposal that it needed to obtain Colorado River water for its growing development. In return, the states requested a guaranteed amount of water from the river for their own future development. This agreement between the states resulted in the Colorado River Compact in 1922, which Arizona did not ratify until 1944.

Under the Colorado River Compact, it was agreed that the upper and lower Colorado River Basin would each receive 7.5 maf. It was also agreed that the lower basin would have the right to increase its beneficial consumptive use by 1 maf annually. All of the states supported the compact except Arizona, which opposed the Compact and refused to sign it. The dispute over the water continued as the Boulder Canyon Project Act was passed on December 21, 1928 by

Congress, which authorized the construction of the Boulder Dam (now Hoover Dam). However, the one stipulation was that California must agree to limit its use of Colorado River water to an amount of 4.4 maf. Arizona and California fought over both the Colorado River Compact and the Boulder Canyon Act. Arizona was against the Act and did not want California to have any of their water.

In order to help in settling the dispute, the U.S. Congress made it clear to Arizona that, until they could settle the dispute of water allocation in the lower Colorado River Basin, the state would not receive any support for their water canal system, the Central Arizona Project (CAP), which would later become a controversy in itself. Arizona finally agreed to share its water with California in order to receive funding for the CAP. As a result of the case *Arizona v. California*, which took place in 1964, the Supreme Court decreed that California would receive 4.4 maf of Colorado River water, Arizona would receive 2.8 maf, and Nevada 0.3 maf.

The CAP is a 336-mi-long system of aqueducts, tunnels, pumping plants, and pipelines (Figure 2.3.5) that carries water from the Colorado River at Lake Havasu, through Phoenix, to the San Xavier Indian reservation southwest of Tucson. The main purpose of the CAP was to help Arizona conserve its groundwater supplies by importing surface water from the Colorado River. Figure 2.3.5 shows the layout and major features of the Central Arizona Project (CAP) system.

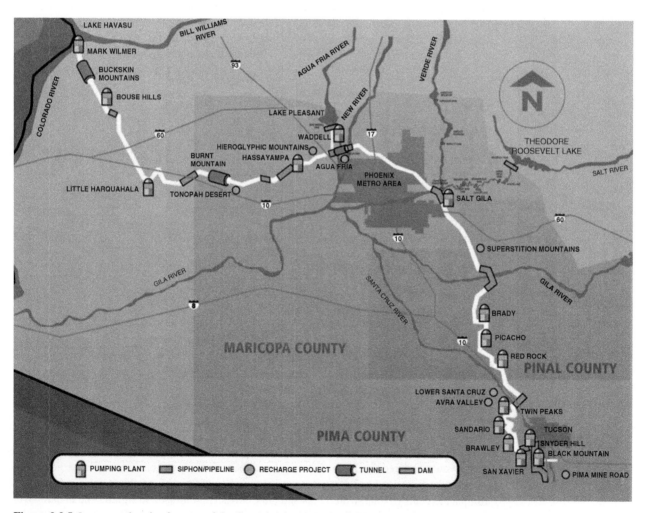

Figure 2.3.5 Layout and major features of the Central Arizona Project (CAP) system.

The CAP was the largest, most expensive, and most politically volatile water development project in U.S. history; it was also the most ambitious basin project that the bureau attempted (Espeland, 1998). Even early on in 1947, the strategy of CAP supporters was to paint CAP as a "rescue" operation. This was the project necessary to replace the "exhausted" groundwater supply in order to save the local economy. By 1963, the CAP was still justified as a "rescue" project; a doubling of the population over the previous 10 years supposedly made the project even more urgent. Economic development was assumed to be driven by agricultural development. The thought was that without more irrigated farmland, urban growth (which reduces irrigated farmland) would be stymied. How did the population grow so fast despite the previous prediction that water supply would limit economic growth?

In 1968, Congress authorized the construction of the CAP under the Colorado River Basin Project Act. The main purpose for the authorization was to assist Arizona in reducing its water deficiencies. By 1971, the first *environmental impact statement* (EIS) on the CAP was written and then finalized in 1972. The 1976 EIS was devoted solely to the Orme Dam, to become the beginning of a series of EISs in the major features of the CAP. In 1971, the *Central Arizona Water Conservation District* (CAWCD) was created to provide a means for Arizona to repay the federal government for the reimbursable costs of construction and to manage and operate the CAP once complete. The construction began in 1973 at Lake Havasu and was completed in 1993. The entire cost of the project was more than $4 billion. Under the Colorado River Basin Project Act, the CAP would be the first to take shortages in the lower Colorado River Basin.

The users of the CAP water fall into three categories. The first category is municipal and industrial. These customers include cities and water utilities which are responsible for treating drinking water and delivering it to residences, commercial buildings, and industries. The next water use category is agricultural. These agricultural users are primarily irrigation districts. The last category is the Indian community. These communities receive water from the CAP under contracts with the federal government. Agriculture has been the main water user in the past; however, due to the increasing development of Arizona, cities will soon become the largest customer for the CAP.

The Future

A recent study at the Scripps Institution of Oceanography has shown that "under current conditions there is a 10% chance live storage in Lakes Mead and Powell will be gone by about 2013 and a 50% chance it will be gone by 2021 if no changes in water allocation from the Colorado River system are made." These results are driven by climate change resulting from global warming, the effects of natural climate variability, and the current operation of the reservoir system. Water planners and managers at all levels have denied any knowledge that such a thing could happen, even though there have been a number of studies suggesting that there will be a decrease in runoff in the southwestern United States due to global warming.

As stated in the Water Science and Technology (2007) report, "Steadily rising population and urban water demands in the Colorado River region will inevitably result in increasingly costly, controversial, and unavoidable trade-off choices to be made by water managers, politicians, and their constituents. These increasing demands are also impeding the region's ability to cope with droughts and water shortages."

2.4   GROUNDWATER SYSTEMS – THE EDWARDS AQUIFER, TEXAS

The Edwards Aquifer, located in south-central Texas (Figure 2.4.1), is a large area of karst terrain. Limestone and dolomites are exposed at the land surface, referred to as the outcrop area or recharge area (see Figures 2.4.1 and 2.4.2). Figure 6.9.2 also points out the outcrop area of the aquifer. Recharge to the aquifer is derived mainly from streams that cross the outcrop area of the aquifer and from direct infiltration of precipitation on the outcrop. The watershed areas that provide recharge

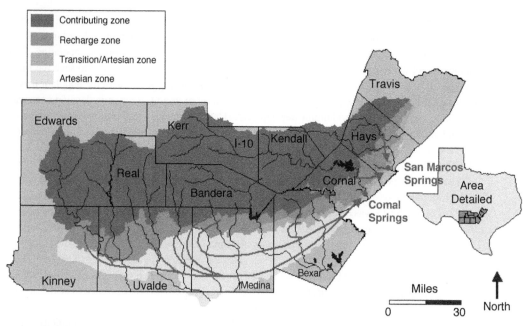

Figure 2.4.1 Edwards Aquifer in south-central Texas showing general flow paths of aquifer. (Courtesy of Greg Eckhardt.)

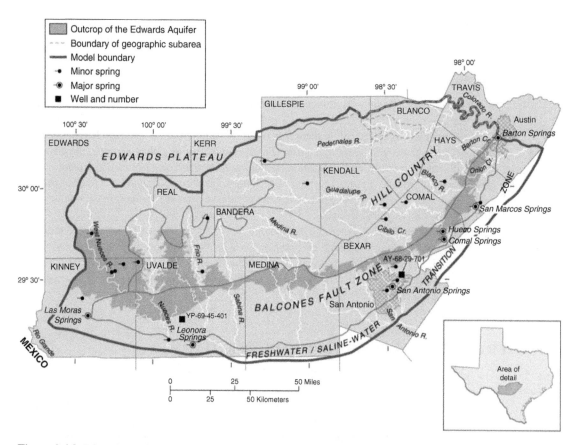

Figure 2.4.2 Edwards Aquifer showing springs (from Kuniansky et al., 2001).

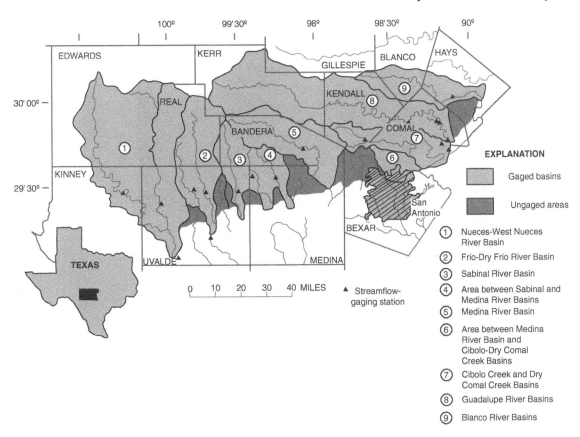

Figure 2.4.3 Map showing gaged basins and ungauged areas (modified from Puente (1978), Figure 1, as presented by Slattery and Thomas (2002)).

to the aquifer are shown in Figure 2.4.3. Estimated annual recharge to the aquifer by river basin is listed in Table 2.4.1.

Discharge from the Edwards Aquifer is from wells and springs. The major discharge from wells is in Bexar, Medina, and Uvalde Counties. The City of San Antonio, Texas obtains all of its water supply from the aquifer. Table 2.4.2 lists the estimated annual discharge from the Edwards Aquifer by county

Table 2.4.1 Estimated Annual Recharge to the Edwards Aquifer by Basin, 1980–2001 (thousands of ac-ft)

Calendar year	Nueces–West Nueces River Basin	Frio-Dry Frio River Basin[1]	Sabinal River Basin[1]	Area between Sabinal and Medina River Basins[1]	Medina River Basin[2]	Area between Medina River Basin and Cibolo–Dry Comal Creek Basin[1]	Cibolo Creek and Dry Comal Creek Basins	Blanco River Basin[1]	Total[3]
1980	58.6	85.6	42.6	25.3	88.3	18.8	55.4	31.8	406.4
1981	205.0	365.2	105.6	252.1	91.3	165.0	196.8	67.3	1448.4
1982	19.4	123.4	21.0	90.9	76.8	22.6	44.8	23.5	422.4
1983	79.2	85.9	20.1	42.9	74.4	31.9	62.5	23.2	420.1
1984	32.4	40.4	8.8	18.1	43.9	11.3	16.9	25.9	197.9
1985	105.9	186.9	50.7	148.5	64.7	136.7	259.2	50.7	1003.3
1986	188.4	192.8	42.2	173.6	74.7	170.2	267.4	44.5	1153.7
1987	308.5	473.3	110.7	405.5	90.4	229.3	270.9	114.9	2003.6

(Continued)

Table 2.4.1   (*Continued*)

Calendar year	Nueces–West Nueces River Basin	Frio-Dry Frio River Basin[1]	Sabinal River Basin[1]	Area between Sabinal and Medina River Basins[1]	Medina River Basin[2]	Area between Medina River Basin and Cibolo–Dry Comal Creek Basin[1]	Cibolo Creek and Dry Comal Creek Basins	Blanco River Basin[1]	Total[3]
1988	59.2	117.9	17.0	24.9	69.9	12.6	28.5	25.5	355.5
1989	52.6	52.6	8.4	13.5	46.9	4.6	12.3	23.6	214.4
1990	479.3	255.0	54.6	131.2	54.0	35.9	71.8	41.3	1123.2
1991	325.2	421.0	103.1	315.2	52.8	84.5	109.7	96.9	1508.4
1992	234.1	586.9	201.1	566.1	91.4	290.6	286.6	228.9	2485.7
1993	32.6	78.5	29.6	60.8	78.5	38.9	90.9	37.8	447.6
1994	124.6	151.5	29.5	45.1	61.1	34.1	55.6	36.6	538.1
1995	107.1	147.6	34.7	62.4	61.7	36.2	51.1	30.6	531.3
1996	130.0	92.0	11.4	9.4	42.3	10.6	14.7	13.9	324.3
1997	176.9	209.1	57.0	208.4	63.3	193.4	144.2	82.3	1134.6
1998	141.5	214.8	72.5	201.4	80.3	86.2	240.9	104.7	1142.3
1999	101.4	136.8	30.8	57.2	77.1	21.2	27.9	21.0	473.5
2000	238.4	123.0	33.1	55.2	53.4	28.6	48.6	34.1	614.5
2001	297.5	126.7	66.2	124.1	90.0	101.5	173.7	89.7	1069.4
Average	121.2	134.1	42.2	107.2	61.9	69.9	105.0	43.1	684.7

[1]Includes recharge from ungaged areas.
[2]Recharge to Edwards Aquifer from the Medina River Basin consists entirely of losses from Medina Lake (Puente, 1978).
[3]Total might not equal the sum of basin values due to rounding.

Source: Slattery and Thomas (2002).

Table 2.4.2   Estimated Annual Discharge from the Edwards Aquifer by County, 1986–2001 (thousands of ac-ft)

Calendar year	Kinney-Uvalde Counties	Medina County	Bexar County	Comal County	Hays County	Total[1]	Well discharge	Spring discharge
1980	151.0	39.9	300.3	220.3	107.9	819.4	491.1	328.3
1981	104.2	26.1	280.7	241.8	141.6	794.4	387.1	407.3
1982	129.2	33.4	305.1	213.2	105.5	786.4	453.1	333.3
1983	107.7	29.7	277.6	186.6	118.5	720.1	418.5	301.6
1984	151.1	46.9	309.7	108.9	85.7	702.3	529.8	172.5
1985	156.9	59.2	295.5	200.0	144.9	856.5	522.5	334.0
1986	91.7[2]	41.9	294.0	229.3	160.4	817.3[2]	429.3	388.1[2]
1987	95.1[2]	15.9	326.6	286.2	198.4	922.0[2]	364.1	558.0[2]
1988	156.7[2]	82.2	317.4	236.5	116.9	909.7[2]	540.0	369.8[2]
1989	156.9	70.5	305.6	147.9	85.6	766.6	542.4	224.1
1990	118.1	69.7	276.8	171.3	94.1	730.0	489.4	240.6
1991	76.6	25.6	315.5	221.9	151.0	790.6	436.3	354.3
1992	76.5	9.3	370.5	412.4	261.3	1,130.2	327.3	802.8
1993	107.5	17.8	371.0	349.5	151.0	996.7	407.3	589.4
1994	95.5	41.1	297.7	269.8	110.6	814.8	424.6	390.2
1995	90.8	35.2	272.1	235.0	127.8	761.0	399.6	361.3
1996	117.6	66.3	286.8	150.2	84.7	705.6	493.6	212.0
1997	29.9[3]	7.0[3]	255.3[3]	243.3	149.2	684.7[3]	300.7[3]	384.0
1998	113.1	51.3	312.8	271.4	169.2	915.9	451.7	464.1

(*Continued*)

Table 2.4.2 (*Continued*)

Calendar year	Kinney-Uvalde Counties	Medina County	Bexar County	Comal County	Hays County	Total[1]	Well discharge	Spring discharge
1999[4]	99.8[5]	48.3	298.3	295.2	142.3	884.0	427.8	456.2
2000	89.1	45.1	283.6	226.1	108.4	752.3	414.8	337.5
2001	68.7	33.9	291.6	327.4	175.3	896.9	367.7	529.1

[1]Total might not equal the sum of county values due to rounding.
[2]Differs from value to the *Edwards Underground Water District Bulletins*, pp. 46–48, table 3, due to correction of an error in the method of computing the Leona Formation underflow.
[3]Does not include irrigation discharge (Bexar, Medina, and Uvalde Counties).
[4]Does not include discharge for domestic supply, stock, and miscellaneous use.
[5]Does not include discharge from Kinney County.

Source: Slattery and Thomas (2002).

and lists the portion from wells and from springs. Note the relation of discharge from springs as compared to well discharge as a function of time. Discharge from the Comal Springs and San Marcos Springs for 2001 was 414,800 ac-ft, accounting for 78 percent of the discharge for that year.

There are many technical, legal, economic, and institutional issues concerning the Edwards Aquifer. Even though the Edwards Aquifer has been studied extensively over the years there are still no solutions or answers to many issues. These include technical, legal, economic, and institutional issues (http://www.edwardsaquifer.net/).

2.5 WATER BUDGETS

2.5.1 What are Water Budgets?

The use of water budgets is very important in the evaluation of water resources sustainability. A water budget, hydrologic budget, or water balance is a measurement of the continuity of the flow of water through a system or control volume. The budget holds true for any time interval and applies to any size area ranging from local-scale areas to regional scale areas or from any drainage area to the earth as a whole. An open system is usually considered for which quantification is a mass balance equation in which the change of storage of water (ΔS) over a time period within the system equals the inputs (I) to the system minus the outputs (O) from the system.

A common application of the water budget (hydrologic budget) is to geographical areas to establish hydrologic characteristics of the area. A watershed or drainage basin or catchment is defined as the area that contributes, on the basis of topography, all the water that flows through a given location of the stream. The water budget for a watershed (Figure 2.5.1) includes the precipitation (P) onto the watershed, the groundwater inflow (G_{in}), the evaporation (E), the evapotranspiration (T), the stream outflow (Q), and the groundwater outflow (G_{out}) from the watershed, all in terms of volume over a time period. Change of storage, ΔS, is then expressed as

$$\Delta S = P + G_{in} - E - T - G_{out} - Q \qquad (2.5.1)$$

EXAMPLE 2.5.1   Determine the net groundwater flow to a lake (which is the groundwater inflow to the lake minus the groundwater outflow) using a water budget for a particular year in which the precipitation was 43 in, the evaporation was 53 in, the surface water inflow was 1 in, the surface outflow was 173 in, and the change in storage was −2 in.

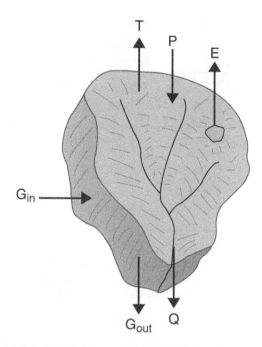

Figure 2.5.1 Watershed showing components of the water budget (hydrologic budget: P = precipitation, G_{in} = groundwater flow inflow, E = evaporation, T = evapotranspiration, Q = streamflow and G_{out} = groundwater outflow, all in terms of volume. Change of storage, ΔS, is $\Delta S = P + G_{in} - E - T - G_{out} - Q$).

SOLUTION

The water budget equation to define the net groundwater inflow to the lake neglecting evapotranspiration is

$$G = G_{in} - G_{out} = \Delta S - P + E - Q_{in} + Q_{out}$$
$$= -2 - 43 + 53 - 1 + 173$$
$$= 180 \text{ in for the year}$$

Figure 2.5.2 illustrates a simple water budget for a groundwater system for predevelopment conditions and development conditions. The possible sources of water entering (recharge) a groundwater system under natural conditions could be (a) the areal recharge from precipitation that percolates through the unsaturated zone to the water table and (b) recharge from losing streams,

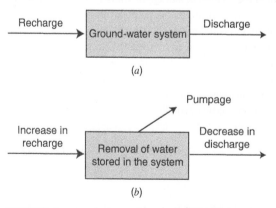

Figure 2.5.2 Diagrams of water budgets for a groundwater system for predevelopment and development conditions. (*a*) Predevelopment water-budget diagram illustrating that inflow equals outflow; (*b*) Water-budget diagram showing changes in flow for a groundwater system being pumped. The sources of water for the pumpage are changes in recharge, discharge, and the amount of water stored. The initial predevelopment values do not directly enter the budget calculation (from Alley et al., 1999).

lakes, and wetlands. The discharge leaving a groundwater system under natural conditions could be (a) discharge to streams, lakes, wetlands, saltwater bodies (bays, estuaries, or oceans), and springs; and (b) groundwater evaporation.

Figure 2.5.3 illustrates the difference in water budget components for predevelopment times and the present for Las Vegas, Nevada. Figure 2.5.4 illustrates the groundwater budgets for predevelopment and development conditions for a part of Nassau and Suffolk Counties, Long Island, New York. Both budgets assume equilibrium conditions with little or no change in storage.

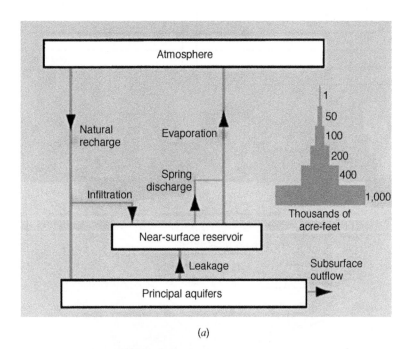

(a)

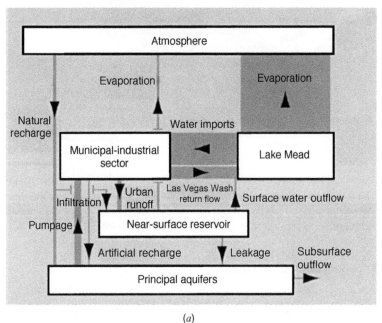

(a)

Figure 2.5.3 Comparison of water budgets for (a) predevelopment period and (b) the present for Las Vegas, Nevada (from Galloway et al., 1999).

OVERALL PREDEVELOPMENT WATER-BUDGET ANALYSIS	
INFLOW TO LONG ISLAND AND HYDROLOGIC SYSTEM	CUBIC FEET PER SECOND
1. Precipitation	2,475
OUTFLOW FROM LONG ISLAND HYDROLOGIC SYSTEM	
2. Evapotranspiration of precipitation	1,175
3. Groundwater discharge to sea	725
4. Streamflow discharge to sea	525
5. Evapotranspiration of groundwater	25
6. Spring flow	25
Total outflow	2,475

GROUNDWATER PREDEVELOPMENT WATER BUDGET ANALYSIS	
INFLOW TO LONG ISLAND GROUNDWATER SYSTEM	CUBIC FEET PER SECOND
7. Groundwater recharge	1,275
OUTFLOW FROM LONG ISLAND GROUNDWATER SYSTEM	
8. Groundwater discharge to streams	500
9. Groundwater discharge to sea	725
10. Evapotranspiration of groundwater	25
11. Spring flow	25
Total outflow	1,275

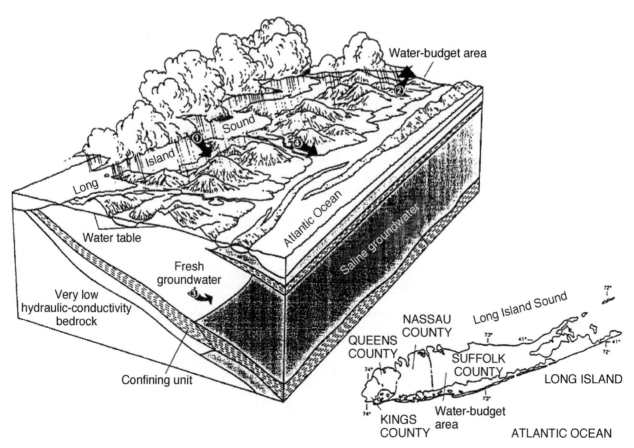

Figure 2.5.4 Groundwater budget for part of Nassau and Suffolk Counties, Long Island, New York. Block diagram of Long Island, New York, and tables listing the overall water budget and groundwater budget under predevelopment conditions. Both water budgets assume equilibrium conditions with little or no change in storage (from Alley et al., 1999).

2.5.2   Water Balance for Tucson, Arizona

To further illustrate the water balance procedure and its importance in water resources sustainability evaluations, Tucson, Arizona, is used. In 1989, Arizona's Groundwater Management Act (GMA) was passed creating five Active Management Areas (AMAs). These AMAs encompassed Arizona's most populous, water-stressed regions. One of these AMAs is the Tucson Active Management Area

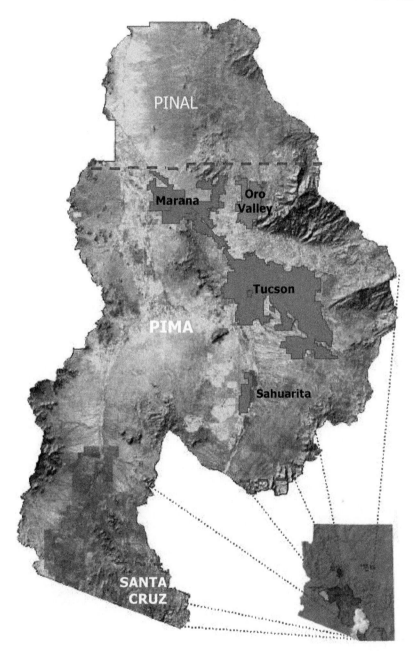

Figure 2.5.5 The Tucson Active Management Area (TAMA).
Source: Arizona Department of Water Resources.

(TAMA), illustrated in Figure 2.5.5, covers 3866 mi^2. Prior to 2001, the Tucson, Arizona metropolitan area relied entirely on groundwater to satisfy the potable water demand. Because there was no engineered effort to replenish the groundwater resources, the groundwater levels declined up to 4 ft per year in major Tucson wells. Agricultural, industrial, and municipal water are the three primary water use categories. Water demand in the TAMA has outgrown the local renewable groundwater supply, so that sustained growth is possible only because Colorado River water is delivered to the Tucson area via the CAP canal (Figure 2.3.2). However, it must be pointed

Table 2.5.1   The Year 1995 TAMA Water Budget (Demand and Supply)

Demand (10^6 m^3/year)		Supply (10^6 m^3/year)		
		CAP water	Effluent	Groundwater
Municipal sector	191.8	0.12	9.50	182.2
Agricultural sector	120.9	0	2.22	118.7
Industrial sector	74.3	0	0.99	73.3
Loss due to evaporation	4.56	0	0	4.56
Totals:	**392**	**0.12**	**12.7**	**378.7**
Groundwater use				378.7
(Less) Net natural recharge				−75.0
(Less) Incidental recharge[1]				−101.5
Groundwater overdraft				**202.2**

[1]Incidental recharge arises from fractional infiltration of water demands in municipal, agricultural, and industrial sectors.
Source: Arnold and Arnold (2009).

out that Arizona has the lowest priority use of CAP water, making it the state most vulnerable to shortages. Table 2.5.1 shows the TAMA water budget for 1995.

Referring to Figure 2.5.6, a detailed water balance for the TAMA groundwater resource has the following form:

$$G = S_{gs} + S_c + (RD_m - W) + I_aD_a + I_iD_i + I_mD_m + I_eW - D_a - D_i - D_m \qquad (2.5.2)$$

where G is the net annual accumulation of groundwater in the TAMA; S_{gs} is the natural rate of groundwater replenishment (75×10^6 m^3/yr); S_c is the volume of CAP water supplied to the TAMA

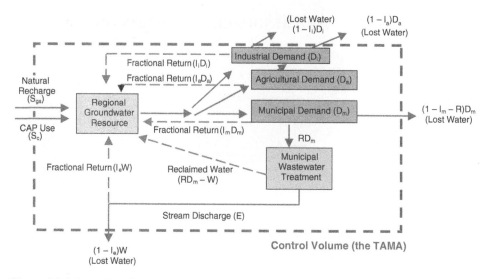

Figure 2.5.6 Water Flow in the Tucson Active Management Area (TAMA). Dotted boundary represents the TAMA control volume used for water balances and sustainability calculations. S_{gs} is the natural rate of groundwater replenishment; S_c is the volume of CAP water supplied to the TAMA; R is the fraction of municipal water demand (D_m) that is recovered, treated and reused or discharged as effluent; D_m, D_i, and D_a are annual municipal, industrial, and agricultural demands; I_m, I_i, and I_a are the fractions of municipal, industrial, and agricultural demands that reenter the ground water as infiltrate; W is the annual volume of wastewater effluent discharged to the Santa Cruz River; and I_e is the fraction of effluent that infiltrates to groundwater in the Santa Cruz River channel (from Arnold and Arnold, 2009).

$(\leq 328.8 \times 10^6 \, \text{m}^3/\text{yr})$; R is the fraction of municipal water demand (D_m) that is recovered, treated, and reused or discharged as effluent (0.44); D_m, D_i, and D_a are annual municipal, industrial, and agricultural demands; I_m, I_i, and I_a are the fractions of municipal, industrial, and agricultural demands that reenter the groundwater as infiltrate (0.04, 0.12, 0.20); W is the annual volume of wastewater effluent discharged to the Santa Cruz River; and I_e is the fraction of effluent that infiltrates to ground water in the Santa Cruz River channel (0.90).

Equation (2.5.2) can be used to recalculate the maximum sustainable rate of water use in the TAMA using steady conditions for a zero net annual accumulation of groundwater, $G = 0$. Referring to the control volume (Figure 2.5.6), the steady condition is achieved when

$$S_{gs} + S_c = (1 - I_i)D_i + (1 - I_a)D_a + (1 - I_m - R)D_m + (1 - I_e)W \qquad (2.5.3)$$

Total water demand is the sum, $D_i + D_a + D_m$, so that under steady (sustainable) conditions,

$$D_i + D_a + D_m = S_{gs} + S_c + I_iD_i + I_iD_a + (I_m + R)D_m + (I_e - 1)W \qquad (2.5.4)$$

To maximize the rate of water use, the TAMA must use its entire CAP entitlement $\left(S_c - S_{c,\,max}\right)$ and recycle all municipal wastewater effluent $(W = 0)$.
Then

$$(1 - I_m - R)D_m = S_{gs} + S_{c,\,max} - (1 - I_i)D_i - (1 - I_a)D_a \qquad (2.5.5)$$

or

$$D_m = \frac{S_{gs} + S_{c,\,max}}{1 - I_m - R} - \frac{(1 - I_i)D_i}{1 - I_m - R} - \frac{(1 - I_a)}{1 - I_m - R}D_a \qquad (2.5.6)$$

in which $R = 0.44$ and $S_{c,\,max} = 328.8 \times 10^6 \, \text{m}^3 \cdot \text{yr}^{-1}$, etc.

2.6 EXAMPLES OF WATER RESOURCES UNSUSTAINABILITY

2.6.1 Aral Sea

The Aral Sea is located in Central Asia between Uzbekistan and Kazakstan (both countries were part of the former Soviet Union) as shown in Figure 2.6.1. The Amu-Darya and the Syr-Darya (*dar'ya* means river in Turkic) flow into the Aral Sea with no outlet from the sea. Over a 30-year-plus time period, water has been diverted from the Amu-Darya and the Syr-Darya to irrigate millions of acres of land for cotton and rice production, which has resulted in a loss of more than 60 percent of the sea's water. The sea has shrunk from over 65,000 km^2 to less than half that size, exposing large areas of the lake bed. From 1973 to 1987, the Aral Sea dropped from fourth to sixth among the world's largest inland seas. The satellite photos in Figure 2.6.2 show the Aral Sea in 1985 and 2003, and Figure 2.6.3 illustrates the decrease in size of the Aral Sea from 1957 to 2000.

The lake's salt concentration increased from 10 percent to more than 23 percent, contributing to the devastation of a once-thriving fishing industry. The local climate reportedly has shifted, with hotter, drier summers and colder, longer winters. With the decline in sea level, salty soil has remained on the exposed lake bed. Dust storms have blown up to 75,000 tons of this exposed soil has annually, dispersing its salt particles and pesticide residues. This air pollution has caused widespread nutritional and respiratory ailments, and crop yields have been diminished by the added salinity, even in some of the same fields irrigated with the diverted water. Additional literature on this subject includes Ellis and Turnley (1990), Ferguson (2003), and Perera (1988, 1993).

The major consequences of the continuous desiccation of the Aral Sea since 1960 are summarized as follows:

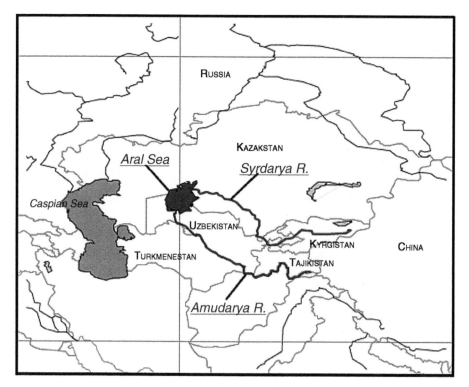

Figure 2.6.1 The Aral Sea Basin (from McKinney, 1996).

- Climatic consequences such as mesoclimatic changes, increase of salt and dust storms, shortening of vegetation period;
- Ecological/economic consequences including degeneration of the delta ecosystems, total collapse of the fishing industry, and decrease of productivity of agricultural fields; and
- Health consequences such as increase in serious diseases, birth defects, and high infant mortality.

People of the region did not make the decision to use the rivers of the Aral Sea basin, but they have certainly suffered the consequences. As stated at the Conference of the Central Asian region ministers, States of Central Asia: Environment Assessment, Aarhus, Denmark, 1998:

> *"The Aral crisis is the brightest example of the ecological problem with serious social and economic consequences, directly or indirectly connected with all the states of Central Asia. Critical situation caused by the Aral Sea drying off was the result of agrarian economy tendency on the basis of irrigated agriculture development and volume growth of irrevocable water consumption for irrigation."*

2.6.2 Mexico City

Now let's look at present day Mexico City, a very large urban center, as another example of water resources unsustainability. Mexico City is the cultural, economic, and industrial center for Mexico. This city is located in the southern part of the Basin of Mexico, which is an extensive, high mountain valley at approximately 2200 m above sea level and surrounded by mountains of volcanic origin with peak altitudes of over 5000 m above sea level.

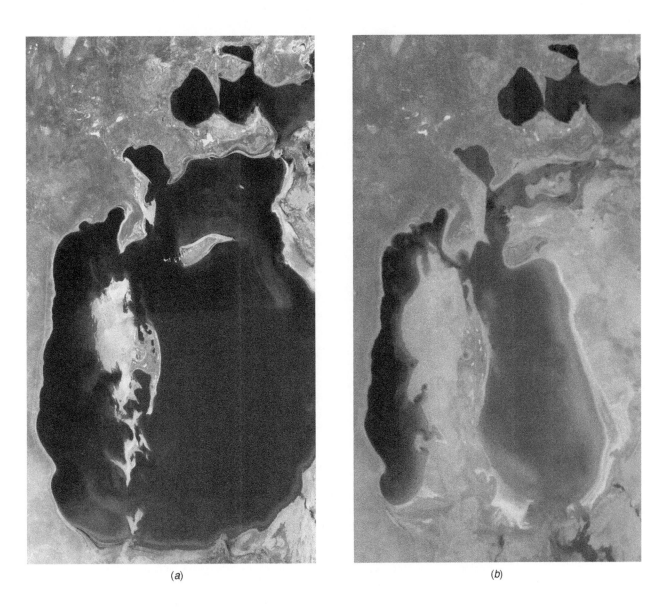

Figure 2.6.2 Comparison of Aral Sea (*a*) 1989 and (*b*) 2003 (NASA).

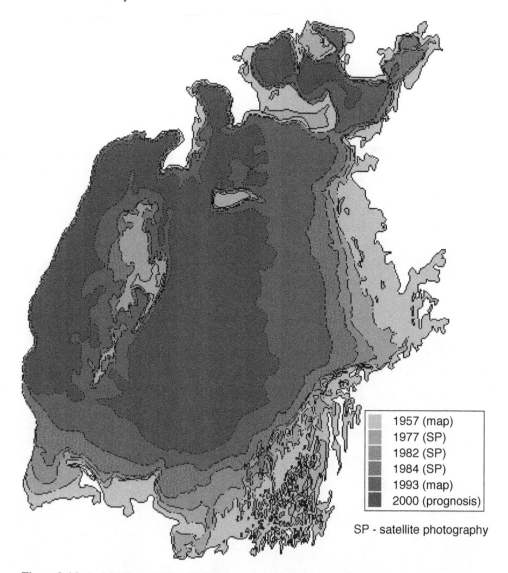

1957 (map)
1977 (SP)
1982 (SP)
1984 (SP)
1993 (map)
2000 (prognosis)

SP - satellite photography

Figure 2.6.3 Aral Sea from 1957 to 2000 from the report "Environment State of the Aral" developed by the International Fund for the Aral Sea (IFAS) and the UN Environment Programme (UNEP) under financial support of the Norway Trust Fund at the World Bank. Coordination from the side of IFAS was held by the Executive Committee of IFAS, and from the side of UNEP–UNEP/GRID-Arendal (http://enrin.grida.no/aral/aralsea/english/arsea/arsea.htm#2).

Beginning in the 14th century, the Aztecs made use of a system of aqueducts to convey spring water from the higher elevations in the southern portion of the Basin of Mexico to their city, Tenochtitlan. This ancient city was built on land reclaimed from the saline Lake Texcoco. The Spaniards defeated the Aztecs in 1520, after which they rebuilt the aqueducts and continued to use the spring water until the mid-1850s. Potable groundwater, under artesian conditions, was discovered in 1846. Over the next century, the increased groundwater extraction and the artificial diversions to drain the valley resulted in the drying up of many of the springs, the draining of lakes, loss of pressure in the aquifer with declining groundwater levels, and the consequent subsidence.

Because Mexico City is located on the valley floor, it has always been subject to flooding. Subsidence has worsened this problem by lowering the land surface of Mexico City below the level of Lake Texcoco, resulting in increased flooding. Drainage systems had to be dug deeper and Lake Texcoco had to be excavated. By 1950, dikes had to be built to confine stormwater flow, and pumping was required to lift drainage water under the city to the level of the drainage canals. By 1953, severe subsidence resulted in the closing of many wells that had to be replaced with new wells.

The Mexico City Metropolitan Area (MCMA) has become a magnet of growth, being the cultural, economic, and industrial center for Mexico, with an estimated population approaching 22 million people. A continual migration of people from rural areas to the city has occurred, with many of the people settling illegally in the urban fringe with the hope of eventually being provided public services. Providing water supply and wastewater services for Mexico City is a formidable challenge. Imagine that the city has the largest population in the world living in an enclosed basin with no natural outflow to the sea. The water supply situation has reached a crisis level, with the continued urban growth and poor system of financing by the government. The consequences include an inability to expand the water supply network to areas that are underserved or not served at all, repair leaks, and provide wastewater treatment. Mexico City cannot meet the water demands of its population.

A case study by Dr. Blanca Jimenez (2008) focuses on water and wastewater management in Mexico City, Mexico, a mega-city experiencing many challenges of urban water management. Mexico City is located in the Mexico Valley, which is a basin of 9600 km^2 located 2240 m above sea level. The present-day water usage is 85.7 m^3/s, of which 48 percent is supplied by the water distribution network, 19 percent is directly pumped from the local aquifer by farmers and industry, and 9 percent is treated wastewater that is reused. Figure 2.6.4 shows the water sources for Mexico City. The first use water (78 m^3/s) comes from different sources: 57 m^3/s from 1965 wells in the local aquifer located mainly south and west of the city; 1 m^3/s from local rivers located in the southern part of the city; 5 m^3/s from an aquifer located in the Lerma region 100 km away and 300 m above Mexico City; and 15 m^3/s from the Cutzamala River located 130 km away and 1100 m below the city.

The water use varies by the social class of the population as presented in Table 2.6.1. Approximately 40 percent of the 62 m^3/s of water distributed to the population is lost due to leakage so that the per capita water use is only 153 L/capita/day instead of 255 L/capita/day.

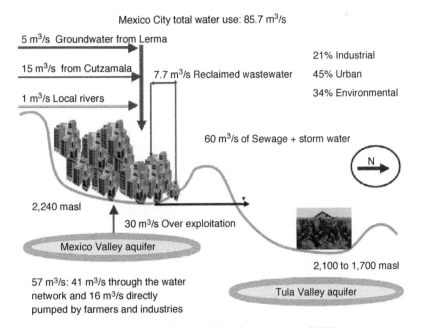

Figure 2.6.4 Water sources for Mexico City (from Jimenez, 2008).

Table 2.6.1   Water Use by Social Class in Mexico City

Social class	Water supply (L/capita/day)	% of the population	Total volume of water demanded (m^3/s)
Low	128	76.5	9.0
Medium	169	18.0	15.7
Medium	399	3.6	1.3
High	567	1.9	1.0
		Total	27.0

Source: DGCOH, 1998, as presented in Jimenez (2008).

The wastewater produced (approximately 67.7 m^3/s on a year-round basis), includes pluvial excess water (collected for only six months) and sewage. Total capacity of public wastewater treatment plants is 15 m^3/s, but only 11 percent (7.7 m^3/s) of the total produced is treated. All of the treated wastewater is reused to fill recreation lakes and canals (54 percent), irrigation (31 percent), industrial cooling (8 percent), diverse commercial uses (5 percent), and recharge (2 percent) (DGCOH, 1998). The system built to drain the excess pluvial water and sewage out of the Mexico Valley is shown in Figure 2.6.5. Three tunnels, the Western Interceptor, the Gran Canal, and the Central Canal were built in 1896, 1989,

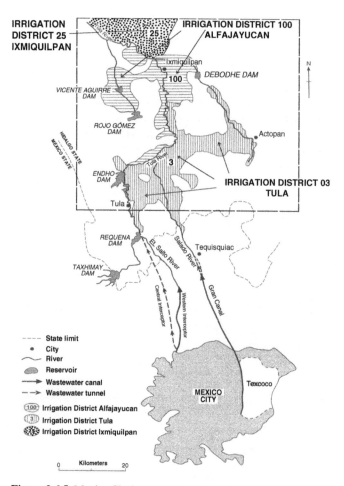

Figure 2.6.5 Mexico City's wastewater disposal drainage system and main components of the irrigation system in the Tula Valley (from Jimenez, 2008).

where μ (mu) is the dynamic viscosity and v is the velocity. The velocity gradient is the time rate of strain. Thus the definition of dynamic viscosity is the ratio of shear stress to the velocity gradient,

$$\mu = \frac{\tau}{dv/dy} \tag{3.1.3}$$

Kinematic viscosity is the ratio of the dynamic viscosity to the density in which the gradient force dimension cancels out in μ/ρ. The Greek symbol v (nu) is used to identify the kinematic viscosity

$$v = \mu/\rho \tag{3.1.4}$$

Kinematic viscosity has been defined because many equations include μ/ρ. Refer to Tables 3.1.1 and 3.1.2 for values of viscosity as a function of temperature.

The shear stress in fluids is involved with the cohesion forces between molecules. Stress applied to fluids causes motion, whereas solids can resist shear stress in a static condition. Considering flow of water in a pipe, the water near the center of the pipe has a greater velocity than the water near the wall. Shear force is increased or decreased in direct proportion to increases or decreases in relative velocity.

Shear stress has units of N/m^2. Dynamic viscosity has units of

$$\mu = \frac{\tau}{\dfrac{dv}{dy}} \equiv \frac{N/m^2}{(m/s)/m} = \frac{N \cdot s}{m^2}$$

Kinematic viscosity has units of

$$v = \frac{\mu}{\rho} \equiv \frac{N \cdot s/m^2}{N \cdot s/m^4} = \frac{m^2}{s}$$

The SI unit for dynamic viscosity is centipoise (cP), in which $1 \, cP = 1 \, N \cdot s/m^2 \times 10^{-3}$. The SI unit for kinematic viscosity is centistoke (cst), in which $1 \, cst = 1 \, m^2/s \times 10^{-6}$.

Ideal fluids are defined as the ones in which viscosity is zero, i.e., there is no friction. Such fluids do not exist in reality but the concept is useful in many types of fluid analysis. *Real fluids* do consider viscosity effects so that shear force exists whenever motion takes place, thus producing fluid friction.

3.1.3   Elasticity

Elasticity (or *compressibility*) is important when we talk about water hammer in the hydraulics of pipe flow. Elasticity of water is related to the amount of deformation (expansion or contraction) induced by a pressure change. Elasticity is characterized by the *bulk modulus of elasticity*, E, which is defined as the ratio of relative change in volume, $d\forall/\forall$, due to a differential change in pressure, dp, so that

$$E = -\frac{dp}{d\forall/\forall} \tag{3.1.5}$$

Also $-d\rho/\rho = d\forall/\forall$, so that

$$E = \frac{dp}{d\rho/\rho} \tag{3.1.6}$$

Refer to Tables 3.1.1 and 3.1.2 for values of the bulk modulus of elasticity as a function of temperature.

3.1.4   Pressure and Pressure Variation

Pressure, p, is the force F acting over an area A, denoted as

$$p = \lim_{\Delta A \to 0} \frac{\Delta F}{\Delta A} = \frac{dF}{dA}$$ (3.1.7)

Pressure at a point is equal in all directions. The pressure at a depth y (neglecting pressure on the surface of water) is

$$p = \gamma y$$ (3.1.8)

Units of pressure are N/m^2 (Pascal), lb/in^2 (psi), lb/ft^2, feet of water, and inches of mercury.

Gauge pressure uses atmospheric pressure as the datum. *Absolute pressure* is the pressure above absolute zero. *Vacuum* refers to pressure less than atmospheric pressure. At absolute zero, pressure is a perfect vacuum. Figure 3.1.1 illustrates the relationship among various pressures.

The *pressure force F* exerted by water on a plane area A is the product of the area and the pressure at its centroid, expressed as

$$F = pA = \gamma y_c A$$ (3.1.9)

where y_c is the vertical depth of the water over the centroid.

For static water, the only variation in pressure is with the elevation in the fluid, i.e.,

$$\frac{dp}{dz} = -\gamma$$ (3.1.10)

where z refers to elevation. Equation (3.1.10) is the basic equation for hydraulic pressure variation with elevation. For water on a horizontal plane, the pressure everywhere on this plane is constant. The greatest possible change in hydrostatic pressure occurs along a vertical path through water.

Considering the specific weight to be constant, equation (3.1.10) can be integrated to obtain

$$p = -\gamma z + \text{constant}$$ (3.1.11)

or

$$\left(\frac{p}{\gamma} + z\right) = \text{constant}$$ (3.1.12)

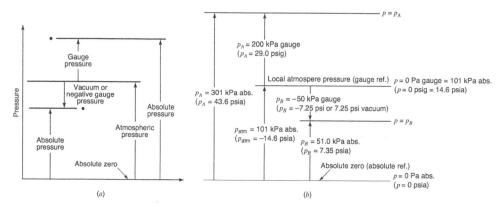

Figure 3.1.1 (*a*) Relationship between various pressures (from Chaudhry (1996)); (*b*) Example of pressure relation (from Roberson & Crowe (1993)).

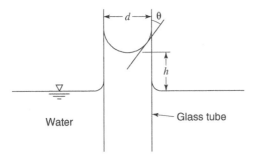

Figure 3.1.2 Capillary action. The effect of surface tension is illustrated for the capillary rise in a small glass tube. θ is the angle of the tangent to the meniscus where it contacts the wall of the tube. Surface tension force acts around the circumference of the tube.

The term $(p/\gamma + z)$ is called the *piezometric head*, which is then constant through any incompressible static fluid. The pressure and elevation at two different points in a static incompressible fluid are then

$$\frac{p_1}{\gamma} + z_1 = \frac{p_2}{\gamma} + z_2 \qquad (3.1.13)$$

3.1.5   Surface Tension

Molecules of water below the surface act on each other by forces that are equal in all directions. Molecules near the surface have a greater attraction for each other. Molecules on the surface are not able to bond in all directions and consequently form stronger bonds with adjacent molecules. The water surface acts like a stretched membrane seeking a minimum possible area by exerting a tension on the adjacent portion of the surface or an object in contact with the water surface. This *surface tension* acts in the plane of the surface as illustrated in Figure 3.1.2 for capillary action. Refer to Tables 3.1.1 and 3.1.2 for values of surface tension as a function of temperature.

3.1.6   Flow Visualization

There are two viewpoints on the motion of fluids, the *Eulerian* viewpoint and the *Lagrangian* viewpoint. The Lagrangian viewpoint focuses on the motion of individual fluid particles and follows these particles for all time. It is more common, however, in hydrologic and hydraulic processes to consider that fluids form a continuum wherein the motion of particles is not traced. This Eulerian viewpoint then focuses on a particular point or control volume in space and considers the motion of fluid that passes through as a function of time.

Streamlines are lines drawn through a fluid field so that the velocity vectors of the fluid at all points on the streamlines are tangent to the streamline at any instant in time. The tangent of the curve at any point along the streamline is the direction of the velocity vector at that point in the flow field. Examples of streamlines are shown in Figure 3.1.3. In the Eulerian viewpoint, the total velocity is expressed as a function of position along a streamline, x, and time, t.

$$\mathbf{V} = \mathbf{V}(x, t) \qquad (3.1.14)$$

A *uniform flow* is defined as one in which the velocity does not change from point to point along any of the streamlines in the flow field. Thus the streamlines are straight and parallel, so that

$$\frac{\partial \mathbf{V}}{\partial x} = 0 \text{ (uniform flow)} \qquad (3.1.15)$$

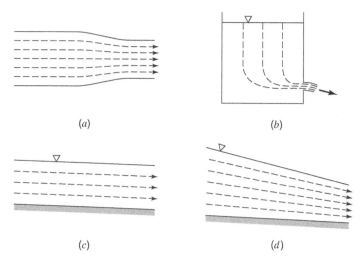

Figure 3.1.3 Streamlines. (*a*) Flow in a conduit; (*b*) Flow from a slot; (*c*) Open-channel flow (uniform); (*d*) Open-channel flow (nonuniform).

When streamlines are not straight there is a directional change in velocity. If they are not parallel, there is a change in speed along the streamlines. Under such circumstances the flow is *nonuniform flow* and

$$\frac{\partial V}{\partial x} \neq 0 \ \text{(nonuniform flow)} \tag{3.1.16}$$

So this flow pattern has streamlines that are curved in space (converging or diverging) as shown in Figure 3.1.3(d) for nonuniform open-channel flow.

The variation in velocity with respect to time at a given point in a flow field is also used to classify flow. *Steady flow* occurs when the velocity at a point in the flow field does not vary in magnitude or direction with respect to time:

$$\frac{\partial V}{\partial t} = 0 \ \text{(steady flow)} \tag{3.1.17}$$

Unsteady flow occurs when the velocity does vary in magnitude or direction at a point in the flow field with respect to time.

3.1.7 Laminar and Turbulent Flow

Turbulent flow is caused by eddies of varying size within the flow that create a mixing action. The fluid particles follow irregular and erratic paths and no two particles have similar motion. Turbulent flow then is irregular with no definite flow patterns. Flow in rivers is a good example of turbulent flow. The index used to relate to turbulence is the *Reynolds number*

$$R_e = \frac{VD\rho}{\mu} = \frac{VD}{\nu} \tag{3.1.18}$$

where D is a characteristic length such as the diameter of a pipe. For pipe flow, the flow is generally turbulent for $R_e > 2000$.

Laminar flow does not have the eddies that cause the intense mixing and therefore the flow is very smooth. The fluid particles move in definite paths and the fluid appears to move by the sliding of laminations of infinitesimal thickness relative to the adjacent layers. The viscous shear of the fluid particles produces the resistance to flow. The resistance to flow varies with the first power of the velocity. For pipe flow, the flow is generally laminar for $R_e < 2000$.

Flow can be *one-, two-,* or *three-dimensional flow,* for which one, two, or three coordinate directions, respectively, are required to describe the velocity and property changes in a flow field.

3.1.8   Discharge

Discharge, or *flow rate,* is the volume rate of flow that passes a given section in a flow stream. The flow velocity v varies across a flow field, as for the example of pipe flow in Figure 3.1.4. The rate of flow through a differential area dA is vdA so that the total volume can be expressed by integrating over the entire flow section as

$$Q = \int_A vdA \tag{3.1.19}$$

Using the mean velocity V, the discharge is defined as

$$Q = AV \tag{3.1.20}$$

By defining an area vector as one that has the magnitude of the area and is oriented normal to the area, then $V\cos\theta dA = \mathbf{V} \cdot d\mathbf{A}$. The discharge is then

$$Q = \int_A \mathbf{V} \cdot d\mathbf{A} \tag{3.1.21}$$

and for a constant (mean) velocity over the cross-sectional area of flow the discharge is

$$Q = \mathbf{V} \cdot \mathbf{A} \tag{3.1.22}$$

The *mass rate of flow* past a flow section is

$$\dot{m} = \int_A \rho vdA \tag{3.1.23a}$$

$$= \rho \int_A vdA \tag{3.1.23b}$$

$$= \rho Q \tag{3.1.23c}$$

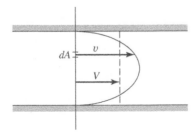

Figure 3.1.4 Velocity distribution in a pipe flow.

3.2   CONTROL VOLUME APPROACH FOR HYDROSYSTEMS

Hydrosystem processes transform the space and time distribution of water in hydrologic systems throughout the hydrologic cycle, in natural and human-made hydraulic systems, and in water resources systems that include both hydrologic and hydraulic systems. The commonality of all hydrosystems is the physical laws that define the flow of fluid in these systems. A consistent mechanism for developing these physical laws is called the *control volume approach*.

The simplified concept of a system is very important in the control volume approach because of the extreme complexity of hydrosystems. Typically a system defined from the fluids viewpoint is defined as a given quantity of mass. A *system* is also a set of connected parts that form a whole. For the present discussion the fluids viewpoint will be used, in which the system has a *system boundary* or *control surface* (CS) as shown in Figure 3.2.1. A control surface is the surface that surrounds the control volume. The control surface can coincide with physical boundaries such as the wall of a pipe or the boundary of a watershed. Part of the control surface may be a hypothetical surface through which fluid flows.

Two properties, *extensive properties* and *intensive properties*, are used in the control volume approach to apply physical properties for discrete masses to a fluid flowing continuously through a control volume. Extensive properties are related to the total mass of the system (control volume), whereas intensive properties are independent of the amount of fluid. The extensive properties are mass m, momentum mV, and energy E. Corresponding intensive properties are mass per unit mass, momentum per unit mass, which is velocity v, and energy per unit mass e. In other words, for an extensive property B, the corresponding intensive property β is defined as the quantity of B per unit mass, $\beta = dB/dm$. Both the extensive and intensive properties can be scalar or vector quantities.

The relationship between intensive and extensive properties for a given system is defined by the following integral over the system:

$$B = \int_{\text{system}} \beta \, dm = \int \beta \rho \, d\forall \tag{3.2.1}$$

where dm and $d\forall$ are the differential mass and differential volume, respectively, and ρ is the fluid density.

The volume rate of flow past a given area A is expressed as

$$Q = \mathbf{V} \cdot \mathbf{A} \tag{3.2.2}$$

where $\mathbf{V}$ is the velocity, directed normal to the area and points outward from the control volume, and $\mathbf{A}$ is the area vector.

For the control volume in Figure 3.2.1 the net flowrate $\dot{Q}$ is

$$\dot{Q} = Q_{\text{out}} - Q_{\text{in}}$$
$$= \mathbf{V}_2 \cdot \mathbf{A}_2 - \mathbf{V}_1 \cdot \mathbf{A}_1$$
$$= \sum_{\text{CS}} \mathbf{V} \cdot \mathbf{A} \tag{3.2.3}$$

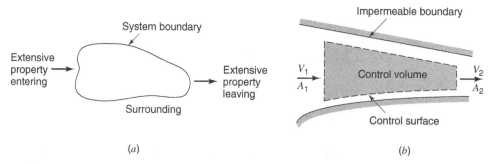

Figure 3.2.1 Control volume approach. (*a*) System and surrounding; (*b*) Control volume as a system.

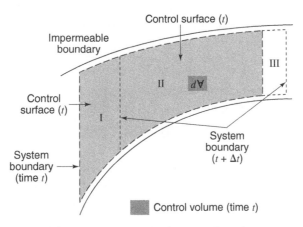

Figure 3.2.2 Control volume at times t and $t + \Delta t$.

In other words, the dot product $\mathbf{V} \cdot \mathbf{A}$ for all flows in and out of a control volume is the net rate of outflow.

The mass rate of flow out of the control volume is

$$\frac{dm}{dt} = \dot{m} = \sum_{\text{CS}} \rho \, \mathbf{V} \cdot \mathbf{A} \tag{3.2.4}$$

The rate of flow of extensive property B is the product of the mass rate and the intensive property:

$$\frac{dB}{dt} = \dot{B} = \sum_{\text{CS}} \beta \rho \, \mathbf{V} \cdot \mathbf{A} \tag{3.2.5}$$

If the velocity varies across the flow section, then it must be integrated across the section, so that the above equation for the rate of flow of extensive property $\dot{B}$ from the control volume becomes

$$\dot{B} = \int_{\text{CS}} \beta \rho \, \mathbf{V} \cdot d\mathbf{A} \tag{3.2.6}$$

Considering the system in Figure 3.2.2, the control volume is defined by the control surface at time $t(\text{I} + \text{II})$ with extensive property B_t. At time $t + \Delta t$ the control volume, defined by the control surface, $(\text{II} + \text{III})$ has moved and has extensive property $B_{t + \Delta t}$. The rate of change of extensive property B is

$$\frac{dB}{dt} = \lim_{\Delta t \to 0} \left[\frac{B_{t + \Delta t} - B_t}{\Delta t} \right] \tag{3.2.7}$$

The mass of the system at time $t + \Delta t$, $m_{\text{sys}, t + \Delta t}$, is

$$m_{\text{sys}, t + \Delta t} = m_{t + \Delta t} + \Delta m_{\text{out}} - \Delta m_{\text{in}} \tag{3.2.8}$$

where $m_{t + \Delta t}$ = mass of fluid within the control volume at time $t + \Delta t$
$\quad\quad\quad\quad \Delta m_{\text{out}}$ = mass of fluid that has moved out of the control volume in time Δt
$\quad\quad\quad\quad \Delta m_{\text{in}}$ = mass of fluid that has moved into the control volume in time Δt

The extensive property of the system at time $t + \Delta t$ is

$$B_{\text{sys}} = B_{\text{CV}, t + \Delta t} + \Delta B_{\text{out}} - \Delta B_{\text{in}} \tag{3.2.9}$$

where $B_{\text{CV}, t + \Delta t}$ = amount of extensive property in the control volume at time $t + \Delta t$
$\quad\quad\quad\quad \Delta B_{\text{out}}$ = amount of extensive property of the system that has moved out of the control volume in time Δt
$\quad\quad\quad\quad \Delta B_{\text{in}}$ = amount of extensive property of the system that has moved into the control volume in time Δt

The time rate of change of extensive property of the system is

$$\frac{dB_{sys}}{dt} = \lim_{\Delta t \to 0} \left[\frac{(B_{CV,t+\Delta t} + \Delta B_{out} - \Delta B_{in}) - B_{CV,t}}{\Delta t} \right] \tag{3.2.10}$$

The expression can be rearranged to yield

$$\frac{dB_{sys}}{dt} = \lim_{\Delta t \to 0} \left[\frac{B_{CV,t+\Delta t} - B_{CV,t}}{\Delta t} \right] + \lim_{\Delta t \to 0} \left[\frac{\Delta B_{out} - \Delta B_{in}}{\Delta t} \right]$$

$$= \left\{ \begin{array}{l} \text{Rate of change with} \\ \text{respect to time of} \\ \text{extensive property} \\ \text{in the control volume} \end{array} \right\} + \left\{ \begin{array}{l} \text{Net flow of} \\ \text{extensive property} \\ \text{from the control} \\ \text{volume} \end{array} \right\} \tag{3.2.11}$$

$$= \frac{dB_{CV}}{dt} + \frac{dB}{dt}$$

The derivative $\dfrac{dB_{CV}}{dt} = \dfrac{d}{dt} \displaystyle\int_{CV} \beta\rho \, d\forall$ and $\dfrac{dB}{dt}$ is defined by equation (3.2.5), so that the *control volume equation for one-dimensional flow* becomes

$$\frac{dB_{sys}}{dt} = \frac{d}{dt} \int_{CV} \beta\rho \, d\forall + \sum_{CS} \beta\rho \, \mathbf{V} \cdot \mathbf{A} \tag{3.2.12}$$

The above equation for the general control volume equation was derived for one-dimensional flow so that the rate of flow of B at each section is $\beta\rho\mathbf{V} \cdot \mathbf{A}$. A more general form for rate of flow of an extensive property considers the velocity as variable across a section. Using equation (3.2.6), then, the *general control volume equation* is expressed as

$$\frac{dB_{sys}}{dt} = \frac{d}{dt} \int_{CV} \beta\rho \, d\forall + \int_{CS} \beta\rho \, \mathbf{V} \cdot d\mathbf{A} \tag{3.2.13}$$

This general control volume equation (also referred to as the *Reynolds transport theorem*) states that the total rate of change of extensive property of a flow is equal to the rate of change of extensive property stored in the control volume, $\dfrac{d}{dt} \displaystyle\int_{CV} \beta\rho \, d\forall$, plus the net rate of outflow of extensive property through the control surface, $\displaystyle\int_{CS} \beta\rho \, \mathbf{V} \cdot d\mathbf{A}$.

Throughout this book the general control volume equation (approach) is applied to develop continuity, energy, and momentum equations for hydrosystem (hydrologic and hydraulic) processes.

3.3 CONTINUITY

In order to write the continuity equation, the extensive property is mass ($B = m$) and the intensive property $\beta = dB/dm = 1$. By the law of conservation of mass, the mass of a system is constant, therefore $dB/dt = dm/dt = 0$. The general form of the continuity equation is then

$$0 = \frac{d}{dt} \int_{CV} \rho \, d\forall + \int_{CS} \rho \, \mathbf{V} \cdot d\mathbf{A} \tag{3.3.1}$$

which is the *integral equation of continuity for an unsteady, variable-density flow*. Equation (3.3.1) can be rewritten as

$$\int_{CS} \rho \, \mathbf{V} \cdot d\mathbf{A} = \frac{d}{dt} \int_{CV} \rho \, d\forall \tag{3.3.2}$$

which states that the net rate of outflow of mass from the control volume is equal to the rate of decrease of mass within the control volume.

For flow with constant density, equation (3.3.2) can be expressed as

$$\int_{CS} \mathbf{V} \cdot d\mathbf{A} = \frac{d}{dt} \int_{CV} d\forall \tag{3.3.3}$$

The continuity equation for flow with a uniform velocity across the flow section and constant density is expressed as

$$\sum_{CS} \mathbf{V} \cdot \mathbf{A} = \frac{d}{dt} \int_{CV} d\forall \tag{3.3.4}$$

For a *constant-density, steady one-dimensional flow*, such as water flowing in a conduit, the velocity is the mean velocity, then

$$\sum_{CS} \mathbf{V} \cdot \mathbf{A} = 0 \tag{3.3.5}$$

For pipe conduit flow we consider a control volume between two locations of the pipe, at sections 1 and 2, then the continuity equation is

$$-V_1 A_1 + V_2 A_2 = 0 \tag{3.3.6a}$$

or

$$V_1 A_1 = V_2 A_2 \tag{3.3.6b}$$

or

$$Q_1 = Q_2 \tag{3.3.6c}$$

For a *constant-density unsteady flow*, consider the integral $\int_{CV} d\forall$ as the volume of fluid stored in a control volume denoted by S, so that

$$\frac{d}{dt} \int_{CV} d\forall = \frac{dS}{dt} \tag{3.3.7}$$

The net outflow is defined as

$$\int_{CS} \mathbf{V} \cdot d\mathbf{A} = \int_{\text{outlet}} \mathbf{V} \cdot d\mathbf{A} + \int_{\text{inlet}} \mathbf{V} \cdot d\mathbf{A} \tag{3.3.8}$$
$$= Q(t) - I(t)$$

Then the integral equation of continuity is determined by substituting equations (3.3.7) and (3.3.8) into equation (3.3.2) to obtain

$$Q(t) - I(t) = -\frac{dS}{dt} \tag{3.3.9}$$

which is more commonly expressed as

$$\frac{dS}{dt} = I(t) - Q(t) \tag{3.3.10}$$

This continuity expression is used extensively in describing hydrologic processes.

EXAMPLE 3.3.1

A river section is defined by two bridges. At a particular time the flow at the upstream bridge is 100 m³/s, and at the same time the flow at the downstream bridge is 75 m³/s. At this particular time, what is the rate at which water is being stored in the river section, assuming no losses?

SOLUTION

Using the continuity equation (3.3.10) yields

$$\frac{dS}{dt} = Q_{up}(t) - Q_{down}(t)$$
$$= 100 \text{ m}^3/\text{s} - 75 \text{ m}^3/\text{s}$$
$$= 25 \text{ m}^3/\text{s}$$

EXAMPLE 3.3.2

A reservoir has the following monthly inflows and outflows in relative units:

Month	J	F	M	A
Inflows	10	5	0	5
Outflows	5	5	10	0

If the reservoir contains 30 units of water in storage at the beginning of the year, how many units of water in storage are there at the end of April?

SOLUTION

The continuity equation (3.3.10) is used to perform a routing of flows into and out of the reservoir. Because the inflow and outflows are for discrete time intervals, the continuity equation (3.3.10) can be reformulated as

$$dS = I(t)dt - Q(t)dt$$

and integrated over time intervals $j = 1, 2, \ldots J$ of each length Δt:

$$\int_{S_{j-1}}^{S_j} dS = \int_{(j-1)\Delta t}^{j\Delta t} I(t)dt - \int_{(j-1)\Delta t}^{j\Delta t} Q(t)dt$$

or

$$S_j - S_{j-1} = I_j - Q_j \text{ for } j = 1, 2, \ldots$$
$$\Delta S_j = I_j - Q_j$$

where I_j and Q_j are the volumes of inflow and outflow for the jth time interval. The cumulative storage is $S_{j+1} = S_j + \Delta S_j$. For the first interval of time,

$$\Delta S_1 = I_1 - Q_1 = 10 - 5 = 5$$

Then $S_2 = S_1 + \Delta S_1 = 30 + 5 = 35$. The remaining computations are:

Time	I_j	Q_j	ΔS_j	S_j
1	10	5	5	30
2	5	5	0	35
3	0	10	−10	25
4	5	0	5	30

3.4 ENERGY

This section uses the first law of thermodynamics along with the control volume approach to develop the energy equation for fluid flow in hydrologic and hydraulic processes. An energy balance for hydrologic and hydraulic processes considers an accounting of all inputs and outputs of energy to and from a system. By the *first law of thermodynamics*, the rate of change of energy, E, with time is the rate at which heat is transferred into the fluid, dH/dt, minus the rate at which the fluid does work on the surroundings, dW/dt, expressed as

$$\frac{dE}{dt} = \frac{dH}{dt} - \frac{dW}{dt} \qquad (3.4.1)$$

The total energy of a fluid system is the sum of the internal energy E_u, the kinetic energy E_k, and the potential energy E_p; thus

$$E = E_u + E_k + E_p \qquad (3.4.2)$$

The extensive property is the amount of energy in the system, $B = E$:

$$B = E_u + E_k + E_p \qquad (3.4.3)$$

and the intensive property is

$$\beta = \frac{dB}{dm} = e = e_u + e_k + e_p \qquad (3.4.4)$$

where e represents the energy per unit mass. Also, the rate of change of extensive property with respect to time is

$$\frac{dB}{dt} = \frac{dE}{dt} = \frac{dH}{dt} - \frac{dW}{dt} \qquad (3.4.5)$$

The *energy balance equation* is now derived by substituting β (equation (3.4.4)) and dB/dt (equation (3.4.5)) into the general control volume equation (3.2.12),

$$\frac{dE}{dt} = \frac{dH}{dt} - \frac{dW}{dt} = \frac{d}{dt} \int_{CV} e\rho d\forall + \sum_{CS} e\rho \, \mathbf{V} \cdot \mathbf{A} \qquad (3.4.6)$$

Next we can replace e by equation (3.3.4):

$$\frac{dH}{dt} - \frac{dW}{dt} = \frac{d}{dt} \int_{CV} (e_u + e_k + e_p)\rho \, d\forall + \sum_{CS} (e_u + e_k + e_p)\rho \, \mathbf{V} \cdot \mathbf{A} \qquad (3.4.7)$$

The kinetic energy per unit mass e_k is the total kinetic energy of mass with velocity V divided by the mass m:

$$e_k = \frac{mV^2/2}{m} = \frac{V^2}{2} \qquad (3.4.8)$$

The potential energy per unit mass e_p is the weight of the fluid $\gamma \forall$ times the centroid elevation z of the mass divided by the mass:

$$e_p = \frac{\gamma \forall z}{m} = \frac{\gamma \forall z}{\rho \forall} = gz \qquad (3.4.9)$$

because $\gamma/\rho = g$.

Now the *general energy equation for unsteady variable density flow* can be written as

$$\frac{dH}{dt} - \frac{dW}{dt} = \frac{d}{dt} \int_{CV} \left(e_u + \frac{1}{2}V^2 + gz \right)\rho d\forall + \sum_{CS} \left(e_u + \frac{1}{2}V^2 + gz \right)\rho \mathbf{V} \cdot \mathbf{A} \qquad (3.4.10)$$

For steady flow, equation (3.4.10) reduces to

$$\frac{dH}{dt} - \frac{dW}{dt} = \sum_{CS}\left(e_u + \frac{1}{2}V^2 + gz\right)\rho \mathbf{V} \cdot \mathbf{A} \tag{3.4.11}$$

The work done by a system on its surroundings can be divided into *shaft work, W_s, and flow work, W_f.* Flow work is the result of pressure force as the system moves through space and shaft work is any other work besides the flow work. In the control volume in Figure 3.2.2 the force on the upstream end of the fluid is $p_1 A_1$ and the distance traveled over time Δt is $l_1 = V_1 \Delta t$. Work done on the surrounding fluid as a result of this force is then the product of the force $p_1 A_1$ in the direction of motion and the distance traveled, $V_1 \Delta t$. The work force on the upstream end is then

$$W_{f_1} = -V_1 p_1 A_1 \Delta t \tag{3.4.12a}$$

and on the downstream end is

$$W_{f_2} = V_2 p_2 A_2 \Delta t \tag{3.4.12b}$$

At the upstream end, a negative sign must be used because the pressure force on the surrounding fluid acts in the opposite direction to the motion of the system boundary. The rate of work at the upstream and downstream ends are, respectively,

$$\frac{dW_{f_1}}{dt} = -V_1 p_1 A_1 \tag{3.4.13}$$

and

$$\frac{dW_{f_2}}{dt} = V_2 p_2 A_2 \tag{3.4.14}$$

The rate of flow work can then be expressed in general terms as

$$\frac{dW_f}{dt} = p\,\mathbf{V} \cdot \mathbf{A} \tag{3.4.15}$$

or for all streams passing through the control volume as

$$\frac{dW_f}{dt} = \sum_{CS} p\,\mathbf{V} \cdot \mathbf{A} = \sum_{CS}\frac{p}{\rho}\,\rho\mathbf{V} \cdot \mathbf{A} \tag{3.4.16}$$

The net rate of work on the system can now be expressed as

$$\frac{dW}{dt} = \frac{dW_s}{dt} + \sum_{CS}\frac{p}{\rho}\,\rho\mathbf{V} \cdot \mathbf{A} \tag{3.4.17}$$

Using equation (3.4.17), the *general energy equation (3.4.10) for unsteady variable density flow* can be expressed as

$$\frac{dH}{dt} - \frac{dW_s}{dt} - \sum_{CS}\frac{p}{\rho}\rho\mathbf{V} \cdot \mathbf{A} = \frac{d}{dt}\int_{CV}\left(e_u + \frac{1}{2}V^2 + gz\right)\rho\,d\forall + \sum_{CS}\left(e_u + \frac{1}{2}V^2 + gz\right)\rho\,\mathbf{V} \cdot \mathbf{A}$$

$$\tag{3.4.18}$$

which can be written as

$$\frac{dH}{dt} - \frac{dW_s}{dt} = \frac{d}{dt}\int_{CV}\left(e_u + \frac{1}{2}V^2 + gz\right)\rho\,d\forall + \sum_{CS}\left(\frac{p}{\rho} + e_u + \frac{1}{2}V^2 + gz\right)\rho\,\mathbf{V} \cdot \mathbf{A} \tag{3.4.19}$$

For steady flow, equation (3.4.19) reduces to

$$\frac{dH}{dt} - \frac{dW_s}{dt} = \sum_{CS} \left(\frac{p}{\rho} + e_u + \frac{1}{2}V^2 + gz \right) \rho\, \mathbf{V} \cdot \mathbf{A} \qquad (3.4.20)$$

EXAMPLE 3.4.1

Determine an expression based upon the energy concept that relates the pressures at the upstream and downstream ends of a nozzle assuming steady flow, neglecting change in internal energy, and assuming $dH/dt = 0$ and $dW_s/dt = 0$.

SOLUTION

Using the energy equation (3.4.20) for steady flow yields

$$\frac{dH}{dt} - \frac{dW_s}{dt} = \sum_{CS} \left(\frac{p}{\rho} + e_u + \frac{1}{2}V^2 + gz \right) \rho\, \mathbf{V} \cdot \mathbf{A}$$

Neglecting dH/dt and dW_s/dt the above energy equation can be expressed as

$$\int_{A_2} \left(\frac{p_2}{\rho} + e_{u_2} + \frac{1}{2}V_2^2 + gz_2 \right) \rho\, V_2\, dA_2 - \int_{A_1} \left(\frac{p_1}{\rho} + e_{u_1} + \frac{1}{2}V_1^2 + gz_1 \right) \rho\, V_1 dA_1 = 0$$

which can be modified to

$$\int_{A_2} \left(\frac{p_2}{\rho} + e_{u_2} + gz_2 \right) \rho\, V_2 dA_2 + \int_{A_2} \frac{\rho\, V_2^3}{2} dA_2 - \int_{A_1} \left(\frac{p_1}{\rho} + e_{u_1} + gz_1 \right) \rho\, V_1\, dA_1 - \int_{A_1} \frac{\rho\, V_1^3}{2} dA_1 = 0$$

For hydrostatic conditions, $\left(\dfrac{p}{\rho} + e_u + gz \right)$ is constant across the system, which allows these terms to be taken outside the integral:

$$\left(\frac{p_2}{\rho} + e_{u_2} + gz_2 \right) \int_{A_2} \rho\, V_2\, dA_2 + \int_{A_2} \frac{\rho\, V_2^3}{2} dA_2 - \left(\frac{p_1}{\rho} + e_{u_1} + gz_1 \right) \int_{A_1} \rho V_1\, dA_1 - \int_{A_1} \frac{\rho\, V_1^3}{2} dA_1 = 0$$

The term $\int \rho V dA$ is the mass rate of flow, $\dot{m}$, and the term $\int \dfrac{\rho\, V^3}{2} dA = \dot{m}\dfrac{V^2}{2}$, so

$$\left(\frac{p_2}{\rho} + e_{u_2} + gz_2 \right) \dot{m} + \dot{m}\frac{V_2^2}{2} - \left(\frac{p_1}{\rho} + e_{u_1} + gz_1 \right) \dot{m} - \dot{m}\frac{V_1^2}{2} = 0$$

Dividing through by $\dot{m}g$ and rearranging yields

$$\frac{p_1}{\rho g} + \frac{e_{u_1}}{g} + z_1 + \frac{V_1^2}{2g} = \frac{p_2}{\rho g} + \frac{e_{u_2}}{g} + z_2 + \frac{V_2^2}{2g}$$

$\gamma = \rho g$ and rearranging yields

$$\frac{p_1}{\gamma} + \frac{V_1^2}{2g} + z_1 = \frac{p_2}{\gamma} + \frac{V_2^2}{2g} + z_2 + \frac{e_{u_2} - e_{u_1}}{g}$$

Neglecting changes in internal energy, $(e_{u_2} - e_{u_1})/g = 0$

$$\frac{p_1}{\gamma} + \frac{V_1^2}{2g} + z_1 = \frac{p_2}{\gamma} + \frac{V_2^2}{2g} + z_2$$

Assuming the control volume is horizontal, $z_1 = z_2$, then

$$\frac{p_1}{\gamma} + \frac{V_1^2}{2g} = \frac{p_2}{\gamma} + \frac{V_2^2}{2g}$$

This energy equation relates the pressures assuming steady flow, $z_1 = z_2$, neglecting change of internal energy in the fluid and assuming $dH/dt = 0$ and $dW_s/dt = 0$.

EXAMPLE 3.4.2

For a nozzle, determine the pressure change through the nozzle between the upstream and downstream end of the nozzle. Assume steady flow, neglect changes in internal energy of the fluid, assume $dH/dt = 0$ and $dW_s/dt = 0$, and say that the nozzle is horizontal. Assume the temperature is 20°C. The velocities at the entrance and exit are $V_2 = 2.55$ m/s and $V_1 = 1.13$ m/s, respectively.

SOLUTION

Using the energy equation derived in example 3.4.1 yields

$$\frac{p_1}{\gamma} + \frac{V_1^2}{2g} = \frac{p_2}{\gamma} + \frac{V_2^2}{2g}$$

$$p_1 - p_2 = \left(V_2^2 - V_1^2\right)\frac{\gamma}{2g}$$

$$= \left[(2.55)^2 - (1.13)^2\right] \times \frac{9.79 \text{ kN/m}^3}{2 \times 9.81 \text{ m/s}^2}$$

$$= (5.226 \text{ m}^2/\text{s}^2)(0.499 \text{ kN s}^2/\text{m}^4)$$

$$= 2.608 \text{ kN/m}^2 = 2.608 \text{ kPa} = 2608 \text{ Pa}$$

The pressure change is a pressure decrease of 2608 Pa.

3.5 MOMENTUM

In order to derive the general momentum equation for fluid flow in a hydrologic or hydraulic system, we use the control volume approach along with Newton's second law. *Newton's second law* states that the summation of all external forces on a system is equal to the rate of change of momentum of the system

$$\sum \mathbf{F} = \frac{d(\text{momentum})}{dt} \tag{3.5.1}$$

To apply the control volume approach the extensive property is momentum, $B = m\mathbf{v}$, and the intensive property is the momentum per unit mass, $\beta = d(m\mathbf{v})/dt$, so

$$\sum \mathbf{F} = \frac{d(m\mathbf{v})}{dt} \tag{3.5.2}$$

A lowercase $\mathbf{v}$ is used to denote that this velocity is referenced to the inertial reference frame and to distinguish it from $\mathbf{V}$.

Using the general control volume equation (3.2.13),

$$\frac{dB_{\text{sys}}}{dt} = \frac{d}{dt} \int_{CV} \beta\rho \, d\forall + \int_{CS} \beta\rho \mathbf{V} \cdot d\mathbf{A} \tag{3.2.13}$$

and from equation (3.5.2) then

$$\sum \mathbf{F} = \frac{d}{dt} \int_{CV} \mathbf{v}\rho \, d\forall + \int_{CS} \mathbf{v}\rho \, \mathbf{V} \cdot d\mathbf{A} \tag{3.5.3}$$

which is the *integral momentum equation for fluid flow*. For steady flow, equation (3.5.3) reduces to

$$\sum \mathbf{F} = \int_{CS} \mathbf{v}\rho \, \mathbf{V} \cdot d\mathbf{A} \tag{3.5.4}$$

When a uniform velocity occurs in the stream crossing the control surface, the integral momentum equation is

$$\sum \mathbf{F} = \frac{d}{dt}\int_{CV} \mathbf{v}\,\rho\,d\forall + \sum_{CS} \mathbf{v}\rho\,\mathbf{V}\cdot\mathbf{A} \tag{3.5.5}$$

The momentum can be written for the coordinate directions x, y, and z in the Cartesian coordinate system as

$$\sum F_x = \frac{d}{dt}\int_{CV} v_x\rho\,d\forall + \sum_{CS} v_x(\rho\mathbf{V}\cdot\mathbf{A}) \tag{3.5.6}$$

$$\sum F_y = \frac{d}{dt}\int_{CV} v_y\rho\,d\forall + \sum_{CS} v_y(\rho\mathbf{V}\cdot\mathbf{A}) \tag{3.5.7}$$

$$\sum F_z = \frac{d}{dt}\int_{CV} v_z\rho\,d\forall + \sum_{CS} v_z(\rho\mathbf{V}\cdot\mathbf{A}) \tag{3.5.8}$$

For a steady flow the time derivative in equation (3.5.6) drops out, yielding

$$\sum \mathbf{F} = \sum_{CS} \mathbf{v}\rho\,\mathbf{V}\cdot\mathbf{A} \tag{3.5.9}$$

For a steady flow in which the cross-sectional area of flow does not change along the length of the flow, $\sum_{CS} v_\rho\mathbf{V}\cdot\mathbf{A} = 0$ (referred to as uniform flow), equation (3.5.9) reduces to

$$\sum \mathbf{F} = 0 \tag{3.5.10}$$

3.6   PRESSURE AND PRESSURE FORCES IN STATIC FLUIDS

In section 3.1.4, pressure, absolute pressure, gauge pressure, piezometric head, and pressure force were defined. This section extends that conversation to hydrostatic forces on submerged surfaces and buoyancy.

3.6.1   Hydrostatic Forces

Hydraulic engineers have many engineering applications in which they have to compute the force being exerted on submerged surfaces. The hydrostatic force on any submerged plane surface is equal to the product of the surface area and the pressure acting at the centroid of the plane surface. Consider the force on the plane surface shown in Figure 3.6.1. This plane surface can be divided into an infinite number of differential horizontal planes with width dy and area dA. The distance to the incremental area from the axis O–O is y. The pressure on dA is $p = \gamma y \sin\theta$ so that the force dF is $dF = pdA = \gamma y \sin\theta\, dA$. The force on the entire submerged plane is obtained by integrating the differential force on the differential area:

$$F = \int_A \gamma y \sin\theta\, dA \tag{3.6.1a}$$

$$= \gamma \sin\theta \int_A y\, dA \tag{3.6.1b}$$

$$= \gamma \sin\theta\, y_c A \tag{3.6.1c}$$

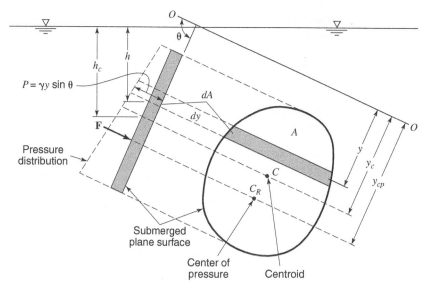

Figure 3.6.1 Hydrostatic pressure on a plane surface.

where γ and $\sin \theta$ are constants. The integral $\int_A y \, dA$ is by definition the first moment of the area and

$\int_A y \, (dA/A) = y_c$ is the distance from the O–O axis to the centroid (center of gravity) of the submerged plane. The vertical distance to the centroid can be defined as $h_c = y_c \sin \theta$, so that the force on the submerged plane is

$$F = \gamma h_c A \tag{3.6.2}$$

Engineers are normally interested in the forces that are in excess of the ambient atmospheric pressures. Keep in mind that atmospheric pressure, for most applications, acts on both sides of the submerged surface so that gauge pressure is of importance.

Even though pressure forces acting on a submerged surface are distributed throughout the surface, engineers are interested in the location of the *center of pressure*, which is the point on the submerged surface where the resultant force acts. The moment equation is

$$y_{cp}F = \int y \, dF \tag{3.6.3}$$

where $dF = pdA$, so

$$y_{cp}F = \int_A yp \, dA \tag{3.6.4}$$

and $p = \gamma y \sin \theta$, so

$$y_{cp}F = \int_A \gamma y^2 \sin \theta \, dA = \gamma \sin \theta \int_A y^2 dA \tag{3.6.5}$$

The integral $\int_A y^2 dA = I_o$ is the *moment of inertia* (*moment of the area*), with respect to an axis formed by the interaction of the plane containing the surface and the free surface. This can also be expressed with respect to the horizontal centroidal axis of the area by the *parallel axis theorem* as

$$I_o = \bar{I} + y_c^2 A \tag{3.6.6}$$

Equation (3.6.5) can now be expressed as

$$y_{cp}F = \gamma \sin\theta \, I_o = \gamma \sin\theta \, (\bar{I} + y_c^2 A) \tag{3.6.7}$$

Substituting equation (3.6.1c) and solving for y_{cp} yields

$$y_{cp} = y_c + \frac{\bar{I}}{y_c A} \tag{3.6.8}$$

The vertical distance to the center of pressure h_{cp} is then

$$h_{cp} = y_{cp} \sin\theta \tag{3.6.9}$$

EXAMPLE 3.6.1

Derive the expression for the depth to the center of pressure y_{cp} for a rectangular area $(b \times h)$ vertically submerged with the long side (h) at the liquid surface.

SOLUTION

Using equation (3.6.8),

$$y_{cp} = y_c + \frac{\bar{I}}{y_c A}$$

where $y_c = h/2$, $\bar{I} = bh^3/12$, and $A = bh$, we get

$$y_{cp} = \frac{h}{2} + \frac{bh^3/12}{\left(\frac{h}{2}\right)(bh)} = \frac{h}{2} + \frac{h}{6} = \frac{4}{6}h = \frac{2}{3}h$$

EXAMPLE 3.6.2

Determine the hydrostatic force and the location of the center of pressure on the 25 m long dam shown in Figure 3.6.2. The face of the dam is at an angle of 60°. Assume 20°C.

SOLUTION

The diagram in Figure 3.6.2 shows the pressure distribution. Using equation (3.6.2), $h_c = 2.5$ m and $A = (25\text{ m} \times 5)/\sin 60°$, so the hydrostatic force is

$$F = \gamma h_c A = \left(9.79 \frac{\text{kN}}{\text{m}^3}\right)(2.5\text{ m})\left(25\text{ m} \times \frac{5}{\sin 60°}\text{ m}\right)$$

$$= 3.532\text{ kN}$$

The center of pressure is at 2/3 of the total water depth, $(2/3) \times 5 = 3.33$ m.

EXAMPLE 3.6.3

Consider a vertical rectangular gate $(b = 4\text{ m and } h = 2\text{ m})$ that is vertically submerged in water so that the top of the gate is 4 m below the surface of the water (as shown in Figure 3.6.3). Determine the total resultant force on the gate and the location of the center of pressure.

SOLUTION

Use the free-body diagram in Figure 3.6.3. The total resultant force is computed using equation (3.6.2), $F = \gamma h_c A$, where $h_c = 4 + (2/2) = 5$ m:

$$F = (9.79\text{ kN/m}^3)(5\text{ m})(4 \times 2\text{ m}^2) = 396.1\text{ kN}$$

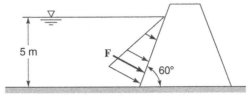

Figure 3.6.2 Hydrostatic force on dam for example 3.6.2.

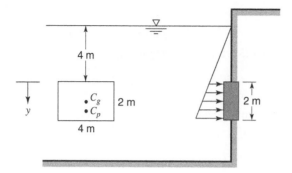

Figure 3.6.3 Vertical rectangular gate for example 3.6.3.

The location of the center of pressure is computed using equation (3.6.8):

$$y_{cp} = y_c + \frac{\bar{I}}{y_cA} = 5 + \frac{\dfrac{4 \times 2^3}{12}}{5 \times (4 \times 2)}$$

$$= 5.067 \, \text{m}$$

Alternatively, this problem can be solved by the simple integration $F = \displaystyle\int_0^2 \gamma \, hdA$, where $dA = 4 \, dy$:

$$F = \int_0^2 \gamma hdA = \int_0^2 (9.79)(4+y)(4dy)$$

$$= 39.16\left[4y + \frac{y^2}{2}\right]_0^2 = 39.16\left[4 \times 2 + \frac{2^2}{2}\right]$$

$$= 391.6 \, \text{kN}$$

EXAMPLE 3.6.4   Consider an inclined rectangular gate with water on one side as shown in Figure 3.6.4. Determine the total resultant force acting on the gate and the location of the center of pressure.

SOLUTION   To determine the total resultant force, $F = \gamma h_c A$, where $h_c = 5 + 1/2(4 \cos 60°)$, so that

$$F = (62.4)\left[5 + \frac{1}{2}(4 \cos 60°)\right](4 \times 6) = 8,986 \, \text{lb}$$

The location of the center of pressure is

$$y_{cp} = y_c + \frac{\bar{I}}{y_cA}$$

where $y_c = \dfrac{5}{\cos 60°} + \dfrac{4}{2} = 12 \, \text{ft}$, $\bar{I} = bh^3/12 = 6 \times 4^3/12 = 32 \, \text{ft}^4$ and $A = 6 \times 4 = 24 \, \text{ft}^2$:

$$y_{cp} = 12 + \frac{32}{12 \times 24} = 12.11 \, \text{ft}$$

Using equation (3.6.9), $h_{cp} = y_{cp} \sin \theta = 12.11(\sin 30°) = 6.06 \, \text{ft}$

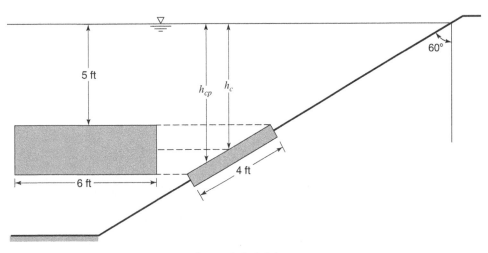

Figure 3.6.4 Inclined rectangular gate for example 3.6.4.

3.6.2 Buoyancy

The submerged body in Figure 3.6.5 is acted upon by gravity and the pressure of the surrounding fluid. On the upper surface of the submerged body, the vertical force is F_y and is equal to the weight of the volume ABCD above the surface. The vertical component of force F_y' on the bottom is the weight of the volume of fluid ABCED. The difference between the two volumes ABCD and ABCED is the volume of the submerged body. Applying the momentum principle, from equation (3.5.7) we get

$$\sum F_y = 0 \tag{3.6.10}$$

The buoyant force F_b is the weight of the volume of fluid DCE and is equal to the weight of the volume of fluid displaced, so that

or

$$F_b - F_y' + F_y = 0 \tag{3.6.11a}$$

$$F_b = F_y' - F_y \tag{3.6.11b}$$

Archimedes' principle (about 250 B.C.) states that the weight of a submerged body is reduced by an amount equal to the weight of liquid displaced by the body. This principle may be viewed as the difference of vertical pressure forces on the two surfaces DC and DEC. Floating bodies are partially submerged due to the balance of the body weight and buoyancy force.

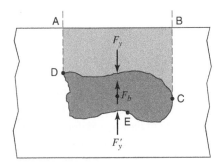

Figure 3.6.5 Forces on a submerged body. Buoyant force, F_b, passes through the centroid of the displaced volume and acts through a point called the center of buoyancy.

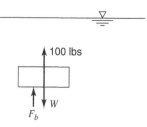

Figure 3.6.6 Free-body diagram for example 3.6.6.

EXAMPLE 3.6.5

A metal block weighs 400 N in air, but when completely submerged in water it weighs 250 N. What is the volume of the metal block?

SOLUTION

Essentially the buoyant force F_b is equal to the weight of water displaced by the metal block, i.e.,

$$F_b = 400\,\text{N} - 250\,\text{N} = 150\,\text{N}$$

The weight $W = (9.79)(1000)\,\forall$, where $\forall$ is the volume.

$$150 = (9.79)(1000)\,\forall$$
$$\forall = 0.0153\,\text{m}^3$$

EXAMPLE 3.6.6

An object is 1 ft thick by 1 ft wide by 2 ft long. It weighs 100 lb at a depth of 10 ft. What is the weight of the object in air and what is its specific gravity?

SOLUTION

Use the free-body diagram in Figure 3.6.6. The summation of forces acting on the object in the vertical direction is

$$\sum F_y = 100 + F_b - W = 0$$

where F_b is the buoyant force and W is the weight of the object.

$$F_b = \left(62.4\,\frac{\text{lb}}{\text{ft}^3}\right)(1\ \text{ft})(1\ \text{ft})(2\ \text{ft})$$
$$= 124.8\,\text{lb}$$
$$100 + 124.8 - W = 0$$
$$W = 224.8\,\text{lb}$$

The specific gravity is $224.8/124.8 = 1.8$.

3.7   VELOCITY DISTRIBUTION

We discussed in section 3.1.8 that the actual velocity varies throughout a flow section (see Figure 3.1.4 for pipe flow as an illustration). Figure 3.7.1 illustrates velocity profiles in various open-channel flow sections. As a result of these nonuniform velocity distributions in pipe flow and open-channel flow, the velocity head is generally greater than the value computed according to $V^2/2g$ where V is the mean velocity. When using the energy principle, the true velocity head is expressed as $\alpha V^2/2g$, where α is a *kinetic energy correction factor*. Chow (1959) also referred to α as an energy coefficient or *Coriolis coefficient*.

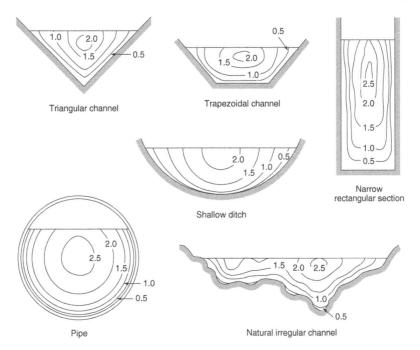

Figure 3.7.1 Typical curves of equal velocity in various channel sections (from Chow (1959)).

Consider the velocity distribution shown in Figure 3.1.4. The mass of fluid flowing through an area dA per unit time is $(\gamma/g)vdA$, where v is the velocity through area dA. The flow of kinetic energy per unit time through this area is $(\gamma/g)vdA(v^2/2) = (\gamma/2g)v^3dA$. For a known velocity distribution, the total kinetic energy flowing through the section per unit time is

$$\begin{array}{c}\text{Total kinetic energy}\\ \text{(per unit time)}\end{array} = \frac{\gamma}{2g}\int_A v^3\,dA \tag{3.7.1}$$

Using the mean flow velocity V and the coefficient α, the total energy per unit weight is $\alpha V^2/2g$; because the flow across the entire section is γAV. The total kinetic energy transmitted is

$$\begin{aligned}\begin{array}{c}\text{Total kinetic energy}\\ \text{(per unit time)}\end{array} &= (\gamma AV)\left(\alpha\frac{V^2}{2g}\right)\\ &= \gamma\alpha A\frac{V^3}{2g}\end{aligned} \tag{3.7.2}$$

From equations (3.7.1) and (3.7.2) we get

$$\frac{\gamma}{2g}\int_A v^3\,dA = \gamma\alpha A\frac{V^3}{2g} \tag{3.7.3}$$

and we then solve for the kinetic energy correction factor:

$$\alpha = \frac{1}{AV^3}\int_A v^3\,dA \tag{3.7.4}$$

The value of α for flow in circular pipes flowing full with a parabolic velocity distribution is equal to 2 for laminar flow and normally ranges from 1.03 to 1.06 for turbulent flow. Because α is not known precisely, it is not commonly used in pipe flow calculations, and the kinetic energy of fluid per unit weight is $V^2/2g$. The values of α for open-channel flow varies by the type of channel flow. For example, in regular channels, flumes, and spillways, α ranges between 1.10 and 1.20, and for river valleys and areas α ranges between 1.5 and 2.0 with an average of 1.75 (see Chow (1959)).

The nonuniform distribution of velocity also affects the computation of momentum in open-channel flow. The corrected momentum of water passing through a channel section per unit time is

$$\begin{aligned}\text{Total momentum} &= \beta_m \frac{\gamma}{g} QV \\ \text{(per unit time)} \\ &= \beta_m \frac{\gamma}{g} AV^2\end{aligned} \qquad (3.7.5)$$

where β_m is the *momentum correction factor*, also called the *momentum coefficient* or *Boussinesq coefficient* by Chow (1959). The momentum of water passing through an elemental area dA per unit time is the product of the mass per unit time $(\gamma/g)v\,dA$ and the velocity v, which is $(\gamma/g)v^2\,dA$. The total momentum of fluid per unit time is

$$\begin{aligned}\text{Total momentum} \\ \text{(per unit time)}\end{aligned} = \frac{\gamma}{g}\int_A v^2\,dA \qquad (3.7.6)$$

From equations (3.7.5) and (3.7.6), we get

$$\beta_m \frac{\gamma}{g} AV^2 = \frac{\gamma}{g}\int_A v^2\,dA \qquad (3.7.7)$$

Solving for the momentum correction factor β_m yields

$$\beta_m = \frac{1}{AV^2}\int_A v^2\,dA \qquad (3.7.8)$$

According to Chow (1959), the value of β_m for fairly straight prismatic channels varies between 1.01 to 1.12 and for river valleys β_m varies between 1.17 and 1.33.

PROBLEMS

3.1.1 If 10 m^3 of a liquid weighs 60 kN, calculate its specific weight, density, and specific gravity.

3.1.2 The specific gravity of a certain oil is 0.7. Calculate its specific weight and density in SI and English units.

3.1.3 What is the unit of force in SI units? What is it identical to?

3.1.4 What is the unit of mass in English units? What is it identical to?

3.1.5 A rocket weighing 9810 N on earth is landed on the moon where the acceleration due to gravity is approximately one-sixth that at the earth's surface. What are the mass of the rocket on the earth and the moon, and the weight of the rocket on the moon?

3.1.6 The summer temperature in Arizona often exceeds 110°F. What is the Celsius equivalent temperature?

3.1.7 What will be the percentage change in the density of water if the temperature changes from 20°C to (a) 50°C and (b) 80°C?

3.1.8 The specific gravity and kinematic viscosity of a certain liquid are 1.5 and 4×10^{-4} m^2/s, respectively. What is its dynamic viscosity?

3.1.9 If the dynamic viscosity of a liquid is 2.09×10^{-5} lb $\cdot$ s/ft^2, what is its dynamic viscosity in N $\cdot$ s/m^3 and centipoises?

3.1.10 If the kinematic viscosity of an oil is 800 centistokes, what is its kinematic viscosity in m^2/s and ft^2/s?

3.1.11 Derive the conversion of stokes to ft^2/s.

3.1.12 A liquid has a density of 700 kg/m^3 and a dynamic viscosity of 0.0042 $kg/m \cdot s$. What is its kinematic viscosity in m^2/s, ft^2/s, centistokes, and stokes?

3.1.13 What is the change in volume of 10 ft^3 of water at 50°F when it is subjected to a pressure increase of 200 psi?

3.1.14 What is the bulk modulus of elasticity of a liquid if its volume is reduced by 0.05% by the application of a pressure of 160 psi?

3.1.15 Determine the bulk modulus of elasticity of a liquid by using the following data: The volume is 3000 cm^3 (= 3 L) at 2 MPa, and the volume is 2750 cm^3 at 3 MPa.

3.1.16 The density of sea water is 1025 kg/m^3 at the ocean surface. The bulk modulus of elasticity of sea water is 234×10^7 Pa. What is the pressure at this depth, if the change in density between the surface and that depth is 1.5?

3.1.17 Suppose it is desired to reduce a given volume of water by 1% by increasing the pressures at two different temperatures, 50°F and 100°F. Compare the changes in the pressures applied at these temperatures. Calculate the percentage of the pressure change at 100°F with reference to that at 50°F.

3.1.18 A fluid pressure is 10 kPa above standard atmospheric pressure (101.3 kPa). Express the pressure as gauge and absolute.

3.1.19 If the pressure 15 ft below the free surface of a liquid is 30 psi, calculate its specific weight and specific gravity.

3.1.20 If the absolute pressure at a bottom of an open tank filled with oil (specific gravity = 0.8) is 200 kPa, what is the depth of oil in the tank?

3.1.21 A fluid pressure is 5 psi below the standard atmospheric pressure (14.7 psi). Express the pressure as vacuum and absolute.

3.1.22 Express the standard atmospheric pressure (14.7 psi) as feet of water and inches of mercury.

3.1.23 A Bourdon gauge reads a vacuum of 10 in of mercury (specific gravity = 13.6) when the atmospheric pressure is 14.3 psi. Calculate the corresponding absolute pressure.

3.1.24 What is the height of the column of a water barometer for an atmospheric pressure of 100 kPa, if the water is at 10°C, 50°C, and 100°C?

3.1.25 Compare the pressure forces exerted on the face of two dams that hold a fresh water of 1000 kg/m^3 density and a salty water of 1030 kg/m^3. Assume the faces of both dams are vertical.

3.1.26 Determine the equation for calculating the capillary rise h in the tube shown in Figure 3.1.2 in terms of θ, σ, γ, and τ.

3.1.27 Derive the relation between the gauge pressure p inside a spherical droplet of a liquid and the surface tension.

3.1.28 Derive the relation between the gauge pressure p inside the spherical bubble and the surface tension.

3.1.29 Calculate the pressure inside the spherical droplet of water having a diameter of 0.5 mm at 25°C if the pressure outside the droplet is standard atmospheric pressure (101.3 kPa).

3.1.30 What is the capillary rise of water in a glass tube having a diameter of 0.02 in at 70°F (take $\theta = 0$)?

3.1.31 Calculate the capillary depression of mercury in a glass tube having a diameter of 0.03 in at 68°F (take $\theta = 140$). The surface tension of mercury is 0.03562 lb/ft, and its specific gravity is 13.57. Take $\gamma_w = 62.3$ lb/ft^3 at 68°F.

3.1.32 Calculate the force necessary to lift a thin wire ring of 10-mm diameter from a water surface at 10°C.

3.1.33 Heavy fuel oil flows in a pipe that has a 6-in diameter with a velocity of 7 ft/s. The kinematic viscosity is 0.00444 ft^2/s. What is the Reynolds number?

3.1.34 The discharge of water in a 20-cm diameter pipe is 0.02 m^3/s. Calculate the velocity, Reynolds number, and mass rate of flow. Assume the temperature is 15°C.

3.1.35 Water flows in a rectangular channel having a slope of 25°C at a mean velocity of 3 ft/s. The depth of flow that is measured along a vertical line is 2 ft. Calculate the discharge if the width of the channel is 1 ft.

3.1.36 Water flows in a 10-cm diameter pipe at 500 kg/min. Calculate the discharge and mean velocity if the temperature is 20°C.

3.1.37 Water flows into a weigh tank for 30 min. Calculate the increase in the weight of the tank, if the discharge is 1.5 ft^3/s. The temperature is 50°F.

3.1.38 The hypothetical velocity distribution in a horizontal, rectangular, open channel is $v = 0.3\,y$ (m/s), where v is the velocity at a distance y (m) above the bottom of the channel. If the vertical depth of flow is 2 m and the width of the channel is 4 m, what are the maximum velocity, the discharge, and the mean velocity?

3.1.39 Assume water flows full in a 10-mm piezometer pipe at a temperature of 20°C. What is the approximate maximum discharge for which a laminar flow may be expected?

3.3.1 It is required to reduce a pipe of diameter 8 in to a minimum-diameter pipe that allows the downstream velocity not to exceed twice the upstream velocity. Determine the diameter of the pipe. Assume smooth transition.

3.4.1 Water flows through a pipe of diameter 3 in. If it is desired to use another pipe for the same flow rate such that the velocity head in the second pipe is four times the velocity head in the first pipe, determine the diameter of the pipe.

3.5.1 If the pipeline in problem 3.3.1 is horizontal, what is the proportion of the potential energy head at the upstream cross-section that is changed to kinetic energy head at the downstream cross-section? Determine the answer in terms of the discharge.

3.6.1 Derive an expression for the depth to the center of pressure for a triangle of height h and base b that is vertically submerged in water with the vertex at the water surface.

3.6.2 Derive an expression for the depth to the center of pressure for a triangle of height h and base b that is vertically submerged in water with the vertex a distance x below the water surface.

3.6.3 Determine the magnitude and the location of the hydrostatic force on the 2-m by 4-m vertical rectangular gate shown in Figure 3.6.3. The top of the gate is 6 m below the water surface.

3.6.4 Suppose a vertical flat plate supports water on one side and oil of specific gravity 0.86 on the other side, as shown in Figure P3.6.4. How deep should the oil be so that there is no net horizontal force on the plate? Calculate the moments of the pressure forces about the base of the plate. Are the magnitudes of the moments equal? Why?

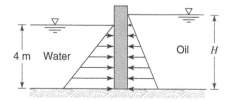

Figure P3.6.4 Vertical flat plate for problem 3.6.4.

3.6.5 Suppose a steel material of specific gravity of 7.8 is attached to a wood of specific gravity 0.8 as shown in Figure P3.6.5. If it is required that the material does not sink or rise when left in static water, what should be the proportion of the volume of the steel to that of the wood in Figure P3.6.5?

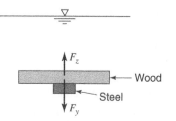

Figure P3.6.5 Problem 3.6.5 system.

3.6.6 Rework example 3.6.4 if the top of the inclined rectangular gate is 3 ft below the water surface.

3.7.1 Figure P3.7.1 shows a compound open channel cross-section. Determine the energy correction factor α. Assume uniform velocities within the subsections.

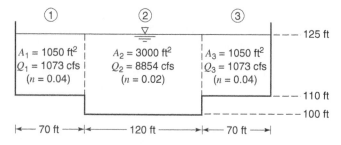

Figure P3.7.1 Compound open-channel cross-section for problem 3.7.1.

3.7.2 Determine the momentum correction factor β for problem 3.7.1.

REFERENCES

Chaudhry, H., "Principles of Flow of Water," Chapter 2 in *Water Resources Handbook*, edited by L. W. Mays, McGraw-Hill, New York, 1996.

Chow, V. T., *Open-Channel Hydraulics*, McGraw-Hill, New York, 1959.

Crowe, C. T., D. F. Elger, B. C. Williams, and J. A. Roberson, *Engineering Fluid Mechanics*, 9th edition, John Wiley & Sons, Inc., New York, 2009.

Finnemore, E. J., and J. B. Franzini, *Fluid Mechanics*, 10th edition, McGraw-Hill, New York, 2002.

Fox, R. W., P. J. Pritchard, and A. T. McDonald, *Introduction to Fluid Mechanics*, 7th edition, John Wiley & Sons, Inc., New York, 2009.

Munson, B. R., D. F. Young, and T. H. Okiish, *Fundamentals of Fluid Mechanics*, 4th edition. John Wiley and Sons, Inc., New York, 2002.

Roberson, J. A., and C. T. Crowe, *Engineering Fluid Mechanics*, Houghton Mifflin, Boston, 1993.

Chapter 4

Hydraulic Processes: Pressurized Pipe Flow

4.1 CLASSIFICATION OF FLOW

Hydraulics is typically defined as the study of liquid (water) flow in pipes and open channels, referred to as pipe flow and open-channel flow, respectively. Pipe flow and open-channel flow are similar in many ways but have one major difference. *Open-channel flow* occurs when there is a free surface, whereas pipe flow does not have a free surface. *Pipe flow* here refers to pressurized flow in pipes as long as there is not a free surface. Open-channel flow can occur in pipes. Analogies can be made between pipe flow and groundwater flow in confined aquifers. Also, groundwater flow in unconfined (water table) conditions is analogous to open-channel flow. The major difference is the geometry of the flow paths in groundwater flow as compared to pipe or open-channel flow. Because of the many varied-flow paths that occur in groundwater flow, the macroscopic average of the liquid (water) and medium properties is used. The control volumes for these three types of flow are illustrated in Figure 4.1.1.

Both pipe flow and open-channel flow are expressed in terms of the discharge Q, cross-sectional averaged velocity V, and cross-sectional area of flow A. They are related mathematically as

$$Q = AV \tag{4.1.1}$$

and

$$A = \int_0^y B \, dy \tag{4.1.2}$$

where y is the depth of flow and B is the channel width, so that

$$Q = \int_A v \, dA \tag{4.1.3}$$

where v is the local point velocity along the direction normal to A.

For groundwater flow, the discharge is expressed as

$$Q = Aq \tag{4.1.4}$$

where A is the total cross-sectional area including the space occupied by the porous medium and q is the *Darcy flux* or volumetric flow rate per unit area of porous medium Q/A, also sometimes called the *specific discharge* or *Darcy velocity*. This velocity assumes that flow occurs through the entire cross-section of the porous medium without regard to solids and pores. *Darcy's law* states that flow

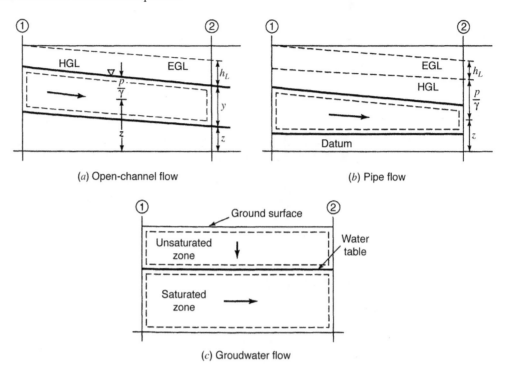

(a) Open-channel flow

(b) Pipe flow

(c) Groudwater flow

Figure 4.1.1 Control volumes for open-channel flow, pipe flow, and groundwater flow.

through a porous medium is proportional to the headloss and inversely proportional to the length of the flow path, $Q \sim dh/dL$. Darcy's flux is expressed as

$$q = -K\frac{dh}{dL} \tag{4.1.5}$$

and

$$Q = -KA\frac{dh}{dL} \tag{4.1.6}$$

where K is the saturated hydraulic conductivity (L/T) or units of velocity.

The criterion to differentiate viscosity effects, i.e., turbulent or laminar flow, is the *Reynolds number*, R_e:

$$R_e = \frac{VD}{v} = \frac{V4R}{v} \tag{4.1.7}$$

where D is a characteristic length such as pipe diameter, R is the *hydraulic radius* defined as the cross-sectional area of flow A divided by the wetted perimeter, P, of A, and v is the kinematic viscosity of the fluid. For full pipe flow, $R = A/P = \pi D^2/4/\pi D = D/4$. The critical number is around 2000 for pressurized pipe flow and 500 for open-channel flow. Below this critical value is laminar flow and above this number is turbulent flow. For groundwater flow the Reynolds number can be expressed with the velocity being the Darcy flux, q, and the effective grain size (d_{10}) is used for D. The *effective grain size* (d_{10}) is used to indicate that 10% (by weight) of a sample has a diameter smaller than d_{10}. Darcy's law is typically valid for Reynolds number $R_e < 1$ and does not depart seriously until $R_e = 10$, which can be thought of as the upper limit for Darcy's law.

The criterion to differentiate gravity effects, i.e., subcritical or supercritical flow, is the *Froude number:*

$$F_r = \frac{V}{\sqrt{gD_h}} \tag{4.1.8}$$

Table 4.1.1   Flow Classification

Classification criteria	Flow classification	Book section defined
Time	Steady flow	3.1.6
	Unsteady flow	3.1.6
Space	Uniform flow	3.1.6
	Nonuniform flow	3.1.6
	Gradually varied	5.3
	Rapidly varied	5.3
Viscosity (Reynolds number)	Laminar	3.1.7
	Turbulent	3.1.7
Gravity (Froude number)	Subcritical	5.2
	Critical	5.2
	Supercritical	5.2

where D_h is the *hydraulic depth* defined as the cross-sectional area of flow divided by the top width of flow. The critical value of the Froude number is 1, $F_r = 1$ for critical flow, i.e., when $F_r < 1$ the flow is subcritical and when $F_r > 1$ the flow is supercritical. Table 4.1.1 lists the flow classification for the criteria of time, space, viscosity, and gravity.

EXAMPLE 4.1.1   Water flows in a trapezoidal channel having a bottom width of 15 ft and side slopes of 1 vertical to 1.5 horizontal at a rate of 200 cfs. Is the flow a subcritical or a supercritical flow if the depth of the flow is 2 ft?

SOLUTION   To find out whether the flow is subcritical or supercritical, the Froude number should be calculated by equation (4.1.8):

$$F_r = \frac{V}{\sqrt{gD_h}}$$

The top width of the channel T is

$$T = 15 + 2 \times (2 \times 1.5) = 21$$

To determine the mean velocity, cross-sectional area must be computed:

$$A = 2 \times (15 + 21)/2 = 36 \text{ ft}^2$$

and from equation (4.1.1)

$$V = Q/A$$
$$V = 200/36 = 5.56 \text{ ft/s}$$

the hydraulic depth is

$$D_h = A/T = 36/21 = 1.71 \text{ ft}$$

Then

$$F_r = \frac{5.56 \text{ ft/s}}{\sqrt{32.2 \text{ ft/s}^2 \times 1.71 \text{ ft}}} = 0.75$$

F_r is < 1, so the flow is subcritical flow.

4.2 PRESSURIZED (PIPE) FLOW

4.2.1 Energy Equation

In section 3.4, the general energy equation for steady fluid flow was derived as equation (3.4.20). Using equation (3.4.20) and considering pipe flow between sections 1 and 2 (Figure 4.1.1b), the energy equation for pipe flow is expressed as

$$\frac{dH}{dt} - \frac{dW_s}{dt} = \int_{A_2} \left(\frac{p_2}{\rho} + e_{u_2} + \frac{1}{2}V_2^2 + gz_2 \right)\rho V_2 dA_2 - \int_{A_1} \left(\frac{p_1}{\rho} + e_{u_1} + \frac{1}{2}V_1^2 + gz_1 \right)\rho V_1 dA_1$$

$$(4.2.1)$$

which can be modified to

$$\frac{dH}{dt} - \frac{dW_s}{dt} = \int_{A_2} \left(\frac{p_2}{\rho} + e_{u_2} + gz_2 \right)\rho V_2 dA_2 + \int_{A_2} \frac{\rho V_2^3}{2} dA_2 - \int_{A_1} \left(\frac{p_1}{\rho} + e_{u_1} + gz_1 \right)\rho V_1 dA_1 - \int_{A_1} \frac{\rho V_1^3}{2} dA_1$$

$$(4.2.2)$$

Flow is uniform at sections 1 and 2 and therefore hydrostatic conditions prevail across the section. For hydrostatic conditions, $p/\rho + e_u + gz$ is constant across the system. This allows the term $(p/\rho + e_u + gz)$ to be taken outside the integral, so that (4.2.2) can be expressed as

$$\frac{dH}{dt} - \frac{dW_s}{dt} = \left(\frac{p_2}{\rho} + e_{u_2} + gz_2 \right) \int_{A_2} \rho V_2 dA_2 + \int_{A_2} \frac{\rho V_2^3 dA_2}{2} - \left(\frac{p_1}{\rho} + e_{u_1} + gz_1 \right) \int_{A_1} \rho V_1 dA_1 - \int_{A_1} \frac{\rho V_1^3}{2} dA_1$$

$$(4.2.3)$$

The term $\int \rho V \, dA = \dot{m}$ is the mass rate of flow and $\int_A (\rho V^3/2) dA = (\rho V^3/2)A = \dot{m}(\rho V^3/2)$. Substituting these definitions, equation (4.2.3) now becomes

$$\frac{dH}{dt} - \frac{dW_s}{dt} = \left(\frac{p_2}{\rho} + e_{u_2} + gz_2 \right)\dot{m} + \dot{m}\frac{V_2^2}{2} - \left(\frac{p_1}{\rho} + e_{u_1} + gz_1 \right)\dot{m} - \dot{m}\frac{V_1^2}{2} \qquad (4.2.4)$$

Dividing through by $\dot{m}$ and rearranging then yields

$$\frac{1}{\dot{m}}\left(\frac{dH}{dt} - \frac{dW_s}{dt} \right) + \frac{p_1}{\rho} + e_{u_1} + gz_1 + \frac{V_1^2}{2} = \frac{p_2}{\rho} + e_{u_2} + gz_2 + \frac{V_2^2}{2} \qquad (4.2.5)$$

The shaft work term $(d\,W_s/dt)$ can be the result of a turbine (W_T) or a pump (W_P) in the system, so that the shaft work term can be expressed as

$$\frac{dW_s}{dt} = \frac{dW_T}{dt} - \frac{dW_P}{dt} \qquad (4.2.6)$$

The minus sign occurs because a pump does work on the fluid and conversely a turbine does shaft work on the surroundings. The term dW/dt has the units of power, which is work per unit time. Substituting equation (4.2.6) into (4.2.5) and dividing through by g results in

$$\left(\frac{1}{\dot{m}g} \right)\frac{dW_P}{dt} + \frac{p_1}{g} + z_1 + \frac{V_1^2}{2g} = \left(\frac{1}{\dot{m}g} \right)\frac{dW_T}{dt} + \frac{p_2}{g} + z_2 + \frac{V_2^2}{2g} + \left[\frac{e_{u_2} - e_{u_1}}{g} - \frac{1}{\dot{m}g}\frac{dH}{dt} \right] \qquad (4.2.7)$$

Each term in the above equation has the dimension of length. The following are defined:

Head supplied by pumps:
$$h_P = \left(\frac{1}{\dot{m}g} \right) \frac{dW_P}{dt} \tag{4.2.8}$$

Head supplied by turbines:
$$h_T = \left(\frac{1}{\dot{m}g} \right) \frac{dW_T}{dt} \tag{4.2.9}$$

Headloss (loss of mechanical energy due to viscous stress):
$$h_L = \left[\left(\frac{e_{u_2} - e_{u_1}}{g} \right) - \left(\frac{1}{\dot{m}g} \right) \frac{dH}{dt} \right] \tag{4.2.10}$$

The term $((e_{u_2} - e_{u_1})/g)$ represents finite increases in internal energy of the flow system because some of the mechanical energy is converted to thermal energy through viscous action between fluid particles. The term $-(1/\dot{m}g)dH/dt$ represents heat generated through energy dissipation that escapes the flow system.

Using h_P, h_T, and h_L, the energy equation (4.2.7) is now expressed as

$$\frac{p_1}{\gamma} + z_1 + \frac{V_1^2}{2g} + h_P = \frac{p_2}{\gamma} + z_2 + \frac{V_2^2}{2g} + h_T + h_L \tag{4.2.11}$$

This energy equation is expressed with the velocity representing the mean velocity; p/γ is referred to as the *pressure head* and $V^2/2g$ is referred as the *velocity head*.

Because the velocity is not actually uniform over a cross-section, an *energy correction factor* is typically introduced, which is defined as (see section 3.7)

$$\alpha = \frac{\int_A v^3 dA}{V^3 A} \tag{4.2.12}$$

where v is the velocity at any point in the section. Using α_1 and α_2 for sections 1 and 2, the energy equation can be expressed as

$$\frac{p_1}{\gamma} + z_1 + \alpha_1 \frac{V_1^2}{2g} + h_P = \frac{p_2}{\gamma} + z_2 + \alpha_2 \frac{V_2^2}{2g} + h_T + h_L \tag{4.2.13}$$

Because α has a value very near 1 for pipe flow, it is typically omitted.

EXAMPLE 4.2.1

The flow of water in a horizontal pipe of constant cross-section has a pressure gauge at location 1 with a pressure of 100 psig and a pressure gauge at location 2 with a pressure of 75 psig. What is the headloss between the two pressure gauges?

SOLUTION

For the energy equation (4.2.11), $h_P = 0$, $h_T = 0$, $z_1 = z_2$, and because the pipe size is constant $V_1^2/2g = V_2^2/2g$; then (4.2.11) reduces to

$$h_L = \frac{p_1 - p_2}{\gamma} = \frac{\left(100 \text{ lb/in}^2 - 75 \text{ lb/in}^2 \right) \times 144 \text{ in}^2/\text{ft}^2 \right)}{62.4 \text{ lb/ft}^3} = \frac{25 \times 144}{62.4}$$

$$h_L = 57.7 \text{ ft}$$

EXAMPLE 4.2.2

For the simple pipe system shown in Figure 4.2.1, the pressures are $p_1 = 14$ kPa, $p_2 = 12.5$ kPa, and $p_3 = 10$ kPa. Determine the headloss between 1 and 2 and the headloss between 1 and 3. The discharge is 7 L/s.

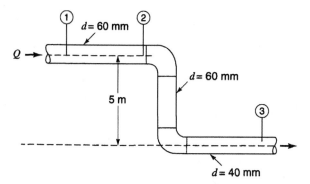

Figure 4.2.1 Pipe system for example 4.2.2.

SOLUTION

The energy equation between 1 and 2 is

$$\frac{p_1}{\gamma} + \frac{V_1^2}{2g} + z_1 = \frac{p_2}{\gamma} + \frac{V_2^2}{2g} + z_2 + h_{L_{1-2}}$$

where $\frac{V_1^2}{2g} = \frac{V_2^2}{2g}$ and $z_1 = z_2$, so

$$\frac{p_1}{\gamma} = \frac{p_2}{\gamma} + h_{L_{1-2}} \text{ and } h_{L_{1-2}} = \frac{p_1 - p_2}{\gamma} = \frac{14 - 12.5}{9.79} = 0.153 \text{ m}$$

The energy equation between 1 and 3 is

$$\frac{p_1}{\gamma} + \frac{V_1^2}{2g} + z_1 = \frac{p_3}{\gamma} + \frac{V_3^2}{2g} + z_3 + h_{L_{1-3}}$$

$$V_1 = Q/A_1 = \frac{(7/1000)}{\left[\frac{\pi(60/1000)^2}{4}\right]} = \frac{0.007}{0.0028} = 2.477 \text{ m/s}$$

$$V_3 = Q/A_3 = \frac{(7/1000)}{\left[\frac{\pi(40/1000)^2}{4}\right]} = \frac{0.007}{0.0013} = 5.385 \text{ m/s}$$

Now using the above energy equation, we get

$$\frac{14}{9.79} + \frac{(2.477)^2}{2 \times 9.81} + 5 = \frac{10}{9.79} + \frac{(5.385)^2}{2 \times 9.81} + 0 + h_{L_{1-3}}$$

$$1.43 + 0.313 + 5 = 1.022 + 1.478 + 0 + h_{L_{1-3}}$$

$$h_{L_{1-3}} = 4.24 \text{ m}$$

EXAMPLE 4.2.3

A *Venturi meter* is a device, as shown in Figure 4.2.2, that is inserted into a pipeline to measure the incompressible flow rate. These meters consist of a convergent section that reduces the diameter to between one-half and one-fourth the pipe diameter, followed by a divergent section. Pressure differences between the position just before the Venturi and at the throat are measured by a differential manometer. Using the energy equation, develop an equation for the discharge in terms of the pressure difference $p_1 - p_2$. Use a coefficient of discharge that takes into account frictional effects.

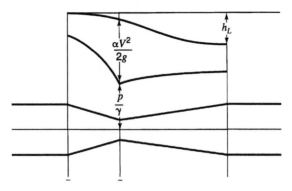

Figure 4.2.2 Venturi tube.

SOLUTION

Write the energy equation between 1 and 2:

$$\frac{p_1}{\gamma} + \frac{V_1^2}{2g} + z_1 = \frac{p_2}{\gamma} + \frac{V_2^2}{2g} + z_2 + h_L$$

For $z_1 = z_2$ and assuming $h_L = 0$, the energy equation reduces to

$$\frac{p_1}{\gamma} + \frac{V_1^2}{2g} = \frac{p_2}{\gamma} + \frac{V_2^2}{2g}$$

$$V_1^2 - V_2^2 = 2g\left(\frac{p_2 - p_1}{\gamma}\right)$$

From continuity, $A_1V_1 = A_2V_2$, so $V_1 = V_2(A_2/A_1)$. Substituting V_1 into the above energy equation yields

$$\left[V_2\frac{A_2}{A_1}\right]^2 - V_2^2 = 2g\left(\frac{p_2 - p_1}{\gamma}\right)$$

$$\left[\left(\frac{A_2}{A_1}\right)^2 - 1\right]V_2^2 = 2g\left(\frac{p_2 - p_1}{\gamma}\right)$$

Multiplying through by (-1) and solving for V_2, we get

$$V_2 = \sqrt{\frac{1}{1 - (A_2/A_1)^2}}\sqrt{2g\left(\frac{p_1 - p_2}{\gamma}\right)}$$

which neglects friction effects. A coefficient of discharge C_d is used in the continuity equation to account for friction effect, so

$$Q = C_dA_2V_2 = C_dA_2\left[\frac{1}{1 - (A_2/A_1)^2}\right]^{1/2}\left[2g\left(\frac{p_1 - p_2}{\gamma}\right)\right]^{1/2} = C_d\frac{A_2}{\sqrt{1 - (A_2/A_1)^2}}\sqrt{2g\frac{p_1 - p_2}{\gamma}}$$

The value of C_d must be determined experimentally.

4.2.2 Hydraulic and Energy Grade Lines

The terms in equation (4.2.13) have dimensions of length and units of feet or meters. The concept of the *hydraulic grade line (HGL)* and the *energy grade line (EGL)* can be used to give a physical

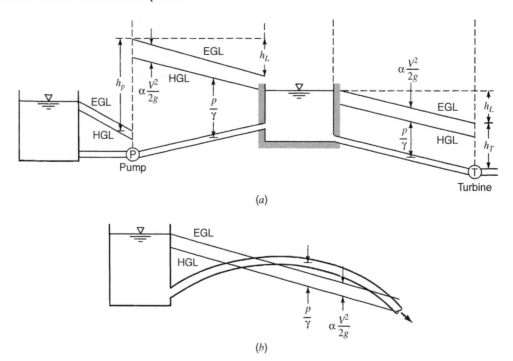

Figure 4.2.3 Energy and hydraulic grade lines. (*a*) EGL and HGL for system with pump and turbine; (*b*) System with subatmospheric pressure (pipe is above the HGL).

relationship to these terms. The HGL is essentially the line p/γ above the center line of a pipe, which is the distance water would rise in a piezometer tube attached to the pipe. The EGL is a distance of $\alpha V^2/2g$ above the HGL. Figure 4.2.3 illustrates the HGL and EGL.

4.3   HEADLOSSES

The energy equation (4.2.13) for pipe flow was derived in the previous section. This equation has a headloss term h_L that was defined as the loss of mechanical energy due to viscous stress. The objective of this section is to discuss the headlosses that occur in pressurized pipe flow.

4.3.1   Shear-Stress Distribution of Flow in Pipes

Consider steady flow (laminar or turbulent) in a pipe (a cylindrical element of fluid) of uniform cross-section as shown in Figure 4.3.1. For a uniform flow, the general form of the integral momentum equation in the *x*-direction is expressed by

$$\sum F_x = \frac{d}{dt} \int_{CV} v_x \rho \, d\forall + \sum_{CS} v_x (\rho \mathbf{V} \cdot \mathbf{A}) \tag{3.5.6}$$

where $\dfrac{d}{dt} \displaystyle\int_{CV} v_x \rho \, d\forall = 0$ because the flow is steady and $\displaystyle\sum_{CS} v_x (\rho \mathbf{V} \cdot \mathbf{A}) = 0$ because there is no flow of extensive property through the control surface so that

$$\sum F_x = 0 \tag{4.3.1}$$

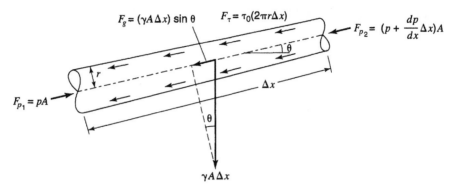

Figure 4.3.1 Cylindrical element of water.

The forces are: (1) the *pressure forces*, $F_{p_1} = pA$ and $F_{p_2} = (p + dp/dx\Delta x)A$; (2) the *gravity force* due to the weight of the water, $F_g = (\gamma A\Delta x)\sin\theta$; and (3) the *shearing force* $F_\tau = \tau(2\pi r\Delta x)$ where τ is the shear stress. The sum of forces is

$$F_{P_1} - F_{P_2} - F_g - F_\tau = 0 \qquad (4.3.2)$$

or

$$pA - \left(p + \frac{dp}{dx}\Delta x\right)A - (\gamma A\Delta x)\sin\theta - \tau(2\pi r\Delta x) = 0 \qquad (4.3.3)$$

Equation (4.3.3) reduces to

$$-\frac{dp}{dx}\Delta xA - (\gamma A\Delta x)\sin\theta - \tau(2\pi r\Delta x) = 0 \qquad (4.3.4)$$

Solving for the shear stress by using $dz = \sin\theta\, dx$ gives

$$\tau = \frac{r}{2}\left[-\frac{d}{dx}(p + \gamma z)\right] \qquad (4.3.5)$$

Equation (4.3.5) indicates that τ is zero at the center of the pipe where $r = 0$ and increases linearly to a maximum at the pipe wall (see Figure 4.3.2). Keep in mind that $p + \gamma z$ is constant across the section because the streamlines are straight and parallel in a uniform flow, so that there is no acceleration of fluid normal to the streamline. In other words, hydrostatic conditions prevail across the flow section, resulting in $p + \gamma z$ being constant across the flow section. The gradient $d(p + \gamma z)/dx$ is therefore negative and constant across the flow section for uniform flow.

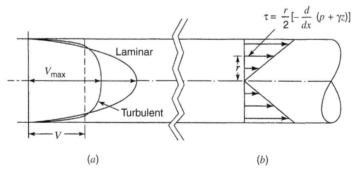

Figure 4.3.2 Distribution of velocity and shear stress for pipe flow. (*a*) Velocity distribution; (*b*) Shear stress.

4.3.2   Velocity Distribution of Flow in Pipes

Laminar Flow

In Chapter 3 the following shear stress equation (3.1.2) was introduced:

$$\tau = \mu \frac{dv}{dy} \qquad (3.1.2)$$

where μ is the dynamic viscosity and dv/dy is the velocity gradient. For our purposes here, the gradient is actually $dv/dy = -dv/dr$, so that using equation (4.3.5) we get

$$\tau = \mu \frac{dv}{dr} = \frac{r}{2}\left[-\frac{d}{dx}(p + \gamma z)\right] \qquad (4.3.6)$$

Integrating this equation by the separation of variables and boundary condition ($r = r_0$, $v = 0$) yields

$$v = \frac{r_0^2 - r^2}{4\mu}\left[-\frac{d}{dx}(p + \gamma z)\right] \qquad (4.3.7)$$

This equation indicates that the velocity distribution for laminar pipe flow is parabolic across the section and has the maximum velocity at the pipe center (refer to Figure 4.3.2). By integrating the velocity across the section using $Q = \int v\,dA$ and using the energy equation, the headloss for laminar flow is

$$h_{L_f} = \frac{32\mu LV}{\gamma D^2} \qquad (4.3.8)$$

where D is the diameter of the pipe.

Turbulent Flow—Smooth Pipes

For a smooth pipe, the following velocity distribution equations are based upon experiment (Roberson and Crowe, 1990):

$$\frac{u}{u_*} = \frac{u_* y}{v} \text{ for } 0 < \frac{u_* y}{v} < 5 \qquad (4.3.9)$$

and

$$\frac{u}{u_*} = 5.75 \log\frac{u_* y}{v} + 5.5 \text{ for } 20 < \frac{u_* y}{v} < 10^5 \qquad (4.3.10)$$

where u is the velocity, y is the distance from the pipe wall, v is the kinematic viscosity, u_* is the shear velocity ($\sqrt{\tau_0/\rho}$), and τ_0 is the shear stress at the pipe wall.

The velocity distribution for turbulent flow can also be approximated using power law formulas of the form $u/u_{\max} = (y/r_0)^m$, where $u_{\max}$ is the velocity at the center of the pipe, r_0 is the pipe radius, and m is an exponent that increases with Reynolds number (see Roberson and Crowe (1990) for values of m).

Turbulent Flow—Rough Pipes

Experimental results on flow in rough pipes indicate the following form of equation for velocity distribution of turbulent flow in rough pipes:

$$\frac{u}{u_*} = 5.75 \log\frac{y}{k} + B \qquad (4.3.11)$$

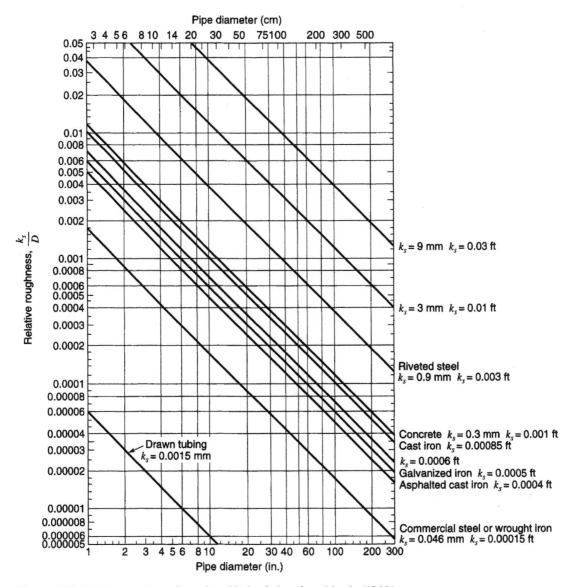

Figure 4.3.3 Relative roughness for various kinds of pipe (from Moody (1944)).

where y is the distance from the wall, k is a measure of the height of the roughness elements, and B is a function of the roughness characteristics. Nikuradse (1933) determined the value of $B = 8.5$, y was measured from the geometric mean of the wall surface, and $k_s = k$ was the sand grain size, so that

$$\frac{u}{u_*} = 5.75 \log \frac{y}{k_s} + 8.5 \tag{4.3.12}$$

Nikuradse's work revealed that (1) for low Reynolds numbers and small sand grains the flow resistance is basically the same as for smooth pipes (roughness elements became submerged in the viscous sublayer) and (2) for high Reynolds numbers the resistance coefficient is a function of only the relative roughness k_s/D, where D is the diameter (the viscous sublayer is very thin so that the roughness elements project into the flow, causing flow resistance from drag of the individual roughness elements). Figure 4.3.3 presents values of relative roughness for different kinds of pipe as a function of pipe diameter.

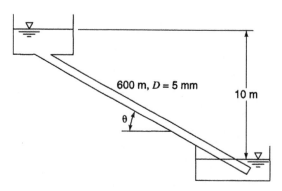

Figure 4.3.4 Pipe system for example 4.3.1.

EXAMPLE 4.3.1

For the system shown in Figure 4.3.4 determine the headloss per unit length of pipe and the discharge, assuming laminar flow $(\mu = 1.002 \times 10^{-3}\ \text{N-s/m}^2)$.

SOLUTION

The headloss equation (4.3.8) for laminar pipe flow is used to determine the velocity of flow in the pipe, where $h_L = 10\ \text{m}$, $D = 5\ \text{mm}$, $\gamma = 9.79\ \text{kN/m}^3$, $\mu = 1.002 \times 10^{-3}\ \text{N-s/m}^2$:

$$V = \frac{\gamma h_L D^2}{32\mu L} = \frac{(9.79\ \text{kN/m}^3 \times 1000)(10\text{m})(5/1000\text{m})^2}{32(1.002 \times 10^{-3}\ \text{N-s/m}^2)(600\text{m})} = 0.125\ \text{m/s}$$

Check the Reynolds number:

$$R_e = \frac{VD}{v} = \frac{VD\rho}{\mu} = \frac{0.125 \times (5/1000) \times 1000}{1.002 \times 10^{-3}} = 6.24 \times 10^2 = 624$$

Flow is laminar. The flowrate is then

$$Q = AV = \left[\frac{\pi(5/1000)^2}{4}\right] \times 0.125\ \text{m/s} = 2.45 \times 10^{-6}\ \text{m}^3/\text{s}$$

$$= 2.45 \times 10^{-6}(1000)(60)$$

$$= 0.147\ \text{L/min}$$

The headloss per unit length of pipe is $(10\ \text{m}/600\ \text{m}) = 0.0167\ \text{m/m}$.

4.3.3 Headlosses from Pipe Friction

Various equations have been proposed to determine the headlosses due to friction, including the Darcy–Weisbach, Chezy, Manning, Hazen–Williams, and Scobey formulas. These equations relate the friction losses to physical characteristics of the pipe and various flow parameters. The Darcy–Weisbach equation is scientifically based and applies to both laminar and turbulent flows. The *Darcy–Weisbach equation* is

$$h_{L_f} = f\frac{L}{D}\frac{V^2}{2g} \tag{4.3.13}$$

where h_{L_f} is the headloss due to pipe friction, f is the dimensionless friction factor, L is the length of the conduit, D is the inside diameter of the pipe, V is the mean flow velocity, and g is the acceleration due to gravity.

The friction factor is a function of the Reynolds number (R_e) and the relative roughness k_s/D, where k_s is the average nonuniform roughness of the pipe. For laminar flow $(R_e < 2000)$ the friction factor is

$$f = \frac{64}{R_e} \tag{4.3.14}$$

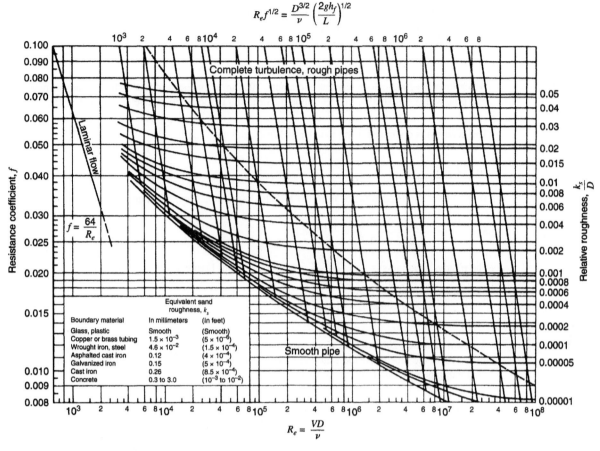

$$R_e f^{1/2} = \frac{D^{3/2}}{\nu}\left(\frac{2gh_f}{L}\right)^{1/2}$$

Figure 4.3.5 Resistance coefficient f versus R_e (from Moody (1944)).

where

$$R_e = \frac{VD}{\nu} \tag{4.3.15}$$

and ν is the kinematic viscosity. For turbulent flow in

$$\textit{Smooth pipe} \quad \frac{1}{\sqrt{f}} = 2\log_{10}\left(R_e\sqrt{f}\right) - 0.8 \text{ for } R_e > 3000 \tag{4.3.16}$$

$$\textit{Rough pipe} \quad \frac{1}{\sqrt{f}} = 2\log_{10}\frac{D}{(k_s)} + 1.14 = 1.14 - 2\log_{10}\left(\frac{k_s}{D}\right) \tag{4.3.17}$$

where D is the pipe diameter. Equations (4.3.16) and (4.3.17) were proposed by von Karman and Prandtl based upon experiments by Nikuradse (1932).

Colebrook and White (1939) proposed the following semi-empirical formula:

$$\frac{1}{\sqrt{f}} = -2.0\log_{10}\left(\frac{k_s/D}{3.7} + \frac{2.51}{R_e\sqrt{f}}\right) \tag{4.3.18}$$

The above equation is asymptotic to both the smooth and rough pipe equations (4.3.16) and (4.3.17) and is valid for the entire nonlaminar range of the Moody diagram.

Moody (1944) developed the *Moody diagram*, shown in Figure 4.3.5, using experimental data on commercial pipes, the Colebrook–White equation, and the Prandtl–Karman experimental data.

Knowing the pipe roughness (relative roughness) and the Reynolds number, the friction factor can be obtained from the Moody diagram. The use of Manning's equation and the Hazen–Williams equation is discussed in Chapter 12.

EXAMPLE 4.3.2

Water flows in a 1000-m long pipeline of diameter 200 mm at a velocity of 5 m/s. The pipeline is a new cast iron pipe with $k_s = 0.00026$ m. Determine the headloss in the pipeline. Use $v = 1.007 \times 10^{-6}$ m^2/s.

SOLUTION

The headloss is computed using the Darcy–Weisbach equation (4.3.13):

$$h_{L_f} = f \frac{L}{D} \frac{V^2}{2g}$$

To determine the friction factor from the Moody diagram the Reynolds number must be computed:

$$R_e = \frac{VD}{v} = \frac{5 \times (200/1000)}{1.007 \times 10^{-6}} = 9.93 \times 10^5$$

The relative roughness is $\dfrac{k_s}{D} = \dfrac{0.00026}{0.200} = 0.0013$. From Figure 4.3.5, $f = 0.021$, so

$$h_{L_f} = 0.021 \frac{1000}{(200/1000)} \frac{(5)^2}{2(9.81)} = 133.8 \text{ m.}$$

EXAMPLE 4.3.3

A 10-cm diameter, 2000-m long pipeline connects two reservoirs open to the atmosphere. What is the discharge in the pipeline if the water surface elevation difference of the reservoirs is 50 m? Assume a smooth pipe and $v = 1.02 \times 10^{-6}$ m^2/s.

SOLUTION

The energy equation between the reservoir surfaces is used to determine the headloss due to friction, which is then used to determine the velocity of flow in the pipeline:

$$\frac{p_1}{\gamma} + \frac{V_1^2}{2g} + z_1 = \frac{p_2}{\gamma} + \frac{V_2^2}{2g} + z_2 + h_{L_f}$$

$$0 + 0 + z_1 = 0 + 0 + z_2 + h_{L_f}$$

$$h_{L_f} = z_1 - z_2 = 50 \text{ m}$$

$$h_{L_f} = f \frac{L}{D} \frac{V^2}{2g}$$

$$f \left(\frac{2000}{10/100} \right) \frac{V^2}{2 \times 9.81} = 50$$

$$f V^2 = 0.0491 \text{ (or } V = 0.2215/\sqrt{f})$$

This must be solved for V using a trial-and-error procedure by assuming f, computing V, then computing the Reynolds number. Assume $f = 0.02$, $V = 0.2215/\sqrt{0.02} = 1.566$ m/s:

$$R_e = \frac{VD}{v} = \frac{1.566(10/100)}{1.02 \times 10^{-6}} = 1.54 \times 10^5$$

From Figure 4.3.5, $f = 0.016$, $V = 0.2215/\sqrt{0.016} = 1.751$ m/s, so

$$R_e = \frac{VD}{v} = \frac{1.751(10/100)}{1.02 \times 10^{-6}} = 1.72 \times 10^5$$

From Figure 4.3.5, $f = 0.016$, which is close enough. The discharge is

$$Q = AV = \frac{\pi (10/100)^2}{4} (1.751) = 0.0137 \text{ m}^3/\text{s.}$$

EXAMPLE 4.3.4

Compute the friction factor for flow having a Reynolds number of 1.37×10^4 and relative roughness $k_s/D = 0.000375$ using the Colebrook–White formula.

SOLUTION

Use equation (4.3.18) $\dfrac{1}{\sqrt{f}} = -2.0 \log_{10} \left(\dfrac{0.000375}{3.7} + \dfrac{2.51}{1.37 \times 10^4 \sqrt{f}} \right)$ and solve using trial and error $f =$ 0.0291. Referring to the Moody diagram (Figure 4.3.5), we would read approximately 0.028 or 0.029.

4.3.4 Form (Minor) Losses

Headlosses are also caused by inlets, outlets, bends, and other appurtenances such as fittings, valves, expansions, and contractions. These losses, referred to as *minor losses, form losses,* or *secondary losses,* are caused by flow separation and the generation of turbulence. Headlosses produced, in general, can be expressed by

$$h_{L_m} = K \frac{V^2}{2g} \tag{4.3.19}$$

where K is the loss coefficient (see Table 4.3.1) and V is the mean velocity. Table 4.3.2 lists loss coefficients for various transitions and fittings. Table 12.2.5 lists recommended energy loss coefficients for valves and Figure 12.2.12 shows diagrams of various types of valves.

For sudden expansions or enlargements, the headlosses can be expressed as

$$h_{L_m} = \frac{(V_1 - V_2)^2}{2g} \tag{4.3.20}$$

For *gradual enlargements,* such as *conical diffusers,* the headloss is expressed as

$$h_{L_m} = K' \frac{(V_1 - V_2)^2}{2g} \tag{4.3.21}$$

The loss of head due to a *sudden contraction* may be expressed as

$$h_{L_m} = K_c \frac{V_2^2}{2g} \tag{4.3.22}$$

where the values of K_c are a function of the diameter ratios D_2/D_1.

Entrance losses are computed using

$$h_{L_m} = K_e \frac{V^2}{2g} \tag{4.3.23}$$

where values of K_e are found in Table 4.3.1.

Exit (or discharge) losses (Table 4.3.1) from the end of a pipe into a reservoir that has a negligible velocity are expressed as

$$h_{L_m} = \frac{V^2}{2g} \tag{4.3.24}$$

EXAMPLE 4.3.5

Water flows from reservoir 1 to reservoir 2 through a 12-in diameter, 600-ft long pipeline as shown in Figure 4.3.6. The reservoir 1 surface elevation is 1000 ft and reservoir 2 surface elevation is 950 ft. Consider the minor losses due to the sharp-edged entrance, the butterfly valve ($\theta = 20°$), the two bends (90° elbow), and the sharp-edged exit. The pipe is galvanized iron with $k_s = 0.0005$ ft. Determine the discharge from reservoir 1 to reservoir 2. For 120°F, $v = 0.609 \times 10^{-5}$ ft²/s.

SOLUTION

The energy equation between 1 and 2 is

$$\frac{p_1}{\gamma} + \frac{V_1^2}{2g} + z_1 = \frac{p_2}{\gamma} + \frac{V_2^2}{2g} + z_2 + h_{L_f} + \sum h_{L_m}$$

Table 4.3.1   Minor Loss Coefficients for Pipe Flow

Type of minor loss	K Loss in terms of $V^2/2g$
Pipe fittings:	
90° elbow, regular	0.21–0.30
90° elbow, long radius	0.14–0.23
45° elbow, regular	0.2
Return bend, regular	0.4
Return bend, long radius	0.3
AWWA tee, flow through side outlet	0.5–1.80
AWWA tee, flow through run	0.1–0.6
AWWA tee, flow split side inlet to run	0.5–1.8
Valves:	
Butterfly valve ($\theta = 90°$ for closed valve)*	
$\theta = 0°$	0.3–1.3
$\theta = 10°$	0.46–0.52
$\theta = 20°$	1.38–1.54
$\theta = 30°$	3.6–3.9
$\theta = 40°$	10–11
$\theta = 50°$	31–33
$\theta = 60°$	90–120
Check valves (swing check) fully open	0.6–2.5
Gate valves (4 to 12 in) fully open	0.07–0.14
1/4 closed	0.47–0.55
1/2 closed	2.2–2.6
3/4 closed	12–16
Sluice gates:	
As submerged port in 12-in wall	0.8
As contraction in conduit	0.5
Width equal to conduit width and without top submergence	0.2
Entrance and exit losses:	
Entrance, bellmouthed	0.04
Entrance, slightly taunted	0.23
Entrance, square edged	0.5
Entrance, projecting	1.0
Exit, bellmouthed	$0.1\left(\dfrac{V_1^2}{2g} - \dfrac{V_2^2}{2g}\right)$
Exit, submerged pipe to still water	1.0

*Loss coefficients for partially open conditions may vary widely. Individual manufacturers should be consulted for specific conditions.

Source: Adapted from Velon and Johnson (1993).

For the reservoir surface $p_1/\gamma = 0$ and $V^2/2g = 0$, so

$$0 + 0 + 1000 = 0 + 0 + 950 + h_{L_f} + \sum h_{L_m}$$
$$50 = h_{L_f} + \sum h_{L_m}$$

where

$$h_{L_f} = \frac{fLV^2}{D2g} = f\frac{600}{(12/12)}\frac{V^2}{2 \times 32.2} = 9.32fV^2$$

The minor losses are

Table 4.3.2 Loss Coefficients for Various Transitions and Fittings

Description	Sketch	Additional data	K		Source
Pipe entrance $h_{l_m} = K_e V^2/2g$		r/d 0.0 0.1 >0.2	K_e 0.50 0.12 0.03		(a)
Contraction $h_{l_m} = K_C V_2^2/2g$		D_2/D_1 0.0 0.20 0.40 0.60 0.80 0.90	K_C $\theta = 60°$ 0.08 0.08 0.07 0.06 0.05 0.04	K_C $\theta = 180°$ 0.50 0.49 0.42 0.32 0.18 0.10	(a)
Expansion $h_{l_m} = K_E V_1^2/2g$		D_1/D_2 0.0 0.20 0.40 0.60 0.80	K_E $\theta = 10°$ 0.13 0.11 0.06 0.03	K_E $\theta = 180°$ 1.00 0.92 0.72 0.42 0.16	(a)
90° miter bend	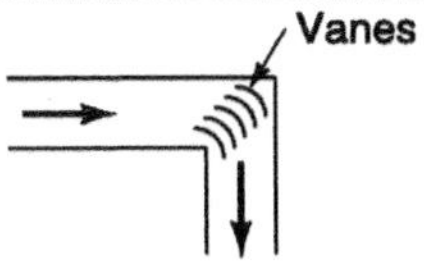	Without vanes	$K_b = 1.1$		(b)
		With vanes	$K_b = 0.2$		(b)
Smooth bend		r/d 1 2 4 6	K_b $\theta = 45°$ 0.10 0.09 0.10 0.12	K_b $\theta = 90°$ 0.35 0.19 0.16 0.21	(c) and (d)
Threaded pipe fittings	Globe valve—wide open Angle valve—wide open Gate valve—wide open Gate valve—half open Return bend Tee 90° elbow 45° elbow		$K_v = 10.0$ $K_v = 5.0$ $K_v = 0.2$ $K_v = 5.6$ $K_b = 2.2$ $K_t = 1.8$ $K_b = 0.9$ $K_b = 0.4$		(b)

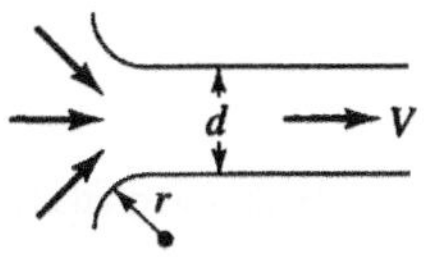

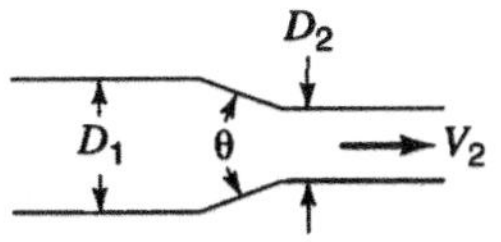

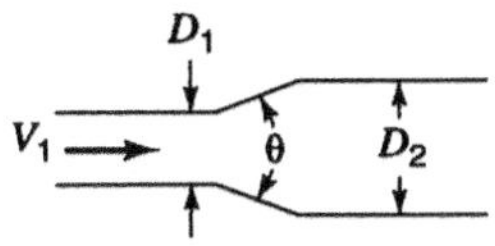

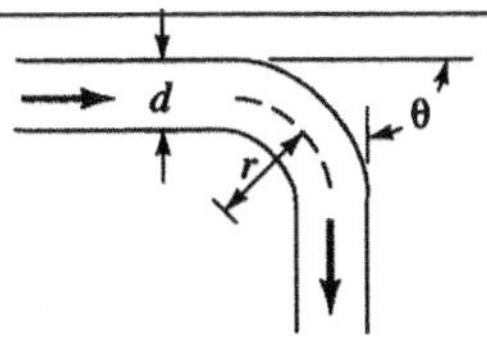

(a) ASHRAE (1977)
(b) Streeter (1961)
(c) Beij (1938)
(d) Idel'chik (1966)

Source: After Roberson et al. (1988).

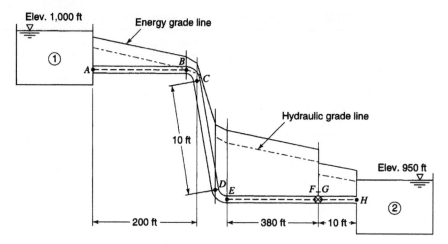

Figure 4.3.6 Pipe system for example 4.3.5.

$$\sum h_{L_m} = h_{L_{\text{entrance}}} + h_{L_{\text{elbow}}} + h_{L_{\text{elbow}}} + h_{L_{\text{globevalve}}} + h_{L_{\text{exit}}}$$

$$= K_e \frac{V^2}{2g} + 2K_{\text{elbow}} \frac{V^2}{2g} + K_{\text{valve}} \frac{V^2}{2g} + \frac{V^2}{2g}$$

$$= (K_e + 2K_{\text{elbow}} + K_{\text{valve}} + 1) \frac{V^2}{2g}$$

where $K_e = 0.5$, $K_{\text{elbow}} = 0.25$, and $K_{\text{valve}} = 1.5$. The energy equation is now expressed as

$$50 = (0.5 + 2 \times 0.25 + 1.5 + 1.0) \frac{V^2}{2g} + 9.32 f\, V^2$$

$$50 = 3.5 \frac{V^2}{2g} + 9.32 f\, V^2$$

$$50 = (0.054 + 9.32 f) V^2$$

$$V = \sqrt{50/(0.054 + 9.32 f)}$$

Assuming fully turbulent flow and using $k_s/D = 0.0005/1 = 0.0005$, we get $f = 0.0165$ from Figure 4.3.5, then

$$V = \sqrt{50/(0.054 + 9.32 \times 0.0165)}$$
$$V = 15.51 \text{ ft/s}$$

Compute $R_e = \dfrac{VD}{v} = \dfrac{15.51 \times 1}{0.609 \times 10^{-5}} = 2.55 \times 10^6$. Referring to Figure 4.3.5 (Moody diagram), we see

that the value of f is OK. Now use the continuity equation to determine Q:

$$Q = AV = [\pi(12/12)^2/4](15.51 \text{ ft/s}) = 12.18 \text{ ft}^3/\text{s}$$

4.4   FORCES IN PIPE FLOW

Change in direction or magnitude of flow velocity of a fluid causes changes in the momentum of the fluid (see Figure 4.4.1). The forces that are required to produce the change in momentum come from the pressure variation within the fluid and from forces transmitted to the fluid from the pipe

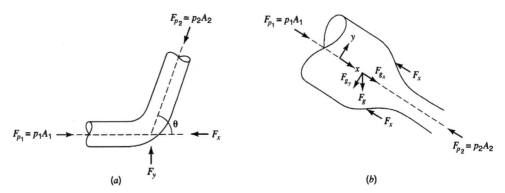

Figure 4.4.1 Forces in pipe flow, (a) Pipe bend; (b) Contraction.

walls. Applying the momentum principle (equation 3.5.6) to the control volumes in Figure 4.4.1, we get

$$\sum F_x = \frac{d}{dt} \int_{CV} v_x \rho\, d\forall + \sum_{CS} v_x \rho \mathbf{V} \cdot \mathbf{A} \tag{3.5.6}$$

where

$$\frac{d}{dt} \int_{CV} v_x \rho\, d\forall = 0$$

because flow is steady and

$$\sum_{CS} v_x \rho \mathbf{V} \cdot \mathbf{A} = \rho V_{x_2} Q - \rho V_{x_1} Q \tag{4.4.1}$$

The momentum entering from the upstream is $-\rho V_{x_1} Q$ and the momentum leaving the control volume is $\rho V_{x_2} Q$. The summation of forces in the x direction for the contraction in Figure 4.4.1b is

$$\sum F_x = p_1 A_1 - p_2 A_2 - F_x + F_{g_x} \tag{4.4.2}$$

where F_{g_x} is the weight of the fluid in the x-direction in the contraction. The momentum principle for the x-direction can be stated using equation (3.5.6) as

$$\sum F_x = \rho V_{x_2} Q - \rho V_{x_1} Q = \rho Q (V_{x_2} - V_{x_1}) \tag{4.4.3}$$

Similar equations can be developed for other directions:

$$\sum F_y = \rho Q (V_{y_2} - V_{y_1}) \tag{4.4.4}$$

EXAMPLE 4.4.1

The nozzle has the dimensions shown in Figure 4.4.2. What is the force exerted on the flange bolts for a flow rate of 0.08 m³/s? The upstream pipe pressure is 75 kPa.

SOLUTION

Using equation (4.4.3) yields

$$\sum F_x = \rho Q (V_{x_2} - V_{x_1})$$

$$F_1 - F_2 - F_{\text{bolt}} = \rho Q (V_{x_2} - V_{x_1})$$

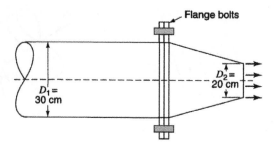

Figure 4.4.2 Nozzle for example 4.4.1.

where $F_1 = p_1A_1$ and $F_2 = p_2A_2 = 0$ (because this end of the nozzle is open to the atmosphere and $p_2 = 0$):

$$p_1A_1 - F_{\text{bolt}} = \rho Q(V_{x_2} - V_{x_1})$$

$$\left(75{,}000 \text{ N/m}^2\right)\left(\frac{\pi(30/100)^2 \text{m}^2}{4}\right) - F_{\text{bolt}} = \left(1000 \text{ kg/m}^3\right)\left(0.08 \text{ m}^3/\text{s}\right)(2.55 \text{ m/s} - 1.13 \text{ m/s})$$

where 1 Pa = 1 N/m^2 and 1 kg/m^3 = 1 N-s^2/m^4, so

$$5298.8 \text{ N} - F_{\text{bolt}} = 113.6 \text{ N}$$

$$F_{\text{bolt}} = 5185 \text{ N(or 1166 lb)}$$

EXAMPLE 4.4.2

Water flows through a horizontal 45° reducing bend shown in Figure 4.4.3, with a 36-in diameter upstream and a 24-in diameter downstream, at the rate of 20 cfs under a pressure of 15 psi at the upstream end of the bend. Neglecting the headloss in the bend, calculate the force exerted by the water on the bend.

SOLUTION

The free-body diagram shown in Figure 4.4.3 is used. To solve this problem, equations (4.4.3) and (4.4.4) will be used:

$$\sum F_x = \rho Q(V_{x_2} - V_{x_1}) \tag{4.4.3}$$

$$\sum F_y = \rho Q(V_{y_2} - V_{y_1}) \tag{4.4.4}$$

The first objective is to determine the velocities in order to apply the energy equation to determine p_2:

$$V_{x_1} = V_1 = Q/A_1 = 20/\pi(3)^2/4 = 2.83 \text{ ft/s}$$
$$V_{y_1} = 0$$
$$V_2 = Q/A = 20/\pi(2)^2/4 = 6.37 \text{ ft/s}$$
$$V_{x_2} = V_{y_2} = V_2 \cos 45° = 6.37(0.707) = 4.50 \text{ ft/s}$$

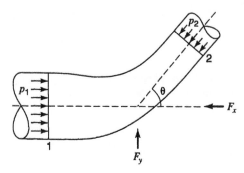

Figure 4.4.3 Reducing bend for example 4.4.2.

Next the energy equation is applied horizontally, $z_1 = z_2 = 0$ and $h_L = 0$:

$$\frac{p_1}{\gamma} + \frac{V_1^2}{2g} = \frac{p_2}{\gamma} + \frac{V_2^2}{2g}$$

$$\frac{15 \text{ lb/in}^2 \times 144 \text{ in}^2/\text{ft}^2}{62.4 \text{ lb/ft}^3} + \frac{(2.83 \text{ ft/s})^2}{2(32.2)\text{ft/s}^2} = \frac{p_2}{\gamma} + \frac{(6.37 \text{ ft/s})^2}{2(32.2)\text{ft/s}^2}$$

$$34.615 + 0.124 = \frac{p_2}{\gamma} + 0.630$$

$$\frac{p_2}{\gamma} = 34.11 \text{ ft}$$

$$p_2 = 34.11 \times \frac{62.4}{144} = 14.78 \text{ lb/in}^2$$

Using equation (4.4.3) yields

$$\sum F_x = p_1 A_1 - p_2 A_2 \cos 45° - F_x = \rho Q(V_{x_2} - V_{x_1})$$

$$F_x = p_1 A_1 - p_2 A_2 \cos 45° - \rho Q(V_{x_2} - V_{x_1})$$

$$F_x = \left(15 \text{ lb/in}^2\right)\left(144 \text{ in}^2/\text{ft}^2\right)\left(\pi \frac{3^2}{4}\text{ft}^2\right) - \left(14.78 \text{ lb/in}^2\right)\left(144 \text{ in}^2/\text{ft}^2\right)\left(\pi \frac{2^2}{4}\text{ft}^2\right)$$

$$- \left[1.94\frac{\text{slugs}}{\text{ft}^3}\right](20 \text{ ft}^3)(4.50 \text{ ft/s} - 2.83 \text{ ft/s})$$

$$F_x = 15,268 - 6686 - 65$$
$$= 8517 \text{ lb}$$

Using equation (4.4.4), we get

$$\sum F_y = F_{y_1} - F_{y_2} + F_y = \rho Q(V_{y_2} - V_{y_1})$$

where $F_{y_1} = 0$ because there is no pressure component in the y direction at 1.

$$- p_2 A_2 \sin 45° + F_y = \rho Q(V_{y_2} - V_{y_1})$$

$$F_y = p_2 A_2 \sin 45° + \rho Q(V_{y_2} - V_{y_1})$$

$$F_y = (14.78)(144)\left(\pi \frac{2^2}{4}\right)(0.707) + (1.94)(20)(4.50 - 0)$$

$$F_y = 4727 + 174$$

$$F_y = 4901 \text{ lb}$$

The resultant force is $F = \sqrt{(8517)^2 + (4901)^2} = 9826$ lb at an angle of $\theta = \tan^{-1}(4901/8517) = 30°$.

4.5   PIPE FLOW IN SIMPLE NETWORKS

4.5.1   Series Pipe Systems

Consider the simple series pipe system in Figure 4.5.1a. Through continuity the discharge is equal in each pipe:

$$Q = Q_1 = Q_2 = Q_3 \tag{4.5.1}$$

and through energy the total headloss is the sum of headlosses in each pipe:

$$h_L = h_{L_1} + h_{L_2} + h_{L_3} \tag{4.5.2}$$

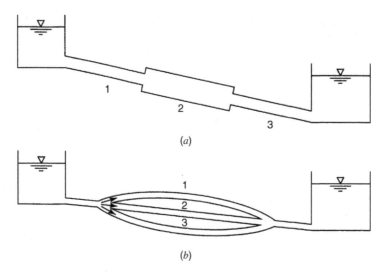

Figure 4.5.1 Pipes in (a) series and (b) parallel.

EXAMPLE 4.5.1

Water flows at a rate of 0.030 m³/s from reservoir 1 to reservoir 2 through three pipes connected in series ($f = 0.025$) as shown in Figure 4.5.2. Neglecting minor losses, determine the difference in water surface elevation.

SOLUTION

Write the energy equation from 1 to 2:

$$z_1 = z_2 + h_{L_f} \qquad \left(\text{or } h_{L_f} = z_1 - z_2\right)$$

$$h_{L_f} = h_{L_A} + h_{L_B} + h_{L_C} = z_1 - z_2$$

Using the Darcy–Weisbach equation (4.3.13)

$$h_{L_f} = f\frac{L}{D}\frac{V^2}{2g}$$

the energy equation is

$$z_1 - z_2 = 0.025\frac{1000}{[200/1000]}\frac{V_A^2}{2(9.81)} + 0.025\frac{1500}{[180/1000]}\frac{V_B^2}{2(9.81)} + 0.025\frac{2000}{[220/1000]}\frac{V_C^2}{2(9.81)}$$

$$z_1 - z_2 = 6.37V_A^2 + 10.62V_B^2 + 11.58V_C^2$$

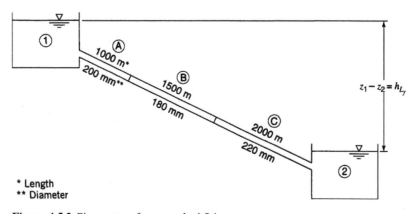

* Length
** Diameter

Figure 4.5.2 Pipe system for example 4.5.1.

Use continuity to determine the velocities:

$$V_A = Q/A_A = 0.03 \bigg/ \frac{\pi(200/1000)^2}{4} = 0.955 \text{ m/s}$$

$$V_A = Q/A_B = 0.03 \bigg/ \frac{\pi(180/1000)^2}{4} = 1.180 \text{ m/s}$$

$$V_C = Q/A_C = 0.03 \bigg/ \frac{\pi(220/1000)^2}{4} = 0.790 \text{ m/s}$$

$$
\begin{aligned}
z_1 - z_2 &= 6.37(0.955)^2 + 10.62(1.180)^2 + 11.58(0.790)^2 \\
&= 5.81 + 14.79 + 7.22 \\
&= 27.82 \text{ m}
\end{aligned}
$$

4.5.2   Parallel Pipe Systems

Consider the simple parallel pipe system in Figure 4.5.1b. Through continuity the total flow is the sum of flow in each of the pipes:

$$Q = Q_1 + Q_2 + Q_3 \tag{4.5.3}$$

Through energy, the flow distribution in the parallel pipes is such that the headloss in each pipe is equal:

$$h_L = h_{L_1} = h_{L_2} = h_{L_3} \tag{4.5.4}$$

EXAMPLE 4.5.2

The three-pipe system shown in Figure 4.5.3 has the following characteristics:

Pipe	D (in)	L (ft)	f
A	8	1500	0.020
B	6	2000	0.025
C	10	3000	0.030

Find the flowrate of water in each pipe and the pressure at point 3. Neglect minor losses.

SOLUTION

Write the energy equation from 1 to 2:

$$\frac{p_1}{\gamma} + \frac{V_1^2}{2g} + z_1 = \frac{p_2}{\gamma} + \frac{V_2^2}{2g} + z_2 + h_{L_f}$$

$$0 + 0 + 200 = 0 + \frac{V_2^2}{2g} + 80 + h_L$$

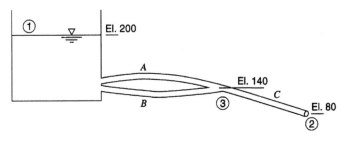

Figure 4.5.3 Pipe system for example 4.5.2.

where V_2 is V_C; $h_{L_f} = h_{L_A} + h_{L_C}$, and using the Darcy–Weisbach equation (4.3.13), we get

$$120 = \frac{V_C^2}{2g} + f_A \frac{L_A}{D_A} \frac{V_A^2}{2g} + f_C \frac{L_C}{D_C} \frac{V_C^2}{2g}$$

Because pipes A and B are parallel, the headloss in A is equal to the headloss in B, so the headloss for B can also be used in the above energy equation instead of for A. This energy equation has two unknowns, V_A and V_C, so that continuity can be used as a second equation:

$$Q_A + Q_B = Q_C$$

$$A_A V_A + A_B V_B = A_C V_C$$

which introduces a third unknown V_A. Because $h_{L_A} = h_{L_B}$, the third equation is

$$f_A \frac{L_A}{D_A} \frac{V_A^2}{2g} = f_B \frac{L_B}{D_B} \frac{V_B^2}{2g}$$

$$0.020 \left(\frac{1500}{8/12} \right) \frac{V_A^2}{2(32.2)} = 0.025 \left(\frac{2000}{6/12} \right) \frac{V_B^2}{2(32.2)}$$

$$0.699 V_A^2 = 1.553 V_B^2$$

$$V_A^2 = 2.221 V_B^2$$

$$V_B = 0.671 V_A$$

So now we have three equations and three unknowns. Using the continuity equation, we get

$$\left[\pi \frac{(8/12)^2}{4} \right] V_A + \left[\pi \frac{(6/12)^2}{4} \right] V_B = \left[\pi \frac{(10/12)^2}{4} \right] V_C$$

$$8^2 V_A + 6^2 V_B = 10^2 V_C$$

substituting

$$V_B = 0.671 V_A$$

$$64 V_A + 36(0.671) V_A = 100 V_C$$

$$88.16 V_A = 100 V_C$$

$$V_A = 1.134 V_C$$

Substitute $V_A = 1.134 V_C$ into the energy equation

$$120 = \frac{V_C^2}{2g} + 0.020 \left(\frac{1500}{8/12} \right) \frac{(1.134 V_C)^2}{2g} + 0.030 \left(\frac{3000}{10/12} \right) \frac{V_C^2}{2g}$$

and solve for V_C:

$$120 = (1 + 57.82 + 108) \frac{V_C^2}{2g} = 166.82 \frac{V_C^2}{2(32.2)}$$

$$V_C = 6.805 \text{ ft/s}$$

The flow rate is then

$$Q = A_C V_C = \left[\pi \frac{(10/12)^2}{4} \right] \times 6.805 = 3.712 \text{ ft}^3/\text{s}$$

The pressure at 3 can be computed using the energy equation from 1 to 3 or from 3 to 2. Using

$$\frac{p_3}{\gamma}+\frac{V_3^2}{2g}+z_3=\frac{p_2}{\gamma}+\frac{V_2^2}{2g}+z_2+h_{L_{3-2}}$$

Because $V_3 = V_2$, the velocity head terms cancel out:

$$\frac{p_3}{\gamma}+140=0+80+h_{L_{3-2}}$$

$$\frac{p_3}{\gamma}=-60+f_C\frac{L_CV_C^2}{D_C2g}$$

$$=-60+0.030\frac{3000}{10/12}\frac{6.805^2}{2(32.2)}$$

$$=-60+77.66$$

$$=17.66$$

and

$$p_3=(62.4)(17.66)=1102\,\text{lb/ft}^2$$
$$=7.65\,\text{lb/in}^2$$

EXAMPLE 4.5.3

The pipe system shown in Figure 4.5.4 connects two reservoirs that have an elevation difference of 20 m. This pipe system consists of 200 m of 50-cm concrete pipe (pipe A), that branches into 400 m of 20-cm pipe (pipe B) and 400 m of 40-cm pipe (pipe C) in parallel. Pipes B and C join into a single 50-cm pipe that is 500 m long (pipe D). For $f = 0.030$ in all the pipes, what is the flow rate in each pipe of the system?

SOLUTION

The objective is to compute the velocity in each pipe. We know that $V_A = V_D$ because they are the same diameter pipe, $h_{L_B} = h_{L_C}$ because pipes B and C are in parallel, and $Q_A = Q_D = Q_B + Q_C$. Express $h_{L_B} = h_{L_C}$ in terms of the velocities

$$f_B\frac{L_B}{D_B}\frac{V_B^2}{2g}=f_C\frac{L_C}{D_C}\frac{V_C^2}{2g}$$

Since $f_B = f_C$ and $L_B = L_C$,

$$\frac{V_B^2}{D_B}=\frac{V_C^2}{D_C}$$

$$\frac{V_B^2}{20/100}=\frac{V_C^2}{40/100}\ \text{or}\ V_B^2=\frac{1}{2}V_C^2\ \text{or}\ V_B=\frac{V_C}{\sqrt2}\ \text{or}\ V_C=\sqrt2\,V_B$$

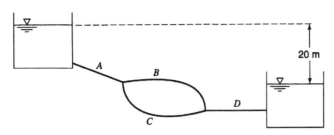

Figure 4.5.4 Pipe system for example 4.5.3.

Using $Q_A = Q_B + Q_C$, we get

$$\left[\frac{\pi(50/100)^2}{4}\right]V_A = \left[\frac{\pi(20/100)^2}{4}\right]V_B + \left[\frac{\pi(40/100)^2}{4}\right]V_C$$

$$50^2 V_A = 20^2 V_B + 40^2 V_C$$

Substituting $V_C = \sqrt{2}\, V_B$ yields

$$2500 V_A = 400 V_B + 1600(\sqrt{2}\, V_B)$$

$$V_A = 1.065 V_B \text{ or } V_B = 0.939\, V_A$$

Next convert the parallel pipes to a single equivalent $D_A = D_D = 50$-cm diameter pipe, with a length of L_E:

$$f\frac{L_B}{D_B}\frac{V_B^2}{2g} = f\frac{L_E}{D_A}\frac{V_A^2}{2g}$$

$$\frac{L_B}{D_B}V_B^2 = \frac{L_E}{D_A}V_A^2$$

$$\frac{400}{(20/100)}V_B^2 = \frac{L_E}{(50/100)}V_A^2$$

$$1000 V_B^2 = L_E V_A^2$$

Substitute $V_B = 0.939\, V_A$:

$$1000(0.939 V_A)^2 = L_E V_A^2$$

$$L_E = 882\text{ m}$$

Write the energy equation from reservoir surface to reservoir surface $\sum h = 20$ m

$$20 = \frac{f(L_A + L_E + L_D)}{(50/100)}\frac{V_A^2}{2g}$$

$$20 = \frac{0.030(200 + 882 + 500)}{(50/100)}\frac{V_A^2}{2(9.81)}$$

$$V_A = 2.033\text{ m/s}, Q_A = \frac{\pi(50/100)^2}{4}(2.033) = 0.399\text{ m}^3/\text{s}$$

Also, $V_A = V_D$, so $Q_D = Q_A$:

$$V_B = 0.939(2.033) = 1.909\text{ m/s}, Q_B = \frac{\pi(20/100)^2}{4}(1.909) = 0.060\text{ m}^3/\text{s}$$

$$Q_C = Q_A - Q_B = 0.399 - 0.060 = 0.339\text{ m}^3/\text{s}.$$

4.5.3 Branching Pipe Flow

Consider the branching pipe system shown in Figure 4.5.5. The following energy equations can be written (neglecting the velocity heads):

$$z_A = z_D + \frac{p_D}{\gamma} + h_{L_{AD}} \tag{4.5.5}$$

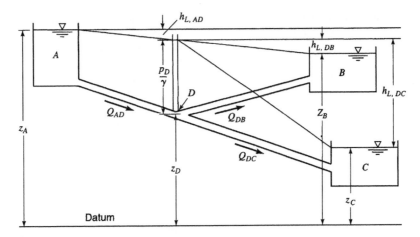

Figure 4.5.5 Branching pipe system.

$$z_B = z_D + \frac{p_D}{\gamma} - h_{L_{DB}} \tag{4.5.6}$$

$$z_C = z_D + \frac{p_D}{\gamma} - h_{L_{DC}} \tag{4.5.7}$$

where the headlosses are defined using the Darcy–Weisbach equation (4.3.13)

$$h_{L_{AD}} = f_{AD} \frac{L_{AD}}{D_{AD}} \frac{V_{AD}^2}{2g} \tag{4.5.8}$$

$$h_{L_{DB}} = f_{DB} \frac{L_{DB}}{D_{DB}} \frac{V_{DB}^2}{2g} \tag{4.5.9}$$

$$h_{L_{DC}} = f_{DC} \frac{L_{DC}}{D_{DC}} \frac{V_{DC}^2}{2g} \tag{4.5.10}$$

The continuity equation is

$$Q_{AD} = Q_{DB} + Q_{DC} \tag{4.5.11}$$

or

$$A_{AD}V_{AD} = A_{DB}V_{DB} + A_{DC}V_{DC} \tag{4.5.12}$$

By substituting the headloss expressions (equations (4.5.8) – (4.5.10)) respectively into equations (4.5.5)–(4.5.7), the three energy equations have four unknowns, p_D/γ, V_{AD}, V_{DB}, and V_{DC}. The continuity equation (4.5.12) provides the fourth equation to solve for the four unknowns.

PROBLEMS

4.1.1 Water flows in a pipe of 3/8-in diameter at a temperature of 70°F. The pressures at 1 and 2 (see Figure P4.1.1) are found to be 15 psi and 20 psi, respectively. Determine the direction of flow. What is the minimum discharge above which the flow will not be laminar?

4.1.2 An experiment was done to determine the hydraulic conductivity K of an aquifer. A dye was injected into the aquifer at point 1 (see Figure P4.1.2) and was 36 hours later observed at point 2. The piezometric head difference between points 1 and 2 was observed to be 0.5 m. Determine the hydraulic conductivity of the aquifer. Take the porosity of the aquifer as 0.24.

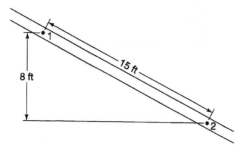

Figure P4.1.1

Figure P4.1.2

4.1.3 The piezometric heads at points 1 and 2 in Figure P4.1.3 are found to be 75 ft and 72.5 ft. If the hydraulic conductivity of the aquifer is 50 ft/day, what is the Darcy flux? Determine the discharge by taking the average thickness of the aquifer as 100 ft.

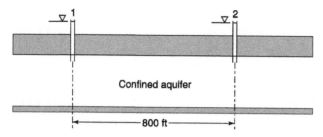

Figure P4.1.3

4.2.1 A Venturi meter with a throat diameter of 150 mm is connected to a pipe of diameter 250 mm to measure the discharge in the pipe. The pressures just upstream of the connection and at the throat were found to be 140 kPa and 80 kPa, respectively. Determine the flow rate in the pipe. Take the coefficient of discharge for the Venturi meter as 0.98.

4.2.2 The pressure difference between the upstream end section and the throat section of a Venturi meter connected to a pipe flow is found to be 12 psi. The diameters of the pipe and the throat of the Venturi meter are 1-7/8 in and 1-1/8 in, respectively. The actual flow in the pipe is 0.353 ft^3/s. Calibrate the Venturi meter for Reynolds numbers at the throat greater than 2×10^6.

4.2.3 Draw (to scale) the hydraulic grade line (HGL) and the energy grade line (EGL) of the system in Figure 4.3.6 (example 4.3.5). Take the length of each pipe as given and neglect the height of the elbows.

4.2.4 Suppose the water fountain in Fountain Hills, Arizona (see Figure P4.2.4), rises vertically to 150 ft above the lake (when operated). Neglecting wind effects and minor losses, determine the velocity at which the water is ejected.

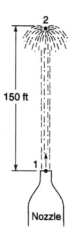

Figure P4.2.4

4.3.1 Develop the expression for the headloss in a pipe for steady, laminar flow of an incompressible fluid (equation (4.3.8)).

4.3.2 Suppose a globe valve ($K = 10$) is present in a pipeline of diameter 300 mm that has a friction factor f of 0.020. What is the equivalent length of this pipe that would cause equal headloss as the globe valve for the same discharge? Repeat this problem for a pipe of 150-mm diameter that has the same friction factor.

4.3.3 Two materials (wrought iron, $k_s = 0.046$ mm, and galvanized iron, $k_s = 0.15$ mm) are being considered for a new pipeline. The expected discharge is 0.15 m^3/s. Both the headloss and cost are sought for. Wrought iron pipe costs 5 cents more than the galvanized iron pipe for every meter length of the pipe. Determine the tradeoff between the cost and the energy head for pipe diameters of 200 mm and 150 mm. Take the temperature as 15°C.

4.3.4 For the pipe system shown in Figure P4.3.4 determine the proportion of each pipe so that the pipe friction loss in each pipe is the same. Assume the same friction factor in all pipes.

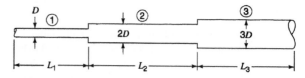

Figure P4.3.4

4.3.5 For the pipe system in problem 4.3.4, compare the expansion losses. Suppose the diameter of the third pipe was twice that of the second. What can you infer by comparing the headlosses at the expansion?

4.4.1 A plate is held against a horizontal water jet in the horizontal plane as shown in Figure P4.4.1. The jet has a diameter of 30 mm and an unknown velocity. A force of 200 N is required to hold the plate in the position. Determine the velocity of flow of the jet just before it hits the plate. What is the discharge?

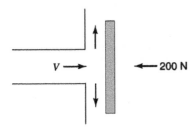

Figure P4.4.1

4.4.2 Suppose the nozzle in example 4.4.1 is connected to the pipe by flange bolts of 1-cm diameter. If the allowable tensile stress in the bolts is 330 N/cm^2, how many bolts are required for a safe connection?

4.4.3 Suppose the horizontal reducing bend in example 4.4.2 has an unknown bend angle (45° in example 4.4.2). What should be this bend angle so that the horizontal component force F_x is three times the vertical component force, F_y, in magnitude?

4.5.1 If the headloss in example 4.5.1 were 15 m, what would be the discharge? Also, determine the velocity in each pipe.

4.5.2 The rate of flow in the pipe system in Figure P4.5.2 is 0.05 m^3/s. The pressure at point 1 is measured to be 260 kPa. All the pipes are galvanized iron with roughness value of 0.15 mm. Determine the pressure at point 2. Take the loss coefficient for the sudden contraction as 0.05 and $v = 1.141 \times 10^{-6}$ m^2/s.

Figure P4.5.2

4.5.3 The pressure difference between points 1 and 2 in the series pipe system in Figure P4.5.3 is 15 psi. All the pipes are galvanized iron with roughness value of 0.0005 ft. The loss coefficient at the sudden contraction is 0.05. Determine the flow rate in the system. The prevailing temperature is 70°F.

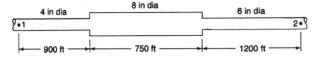

Figure P4.5.3

4.5.4 The pressure at point 1 in the parallel pipe system shown in Figure P4.5.4 is 750 kPa. If the flow rate through the system is 0.50 m^3/s, what is the pressure at point 2? Neglect minor losses. All the pipes are steel with roughness value of 0.046 mm. Also, determine the fraction of the flow in each of the parallel pipes and check your solution. Take $v = 1.141 \times 10^{-6}$ m^2/s.

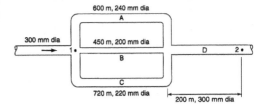

Figure P4.5.4

4.5.5 In problem 4.5.4 above, how far will the water flow before all its energy head is exhausted? Assume pipe D continues horizontally downstream without any other structure.

4.5.6 If the pressure difference between points 1 and 2 in Figure P4.5.6 is 30 psi, what will be the flow rate? The pipes are galvanized iron with $k_s = 0.0005$ ft. Take $v = 1.06 \times 10^{-5}$ ft^2/s and neglect minor losses.

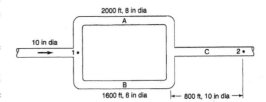

Figure P4.5.6

4.5.7 For the branching pipe system given in Figure P4.5.7, determine the flow to and the elevation of the third reservoir. Neglect minor losses and the velocity heads. The pipes are galvanized iron with $k_s = 0.0005$ ft and $v = 1.06 \times 10^{-5}$ ft^2/s.

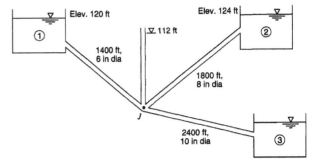

Figure P4.5.7

REFERENCES

ASHRAE, *ASHRAE Handbook, 1977 Fundamentals*, Am. Soc. of Heating, Refrigerating and Air Conditioning Engineers, New York, 1977.

Beij, K. H., "Pressure Losses for Fluid Flow in 90° Pipe Bends," *J. Res. Nat. Bur. Std.*, 21, 1938.

Colebrook, C. F., and C. M. White, "Turbulent Flow in Pipes with Particular Reference to the Transition Region Between Smooth and Rough Pipe Laws," *Institute of Civil Engineers*, London, vol. 11, p. 133, 1939.

Idel'chik, I. E., *Handbook of Hydraulic Resistance Coefficients of Local Resistance and of Friction*, Trans. A. Barouch, Israel Program for Scientific Translation, 1966.

Moody, L. R., "Friction Factors for Pipe Flow," *Trans., Amer. Soc. Mech. Engrs.*, 66, Nov., 1944.

Nikuradse, J., "Gesetzmassigkeiten der turbulenten Stromung in glatten Rohren," *VDI Forschungsheft* 356, 1932.

Nikuradse, J., "Stromungsgesetze in rauhen Rohren," *VDI-Forschungsh*, 361, 1933. Also translated in *NACA Tech.* Memo 1292.

Roberson, J. A., and C. T. Crowe, *Engineering Fluid Mechanics*, Houghton Mifflin, Boston, 1990.

Roberson, J. A., J. J. Cassidy, and M. H. Chaudhry, *Hydraulic Engineering*, Houghton Mifflin, Boston, 1988.

Streeter, V. L. (editor) *Handbook of Fluid Dynamics*, McGraw-Hill, New York, 1961.

Velon, J. P., and T. J. Johnson, "Water Distribution and Treatment," *Davis' Handbook of Applied Hydraulics*, 4th edition, edited by V. I. Zippano and H. Hasen, McGraw-Hill, New York, 1993.

Equating (5.1.16) and (5.1.17) yields

$$C_f \rho \frac{V^2}{2} = \gamma R S_0 \qquad (5.1.18)$$

and solving for the velocity gives

$$V = \sqrt{\frac{2g}{C_f}} \sqrt{R S_0} \qquad (5.1.19)$$

Defining $C = \sqrt{2g/C_f}$, then equation (5.1.19) can be simplified to the well-known *Chezy equation*

$$V = C\sqrt{R S_0} \qquad (5.1.20)$$

where C is referred to as the *Chezy coefficient*.

Robert Manning (1891, 1895) derived the following empirical relation for C based upon experiments:

$$C = \frac{1}{n} R^{1/6} \qquad (5.1.21)$$

where n is the Manning roughness coefficient. Values of n are listed in Table 5.1.1. Values of n for natural channels have been also published by the U.S. Geological Survey (Barnes, 1962). Substituting C from equation (5.1.21) into equation (5.1.20) results in the *Manning equation*

$$V = \frac{1}{n} R^{2/3} S_0^{1/2} \qquad (5.1.22)$$

which is valid for SI units and $S_0 = S_f$.

Table 5.1.1 Values of the Roughness Coefficient n
(Boldface figures are values generally recommended in design)

Type of channel and description	Minimum	Normal	Maximum
A. Closed conduits flowing partly full			
A-1. Metal			
a. Brass, smooth	0.009	**0.010**	0.013
b. Steel			
1. Lockbar and welded	0.010	0.012	0.014
2. Riveted and spiral	0.013	0.016	0.017
c. Cast iron			
1. Coated	0.010	0.013	0.014
2. Uncoated	0.011	0.014	0.016
d. Wrought iron			
1. Black	0.012	0.014	0.015
2. Galvanized	0.013	0.016	0.017
e. Corrugated metal			
1. Subdrain	0.017	0.019	0.021
2. Storm drain	0.021	**0.024**	0.030
A -2. Nonmetal			
a. Lucite	0.008	0.009	0.010
b. Glass	0.009	**0.010**	0.013
c. Cement			
1. Neat, surface	0.010	0.011	0.013
2. Mortar	0.011	0.013	0.015

(Continued)

Table 5.1.1 (*Continued*)

Type of channel and description	Minimum	Normal	Maximum
d. Concrete			
1. Culvert, straight and free of debris	0.010	0.011	0.013
2. Culvert with bends, connections, and some debris	0.011	**0.013**	0.014
3. Finished	0.011	0.012	0.014
4. Sewer with manholes, inlet, etc., straight	0.013	0.015	0.017
5. Unfinished, steel form	0.012	0.013	0.014
6. Unfinished, smooth wood form	0.012	**0.014**	0.016
7. Unfinished, rough wood form	0.015	0.017	0.020
e. Wood			
1. Stave	0.010	0.012	0.014
2. Laminated, treated	0.015	0.017	0.020
f. Clay			
1. Common drainage title	0.011	0.013	0.017
2. Vitrified sewer	0.011	0.014	0.017
3. Vitrified sewer with manholes, inlet, etc.	0.013	0.015	0.017
4. Vitrified subdrain with open joint	0.014	0.016	0.018
g. Brickwork			
1. Glazed	0.011	0.013	0.015
2. Lined with cement mortar	0.012	0.015	0.017
h. Sanitary sewers coated with sewage slimes, with bends and connections	0.012	0.013	0.016
i. Paved invert, sewer, smooth bottom	0.016	0.019	0.020
j. Rubble masonry, cemented	0.018	0.025	0.030
B. Lined or built-up channels			
B-1. Metal			
a. Smooth steel surface			
1. Unpainted	0.011	**0.012**	0.014
2. Painted	0.012	0.013	0.017
b. Corrugated	0.021	0.025	0.030
B-2. Nonmetal			
a. Cement			
1. Neat, surface	0.010	0.011	0.013
2. Mortar	0.011	0.013	0.015
b. Wood			
1. Planed, untreated	0.010	0.012	0.014
2. Planed, creosoted	0.011	0.012	0.015
3. Unplaned	0.011	0.013	0.015
4. Plank with battens	0.012	0.015	0.018
5. Lined with roofing paper	0.010	0.014	0.017
c. Concrete			
1. Trowel finish	0.011	**0.013**	0.015
2. Float finish	0.013	0.015	0.016
3. Finished, with gravel on bottom	0.015	0.017	0.020
4. Unfinished	0.014	0.017	0.020
5. Gunite, good section	0.016	0.019	0.023
6. Gunite, wavy section	0.018	0.022	0.025
7. On good excavated rock	0.017	0.020	—
8. On irregular excavated rock	0.022	0.027	—
d. Concrete bottom float finished with sides of			
1. Dressed stone in mortar	0.015	0.017	0.020
2. Random stone in mortar	0.017	0.020	0.024

Table 5.1.1 (*Continued*)

Type of channel and description	Minimum	Normal		Maximum
3. Cement rubble masonry, plastered	0.016	0.020		0.024
4. Cement rubble masonry	0.020	0.025		0.030
5. Dry rubble or riprap	0.020	0.030		0.035
e. Gravel bottom with sides of				
1. Formed concrete	0.017	0.020		0.025
2. Random stone in mortar	0.020	0.023		0.026
3. Dry rubble or riprap	0.023	0.033		0.036
f. Brick				
1. Glazed	0.011	**0.013**		0.015
2. In cement mortar	0.012	**0.015**		0.018
g. Masonry				
1. Cemented rubble	0.017	0.025		0.030
2. Dry rubble	0.023	0.032		0.035
h. Dressed ashlar	0.013	0.015		0.017
i. Asphalt				
1. Smooth	0.013	0.013		—
2. Rough	0.016	0.016		—
j. Vegetal lining	0.030	—		0.500
C. Excavated or dredged				
a. Earth, straight and uniform				
1. Clean, recently completed	0.016	0.018		0.020
2. Clean, after weathering	0.018	**0.022**		0.025
3. Gravel, uniform section, clean	0.022	0.025		0.030
4. With short grass, few weeds	0.022	0.027		0.033
b. Earth, winding and sluggish				
1. No vegetation	0.023	0.025		0.030
2. Grass, some weeds	0.025	0.030		0.033
3. Dense weeds or aquatic plants in deep channels	0.030	0.035		0.040
4. Earth bottom and rubble sides	0.028	0.030		0.035
5. Stony bottom and weedy banks	0.025	0.035		0.040
6. Cobble bottom and clean sides	0.030	0.040		0.050
c. Dragline-excavated or dredged				
1. No vegetation	0.025	0.028		0.033
2. Light brush on banks	0.035	0.050		0.060
d. Rock cuts				
1. Smooth and uniform	0.025	0.035		0.040
2. Jagged and irregular	0.035	0.040		0.050
c. Channels not maintained, weeds and brush uncut				
1. Dense weeds, high as flow depth	0.050	0.080		0.120
2. Clean bottom, brush on sides	0.040	0.050		0.080
3. Same, highest stage of flow	0.045	0.070		0.110
4. Dense brush, high stage	0.080	0.100		0.140
D. Natural streams				
D-1. Minor streams (top width at flood stage <100 ft)				
a. Streams on plain				
1. Clean, straight, full stage, no rifts or deep pools	0.025	0.030		0.033
2. Same as above, but more stones and weeds	0.030	0.035		0.040
3. Clean, winding, some pools and shoals	0.033	0.040		0.045
4. Same as above, but some weeds and stones	0.035	0.045		0.050
5. Same as above, lower stages, more ineffective slopes and sections	0.040	0.048		0.055

(*Continued*)

Table 5.1.1 (*Continued*)

Type of channel and description	Minimum	Normal	Maximum
6. Same as 4, but more stones	0.045	0.050	0.060
7. Sluggish reaches, weedy, deep pools	0.050	0.070	0.080
8. Very weedy reaches, deep pools, or floodways with heavy stand of timber and underbrush	0.075	0.100	0.150
b. Mountain streams, no vegetation in channel, banks usually steep, trees and brush along banks submerged at high stages			
1. Bottom: gravels, cobbles, and few boulders	0.030	0.040	0.050
2. Bottom: cobbles with large boulders	0.040	0.050	0.070
D-2. Flood plains			
a. Pasture, no brush			
1. Short grass	0.025	0.030	0.035
2. High grass	0.030	0.035	0.050
b. Cultivated areas			
1. No crop	0.020	0.030	0.040
2. Mature row crops	0.025	0.035	0.045
3. Mature field crops	0.030	0.040	0.050
c. Brush			
1. Scattered brush, heavy weeds	0.035	0.050	0.070
2. Light brush and trees, in winter	0.035	0.050	0.060
3. Light brush and trees, in summer	0.040	0.060	0.080
4. Medium to dense brush, in winter	0.045	0.070	0.110
5. Medium to dense brush, in summer	0.070	0.100	0.160
d. Trees			
1. Dense willows, summer, straight	0.110	0.150	0.200
2. Cleared land with tree stumps, no sprouts	0.030	0.040	0.050
3. Same as above, but with heavy growth of sprouts	0.050	0.060	0.080
4. Heavy stand of timber, a few down trees, little undergrowth, flood stage below branches	0.080	0.100	0.120
5. Same as above, but with flood stage reaching branches	0.100	0.120	0.100
D-3. Major streams (top width at flood stage > 100 ft). The *n* value is less than that for minor streams of similar description, because banks offer less effective resistance.			
a. Regular section with no boulders or brush	0.025	—	0.060
b. Irregular and rough section	0.035	—	0.100

Source: Chow (1959).

Manning's equation in SI units can also be expressed as

$$Q = \frac{1}{n} A R^{2/3} S_0^{1/2} \qquad (5.1.23)$$

For V in ft/sec and R in feet (U.S. customary units), equation (5.1.22) can be rewritten as

$$V = \frac{1.49}{n} R^{2/3} S_0^{1/2} \qquad (5.1.24)$$

and equation (5.1.23) can be written as

$$Q = \frac{1.49}{n} A R^{2/3} S_0^{1/2} \qquad (5.1.25)$$

where A is in ft^2 and $S_0 = S_f$. Table 5.1.2 lists the geometric function for channel elements.

Table 5.1.2 Geometric Functions for Channel Elements

Section:	Rectangle	Trapezoid	Triangle	Circle
Area A	$B_w y$	$(B_w + zy)y$	zy^2	$\frac{1}{8}(\theta - \sin \theta)d_o^2$
Wetted perimeter P	$B_w + 2y$	$B_w + 2y\sqrt{1+z^2}$	$2y\sqrt{1+z^2}$	$\frac{1}{2}\theta d_o$
Hydraulic radius R	$\frac{B_w y}{B_w + 2y}$	$\frac{(B_w + zy)y}{B_w + 2y\sqrt{1+z^2}}$	$\frac{zy}{2\sqrt{1+z^2}}$	$\frac{1}{4}\left(1 - \frac{\sin \theta}{\theta}\right)d_o$
Top width B	B_w	$B_w + 2zy$	$2zy$	$\left[\sin\left(\frac{\theta}{2}\right)\right]d_o$ or $2\sqrt{y(d_o - y)}$
$\frac{2dR}{3Rdy} + \frac{1}{A}\frac{dA}{dy}$	$\frac{5B_w + 6y}{3y(B_w + 2y)}$	$\dfrac{(B_w + 2zy)\left(5B_w + 6y\sqrt{1+z^2}\right) + 4zy^2\sqrt{1+z^2}}{3y(B_w + zy)\left(B_w + 2y\sqrt{1+z^2}\right)}$	$\frac{8}{3y}$	$\dfrac{4(2\sin\theta + 3\theta - 5\theta\cos\theta)}{3d_o\theta(\theta - \sin\theta)\sin(\theta/2)}$ where $\theta = 2\cos^{-1}\left(1 - \frac{2y}{d_o}\right)$

Source: Chow (1959) (with additions).

To determine the normal depth (for uniform flow), equation (5.1.23) or (5.1.25) can be solved with a specified discharge. Because the original shear stress τ_0 in equation (5.1.17) is for fully turbulent flow, Manning's equation is valid only for fully turbulent flow. Henderson (1966) presented the following criterion for fully turbulent flow in an open channel:

$$n^6\sqrt{RS_f} \geq 1.9 \times 10^{-13} \quad (R \text{ in feet}) \tag{5.1.26a}$$

$$n^6\sqrt{RS_f} \geq 1.1 \times 10^{-13} \quad (R \text{ in meters}) \tag{5.1.26b}$$

EXAMPLE 5.1.1 An 8-ft wide rectangular channel with a bed slope of 0.0004 ft/ft has a depth of flow of 2 ft. Assuming steady uniform flow, determine the discharge in the channel. The Manning roughness coefficient is $n = 0.015$.

SOLUTION From equation (5.1.25), the discharge is

$$Q = \frac{1.49}{n}AR^{2/3}S_0^{1/2}$$

$$= \frac{1.49}{0.015}(8)(2)\left[\frac{(8)(2)}{8+2(2)}\right]^{2/3}(0.0004)^{1/2}$$

$$= 38.5 \text{ ft}^3/\text{s}$$

EXAMPLE 5.1.2

Solve example 5.1.1 using SI units.

SOLUTION

The channel width is 2.438 m, with a depth of flow of 0.610 m. Using equation (5.1.23), the discharge is

$$Q = \frac{1}{n} A R^{2/3} S_0^{1/2}$$

$$= \frac{1}{0.015} (2.438)(0.610) \left[\frac{(2.438)(0.610)}{2.438 + 2(0.610)} \right]^{2/3} (0.0004)^{1/2}$$

$$= 1.09 \, \text{m}^3/\text{s}$$

EXAMPLE 5.1.3

Determine the normal depth (for uniform flow) if the channel described in example 5.1.1 has a flow rate of 100 cfs.

SOLUTION

This problem is solved using Newton's method wish Q_j defined by equation (5.1.25):

$$Q_j = \frac{1.49}{n} S_0^{1/2} \frac{(B_w y_j)^{5/3}}{(B_w + 2y_j)^{2/3}}$$

$$Q_j = \frac{1.49}{0.015} (0.0004)^{1/2} \frac{(8y_j)^{5/3}}{(8 + 2y_j)^{2/3}} = 1.987 \frac{(8y_j)^{5/3}}{(8 + 2y_j)^{2/3}}$$

Using a numerical method such as Newton's method (see Appendix A), the normal depth is 3.98 ft.

5.1.3 Best Hydraulic Sections for Uniform Flow in Nonerodible Channels

The conveyance of a channel section increases with an increase in the hydraulic radius or with a decrease in the wetted perimeter. Consequently, the channel section with the smallest wetted perimeter for a given channel section area will have maximum conveyance, referred to as the *best hydraulic section* or the cross-section of greatest hydraulic efficiency. Table 5.1.3 presents the geometric elements of the best hydraulic sections for six cross-section shapes. These sections may not always be practical because of difficulties in construction and use of material. The concept of

Table 5.1.3 Best Hydraulic Sections

Cross-section	Area A	Wetted perimeter P	Hydraulic radius R	Top width T	Hydraulic depth D
Trapezoid, half of a hexagon	$\sqrt{3}y^2$	$2\sqrt{3}y$	$\frac{1}{2}y$	$\frac{4}{3}\sqrt{3}y$	$\frac{3}{4}y$
Rectangle, half of a square	$2y^2$	$4y$	$\frac{1}{2}y$	$2y$	y
Triangle, half of a square	y^2	$2\sqrt{2}y$	$\frac{1}{4}\sqrt{2}y$	$2y$	$\frac{1}{2}y$
Semicircle	$\frac{\pi}{2}y^2$	πy	$\frac{1}{2}y$	$2y$	$\frac{\pi}{4}y$
Parabola, $T = 2\sqrt{2}y$	$\frac{4}{3}\sqrt{2}y^2$	$\frac{8}{3}\sqrt{2}y$	$\frac{1}{2}y$	$2\sqrt{2}y$	$\frac{2}{3}y$
Hydrostatic catenary	$1.39586y^2$	$2.9836y$	$0.46784y$	$1.917532y$	$0.72795y$

Source: Chow (1959).

best hydraulic section is only for nonerodible channels. Even though the best hydraulic section gives the minimum area for a given discharge, it may not necessarily have the minimum excavation.

EXAMPLE 5.1.4

Determine the cross-section of greatest hydraulic efficiency for a trapezoidal channel if the design discharge is 10.0 m³/sec, the channel slope is 0.00052, and Manning's $n = 0.025$.

SOLUTION

From Table 5.1.3, the hydraulic radius should be $R = y/2$, so that the width B and area A are

$$B = \frac{2\sqrt{3}y}{3} = 1.155y \quad \left(\text{because } B = \frac{1}{3}P \text{ for half of a hexagon}\right)$$
$$A = \sqrt{3}y^2 = 1.732y^2$$

Manning's equation (5.1.23) is used to determine the depth:

$$Q = \frac{1}{n}AR^{2/3}S_0^{1/2} = \frac{1}{0.025}\left(1.732y^2\right)\left(\frac{y}{2}\right)^{2/3}(0.00052)^{1/2} = 10$$

so

$$\frac{10 \times 0.025 \times 2^{2/3}}{1.732(0.00052)^{1/2}} = y^{8/3}$$

Thus, $y = 2.38$ m, so that $B = 2.75$ m and $A = 9.81$ m².

5.1.4 Slope-Area Method

The slope-area method can be used to estimate the flood discharge through a channel or river reach of length Δx with known cross-sectional areas of flow at the upstream, A_u, and downstream, A_d, ends of the reach. The use of high-water marks from a flood and a survey of the cross sections allow computation of the cross-sectional areas of flow. Manning's equation (5.1.25) can be expressed as

$$Q = K\sqrt{S_o} \tag{5.1.27}$$

where K is the conveyance factor expressed as $K = \frac{1.49}{n}AR^{2/3}$. Conveyance is a measure of the carrying capacity of a channel since it is directly proportional to the discharge Q. The average conveyance factor is the geometric mean of the conveyance factors at the upstream, K_u, and the downstream, K_d, ends of the channel reach, i.e.,

$$\overline{K} = \sqrt{K_u K_d} \tag{5.1.28}$$

The discharge is then expressed as

$$Q = \overline{K}\sqrt{S} \tag{5.1.29}$$

where S is the water slope given as $S = \frac{(z_u - z_d)}{\Delta x}$, z_u and z_d are the water surface elevations at the upstream and downstream ends of the reach, respectively.

Alternatively the friction slope, S_f could be used in equation (5.1.29), $Q = \overline{K}\sqrt{S_f}$, where (Chow, 1959)

$$S_f = \left[(z_u - z_d) - k\left(\alpha_u \frac{V_u^2}{2g} - \alpha_d \frac{V_d^2}{2g}\right)\right]/\Delta x \tag{5.1.30}$$

The difference in water surface elevations is referred to as the fall. The k is a factor to account for a contraction and expansion of a reach. For a contracting reach $V_u < V_d$ so $k = 1.0$ and for an expanding reach $(V_u > V_d)$ so $k = 0.5$. The first approximation would compute the discharge using $Q = \overline{K}\sqrt{S_f}$ with the friction slope computed ignoring the velocity heads. Using the first approximation of Q, the upstream and downstream velocity heads are computed for the next approximation of the friction slope, which is used to compute the second approximation of the discharge. The procedure continues computing the new friction slope using the last discharge approximation to compute the new discharge. This process continues until the discharges approximations do not change significantly.

5.2   SPECIFIC ENERGY, MOMENTUM, AND SPECIFIC FORCE

5.2.1   Specific Energy

The *total head* or *energy head*, H, at any location in an open-channel flow can be expressed as

$$H = y + z + \frac{V^2}{2g} \tag{5.1.6}$$

which assumes that the velocity distribution is uniform (i.e., $\alpha = 1$) and the pressure distribution is hydrostatic (i.e., $p = \gamma y$). Using the channel bottom as the datum (i.e., $z = 0$) then define the total head above the channel bottom as the *specific energy*

$$E = y + \frac{V^2}{2g} \tag{5.2.1}$$

Using continuity $(V = Q/A)$, the specific energy can be expressed in terms of the discharge as

$$E = y + \frac{Q^2}{2gA^2} \tag{5.2.2}$$

Specific energy curves, such as are shown in Figure 5.2.1 and 5.2.2, can be derived using equation (5.2.2).

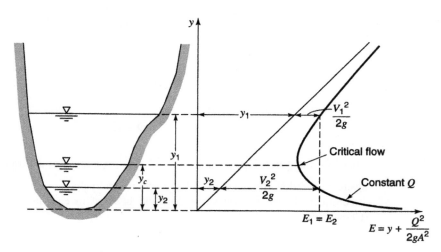

Figure 5.2.1 Specific energy.

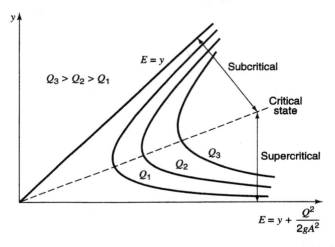

Figure 5.2.2 Specific energy showing subcritical and supercritical flow ranges.

Critical flow occurs when the specific energy is minimum for a given discharge (i.e., $dE/dy = 0$), so that

$$\frac{dE}{dy} = 1 - \frac{Q^2}{gA^3}\frac{dA}{dy} = 0 \qquad (5.2.3)$$

Referring to Figure 5.2.1, the top-width is defined as $T = dA/dy$ so equation (5.2.3) can be expressed as

$$1 - \frac{TQ^2}{gA^3} = 0 \qquad (5.2.4)$$

or

$$\frac{TQ^2}{gA^3} = 1 \qquad (5.2.5)$$

To denote critical conditions use T_c, A_c, V_c, and y_c, so

$$\frac{T_c Q^2}{gA_c^3} = 1 \qquad (5.2.6)$$

or

$$\frac{V_c^2}{g} = \frac{A_c}{T_c} \qquad (5.2.7)$$

Equation (5.2.6) or (5.2.7) can be used to determine the critical depth and/or the critical velocity.

Rearranging equation (5.2.7) yields

$$\frac{V_c^2}{g(A_c/T_c)} = 1 \qquad (5.2.8)$$

The *hydraulic depth* is defined as $D = A/T$ so equation (5.2.7) becomes

$$\frac{V_c^2}{gD_c} = 1 \qquad (5.2.9)$$

or

$$\frac{V_c}{\sqrt{gD_c}} = 1 \tag{5.2.10}$$

This is basically the *Froude number*, F_r, which is 1 at critical flow:

$$F_r = \frac{V}{\sqrt{gD}} \begin{cases} <1 \text{ subcritical flow} \\ =1 \text{ critical flow} \\ >1 \text{ supercritical flow} \end{cases} \tag{5.2.11}$$

Figure 5.2.2 illustrates the range of subcritical flow and the range of supercritical flow along with the location of the critical states. Note the relationship of the specific energy curves and the fact that $Q_3 > Q_2 > Q_1$. Figure 5.2.1 illustrates the alternate depths y_1 and y_2 for which $E_1 = E_2$ or

$$y_1 + \frac{V_1^2}{2g} = y_2 + \frac{V_2^2}{2g} \tag{5.2.12}$$

For a rectangular channel $D_c = A_c/T_c = y_c$, so equation (5.2.10) for critical flow becomes

$$\frac{V_c}{\sqrt{gy_c}} = 1 \tag{5.2.13}$$

If we let q be the flow rate per unit width of channel for a rectangular channel, i.e., $q = Q/B$ where $T = B$, the width of the channel (or $q = Q/T$), then equation (5.2.6) can be rearranged, $T_c Q^2/(gT_c^3 y_c^3) = q^2/(gy_c) = 1$, and solved for y_c to yield

$$y_c = \left(\frac{q^2}{g}\right)^{1/3} \tag{5.2.14}$$

EXAMPLE 5.2.1

Compute the critical depth for the channel in example 5.1.1 using a discharge of 100 cfs.

SOLUTION

Using equation (5.2.13), $V_c = \sqrt{gy_c} = Q/A = 100/8y_c$, so

$$y_c^{3/2} = \frac{100}{8\sqrt{g}} \text{ or } y_c = \left(\frac{100}{8\sqrt{g}}\right)^{2/3} = 1.69 \text{ ft}$$

Alternatively, using equation (5.2.14) yields

$$y_c = \left(\frac{(100/8)^2}{g}\right)^{1/3} = 1.69 \text{ ft}$$

EXAMPLE 5.2.2

For a rectangular channel of 20 ft width, construct a family of specific energy curves for $Q = 0, 50, 100$, and 300 cfs. Draw the locus of the critical depth points on these curves. For each flow rate, what is the minimum specific energy found from these curves?

SOLUTION   The specific energy is computed using equation (5.2.1):

$$E = y + \frac{V^2}{2g} = y + \frac{1}{2g}\left(\frac{Q}{A}\right)^2 = y + \frac{1}{2g}\frac{Q^2}{(20y)^2} = y + \frac{Q^2}{25,760y^2}$$

Computing critical depths for the flow rates using equation (5.2.14) with $q = Q/B$ yields

$Q = 0$:
$$y_c = \sqrt[3]{\frac{Q^2}{B^2 g}} = 0$$

$Q = 50\,\text{cfs}$:
$$y_c = \sqrt[3]{\frac{Q^2}{B^2 g}} = \sqrt[3]{\frac{50^2}{20^2(32.2)}} = 0.58\,\text{ft}$$

$Q = 100\,\text{cfs}$:
$$y_c = \sqrt[3]{\frac{Q^2}{B^2 g}} = \sqrt[3]{\frac{100^2}{20^2(32.2)}} = 0.92\,\text{ft}$$

$Q = 300\,\text{cfs}$:
$$y_c = \sqrt[3]{\frac{Q^2}{B^2 g}} = \sqrt[3]{\frac{300^2}{20^2(32.2)}} = 1.9\,\text{ft}$$

Computed specific energies are listed in Table 5.2.1.

The specific energy curves are shown in Figure 5.2.3. The minimum specific energies are:

$Q = 50\,\text{cfs}$: $E_{min} = 0.868$
$Q = 100\,\text{cfs}$: $E_{min} = 1.379$
$Q = 300\,\text{cfs}$: $E_{min} = 2.868$

Table 5.2.1 Computed Specific Energy Values for Example 5.2.2

Depth, y (ft)	Specific energy, E (ft-lb/lb)			
	$Q = 0$	$Q = 50$	$Q = 100$	$Q = 300$
0.5	0.50	0.89	2.05	14.86
0.6	0.60	0.87	1.68	10.57
0.8	0.80	0.95	1.41	6.41
1.0	1.00	1.10	1.39	4.59
1.2	1.20	1.27	1.47	3.69
1.4	1.40	1.45	1.60	3.23
1.6	1.60	1.64	1.75	3.00
1.8	1.80	1.83	1.92	2.91
2.0	2.00	2.02	2.10	2.90
2.2	2.20	2.22	2.28	2.94
2.4	2.40	2.42	2.47	3.02
2.6	2.60	2.61	2.66	3.13
2.8	2.80	2.81	2.85	3.26
3.0	3.00	3.01	3.04	3.40
3.5	3.50	3.51	3.53	3.79
4.0	4.00	4.01	4.02	4.22
4.5	4.50	4.50	4.52	4.68
5.0	5.00	5.00	5.02	5.14

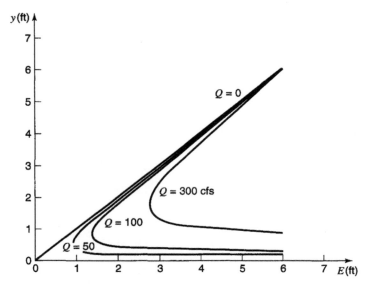

Figure 5.2.3 Specific energy curves for example 5.2.2.

EXAMPLE 5.2.3

A rectangular channel 2 m wide has a flow of 2.4 m^3/s at a depth of 1.0 m. Determine whether critical depth occurs at (a) a section where a hump of $\Delta z = 20$ cm high is installed across the channel bed, (b) a side wall constriction (with no humps) reducing the channel width to 1.7 m, and (c) both the hump and side wall constrictions combined. Neglect headlosses of the hump and constriction caused by friction, expansion, and contraction.

SOLUTION

(a) The computation is focused on determining the critical elevation change in the channel bottom (hump) Δz_{crit} that causes a critical depth at the hump. The energy equation is $E = E_{min} + \Delta z_{crit}$ or $\Delta z_{crit} = E - E_{min}$, where E is the specific energy of the channel flow and E_{min} is the minimum specific energy, which is at critical depth by definition. If $\Delta z_{crit} \leq \Delta z$, then critical depth will occur. Using equation (5.2.2) yields

$$E = y + \frac{Q^2}{2gA^2} = y + \frac{q^2}{2gy^2}$$

which can be solved for q:

$$q = \sqrt{2g(y^2 E - y^3)}$$

Differentiating this equation with respect to y because maximum q and minimum E are equivalent (see Figure 5.2.4) yields

$$\frac{dq}{dy} = \frac{d}{dy}\left[\sqrt{2g(y^2 E - y^3)}\right] = 0$$

$$y_c = \frac{2}{3}E_{min} \quad \text{or} \quad E_{min} = \frac{3}{2}y_c$$

To compute specific energy, use

$$E = y + \frac{q^2}{2gy^2} = 1.0 + \frac{(2.4/2.0)^2}{2(9.81)(1)^2} = 1.073 \text{ m}$$

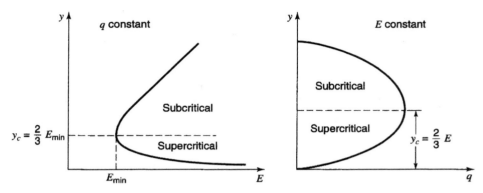

Figure 5.2.4 Specific energy curve and y versus q for constant E.

Next compute E_{min} using $E_{min} = 3/2y_c$, where $y_c = \left(q^2/g\right)^{1/3}$ (equation (5.2.14)):

$$y_c = \left[\frac{(2.4/2.0)^2}{9.81}\right]^{1/3} = 0.528 \text{ m}$$

So $E_{min} = 3/2(0.528 \text{ m}) = 0.792 \text{ m}$. Then $\Delta z_{crit} = E - E_{min} = 1.073 - 0.792 = 0.281 \text{ m}$. In this case $\Delta z = 20 \text{ cm} = 20/100 \text{ m} = 0.2 \text{ m} < \Delta z_{crit} = 0.281 \text{ m}$. Therefore, y_c does not occur at the hump.

(b) The critical depth at the side wall constriction is

$$y_c = \left[\frac{(2.4/1.7)^2}{9.81}\right]^{1/3} = 0.588 \text{ m}$$

Thus $E_{min} = (3/2)y_c = (3/2)(0.588) = 0.882 \text{ m}$. E is computed above as $E = 1.073 \text{ m}$. Because $E_{min} = 0.882 \text{ m} < E = 1.073 \text{ m}$, critical depth does not occur at the constriction. Remember that energy losses are negligible so that the specific energy in the constriction and upstream of the constriction must be equal. For critical flow to occur, the constriction width can be computed as follows: $E_{min} = E = 1.073 \text{ m} = (3/2)y_c$, so that $y_c = 0.715 \text{ m}$. Then using equation (5.2.14), $0.715 = \left[(2.4/B_c)^2/9.81\right]^{1/3}$ and $B_c = 1.267 \text{ m}$.

(c) With both the hump and the side wall constriction, y_c is 0.588 m, so $E_{min} = 0.882 \text{ m}$. Then $\Delta z_{crit} = E - E_{min} = 1.073 - 0.882 = 0.191 \text{ m}$.

Because $\Delta z = 20/100 \text{ m} = 0.20 \text{ m} > \Delta z_{crit} = 0.191 \text{ m}$, critical depth will occur at the hump with a constriction.

5.2.2 Momentum

Applying the momentum principle (equation 3.5.6) to a short horizontal reach of channel with steady flow (Figure 5.2.5), we get

$$\sum F_x = \frac{d}{dt}\int_{cv} v_x\rho \, d\forall + \sum_{cs} v_x\rho \, \mathbf{V} \cdot \mathbf{A} \qquad (2.5.6)$$

where

$$\frac{d}{dt}\int_{cv} v_x\rho \, d\forall = 0 \qquad (5.2.15)$$

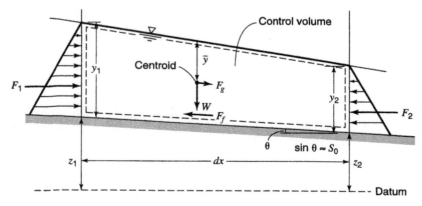

Figure 5.2.5 Application of momentum principle.

The momentum entering from the upstream is $-\rho\beta_1 V_1 Q$ and the momentum leaving the control volume is $\rho\beta_2 V_2 Q$, where β is called the *momentum correction factor* that accounts for the nonuniformity of velocity (equation 3.7.8), so that

$$\sum v_x \rho \, \mathbf{V} \cdot \mathbf{A} = -\rho\beta_1 V_1 Q + \rho\beta_2 V_2 Q \qquad (5.2.16)$$

The forces are

$$\sum F_x = F_1 - F_2 + F_g - F_f' \qquad (5.2.17)$$

The hydrostatic forces are

$$F_1 = \gamma A \bar{y}_1 \qquad (5.2.18)$$

and

$$F_2 = \gamma A \bar{y}_2 \qquad (5.2.19)$$

where $\bar{y}_1$ and $\bar{y}_2$ are the distances to the centroid. The gravity force F_g due to the weight W of the water is $W \sin \theta = \rho g A dx \sin \theta$, where $W = \rho g A dx$. Because the channel slope is small, $S_0 \approx \sin \theta$, and the force due to gravity is

$$F_g = \rho g \bar{A} dx S_0 \qquad (5.2.20)$$

where $\bar{A} = (A_1 + A_2)/2$ is the average cross-sectional area of flow. The external force due to friction created by shear between the channel bottom and sides of the control volume is $-\tau_0 P dx$ where τ_0 is the bed shear stress and P is the wetted perimeter. From equation (5.1.6), $\tau_0 = \gamma R S_f = \rho g (A/P) S_f$. So the friction force is then

$$F_f' = -\rho g A S_f dx \qquad (5.2.21)$$

For our purposes here we will continue to use F_g and F_f'.

Substituting equations (5.2.15) through (5.2.21) into (3.5.6) gives

$$\gamma A_1 \bar{y}_1 - \gamma A_2 \bar{y}_2 + W \sin \theta - F_f' = -\rho\beta_1 V_1 Q + \rho\beta_2 V_2 Q \qquad (5.2.22)$$

which is the *momentum equation for steady state open-channel flow.*

It should be emphasized that in the energy equation the F_f (loss due to friction) is a measure of the internal energy dissipated in the entire mass of water in the control volume, whereas F_f' in the momentum equation measures the losses due to external forces exerted on the water by the wetted perimeter of the control volume. Ignoring the small difference between the energy coefficient α and the momentum coefficient β in gradually varied flow, the internal energy losses are practically identical with the losses due to external forces (Chow, 1959). For uniform flow, $F_g = F_f'$.

Application of the energy and momentum principles in open-channel flow can be confusing at first. It is important to understand the basic differences, even though the two principles may produce identical or very similar results. Keep in mind that *energy is a scalar quantity* and *momentum is a vector quantity* and that *energy considers internal losses* in the energy equation and *momentum considers external resistance* in the momentum equation. The energy principle is simpler and clearer than the momentum principle; however, the momentum principle has certain advantages in application to problems involving high internal-energy changes, such as the hydraulic jump (Chow, 1959), which is discussed in section 5.5 on rapidly varied flow.

5.2.3   Specific Force

For a short horizontal reach (control volume) with $\theta = 0$ and the gravity force $F_g = W \sin \theta = 0$, the external force of friction F_f', can be neglected so $F_f' = 0$ and $F_g = 0$. Also assuming $\beta_1 = \beta_2$, the momentum equation (5.2.22) reduces to

$$\gamma A_1 \bar{y}_1 - \gamma A_2 \bar{y}_2 = -\rho V_1 Q + \rho V_2 Q \tag{5.2.23}$$

Substituting $V_1 = Q/A_1$ and $V_2 = Q/A_2$, dividing through by γ and substituting $1/g = \rho/\gamma$, and then rearranging yields

$$\frac{Q^2}{gA_1} + A_1 \bar{y}_1 = \frac{Q^2}{gA_2} + A_2 \bar{y}_2 \tag{5.2.24}$$

The *specific force F* (Figure 5.2.6) is defined as

$$F = \frac{Q^2}{gA} + A\bar{y} \tag{5.2.25}$$

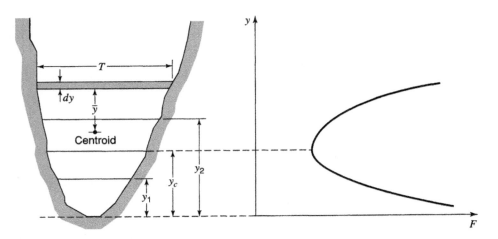

Figure 5.2.6 Specific force curves.

which has units of ft³ or m³. The minimum value of the specific force with respect to the depth is determined using

$$\frac{dF}{dy} = \frac{d\left(\frac{Q^2}{gA}\right)}{dy} + \frac{d(A\bar{y})}{dy} = 0 \qquad (5.2.26)$$

which results in

$$\frac{dF}{dy} = -\frac{TQ^2}{gA^2} + A = 0 \qquad (5.2.27)$$

Refer to Chow (1959) or Chaudhry (1993) for the proof and further explanation of $d(A\bar{y})/dy = A$. Equation (5.2.27) reduces to $-V^2/g + A/T = 0$ where the hydraulic depth $D = A/T$, so

$$\frac{V^2}{g} = D \quad \text{or} \quad \frac{V^2}{gD} = 1 \qquad (5.2.28)$$

which we have already shown is the criterion for critical flow (equation (5.2.9) or (5.2.10)). Therefore, at critical flow the specific force is a minimum for a given discharge.

Summarizing, critical flow is characterized by the following conditions:

- Specific energy is minimum for a given discharge.
- Specific force is minimum for a given discharge.
- Velocity head is equal to half the hydraulic depth.
- Froude number is equal to unity.

Two additional conditions that are not proven here are (Chow, 1959):

- The discharge is maximum for a given specific energy.
- The velocity of flow in a channel of small slope with uniform velocity distribution is equal to the celerity of small gravity waves in shallow water caused by local disturbances.

When flow is at or near the critical state, minor changes in specific energy near critical flow cause major changes in depth (see Figures 5.2.1 or 5.2.2), causing the flow to be unstable. Figure 5.2.7 illustrates examples of locations of critical flow.

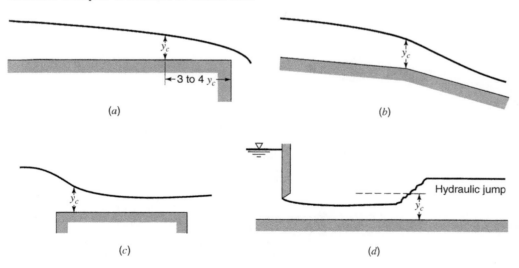

Figure 5.2.7 Example locations of critical flow. (*a*) Critical depth near free overfall; (*b*) Change in grade of channel bottom; (*c*) Flow over a broad-crested weir; (*d*) Flow through hydraulic jump.

EXAMPLE 5.2.4

Compute the specific force curves for the channel and flow rates used in example 5.2.2.

SOLUTION

The specific force values are computed using equation (5.2.25) with the values presented in Table 5.2.2. The curves are plotted in Figure 5.2.8.

Table 5.2.2 Computed Specific Force Curve Values for Example 5.2.4

Depth, y (ft)	Specific force, F (ft^3)			
	$Q = 0$	$Q = 50$	$Q = 100$	$Q = 300$
0.1	0.10	38.92	155.38	1397.62
0.2	0.40	19.81	78.04	699.16
0.4	1.60	11.30	40.42	350.98
0.6	3.60	10.07	29.48	236.52
0.8	6.40	11.25	25.81	181.09
1.0	10.00	13.88	25.53	149.75
1.2	14.40	17.63	27.34	130.86
1.4	19.60	22.37	30.69	119.42
1.6	25.60	28.03	35.30	112.94
1.8	32.40	34.56	41.03	110.04
2.0	40.00	41.94	47.76	109.88
2.2	48.40	50.16	55.46	111.92
2.4	57.60	59.22	64.07	115.83
2.6	67.60	69.09	73.57	121.35
2.8	78.40	79.79	83.95	128.31
3.0	90.00	91.29	95.18	136.58
3.5	122.50	123.61	126.94	162.43
4.0	160.00	160.97	163.88	194.94
4.5	202.50	203.36	205.95	233.56
5.0	250.00	250.78	253.11	277.95

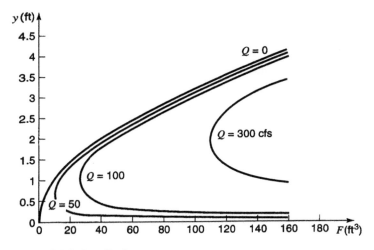

Figure 5.2.8 Specific force curves.

5.3   STEADY, GRADUALLY VARIED FLOW

5.3.1   Gradually Varied Flow Equations

Several types of open-channel flow problems can be solved in hydraulic engineering practice using the concepts of nonuniform flow. The first to be discussed are *gradually varied flow problems* in which the change in the water surface profile is small enough that it is possible to integrate the relevant differential equation from one section to an adjacent section for the change in depth or change in water surface elevation. Consider the energy equation (5.1.6) previously derived for nonuniform flow (Figure 5.1.1) using the control volume approach (with $\alpha_1 = \alpha_2 = 1$):

$$y_1 + z_1 + \frac{V_1^2}{2g} = y_2 + z_2 + \frac{V_2^2}{2g} + h_L \tag{5.1.6}$$

Because $h_L = S_f L = S_f \Delta x$ letting $\Delta y = y_2 - y_1$ and $\Delta z = z_1 - z_2 = S_0 \Delta x$, then equation (5.1.6) can be expressed as

$$S_0 \Delta x + \frac{V_1^2}{2g} = \Delta y + \frac{V_2^2}{2g} + S_f \Delta x \tag{5.3.1}$$

Rearranging yields

$$\Delta y = S_0 \Delta x - S_f \Delta x - \left(\frac{V_2^2}{2g} - \frac{V_1^2}{2g} \right) \tag{5.3.2}$$

and then dividing through by Δx results in

$$\frac{\Delta y}{\Delta x} = S_0 - S_f - \left(\frac{V_2^2}{2g} - \frac{V_1^2}{2g} \right) \frac{1}{\Delta x} \tag{5.3.3}$$

Taking the limit as $\Delta x \to 0$, we get

$$\lim_{\Delta x \to 0} \left(\frac{\Delta y}{\Delta x} \right) = \frac{dy}{dx} \tag{5.3.4}$$

and

$$\lim_{\Delta x \to 0} \left(\frac{V_2^2}{2g} - \frac{V_1^2}{2g} \right) \left(\frac{1}{\Delta x} \right) = \frac{d}{dx} \left(\frac{V^2}{2g} \right) \tag{5.3.5}$$

Substituting these into equation (5.3.3) and rearranging yields

$$\frac{dy}{dx} + \frac{d}{dx} \left(\frac{V^2}{2g} \right) = S_0 - S_f \tag{5.3.6}$$

The second term $\frac{d}{dx} \left(\frac{V^2}{2g} \right)$ can be expressed as $\left[\dfrac{d \left(\frac{V^2}{2g} \right)}{dy} \right] \dfrac{dy}{dx}$, so that equation (5.3.6) can be simplified to

$$\frac{dy}{dx} \left[1 + \frac{d \left(\frac{V^2}{2g} \right)}{dy} \right] = S_0 - S_f \tag{5.3.7}$$

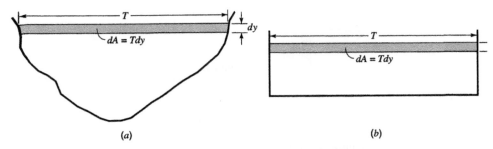

Figure 5.3.1 Definition of top width $(T = dA/dy)$. (*a*) Natural channel; (*b*) Rectangular channel.

or

$$\frac{dy}{dx} = \frac{S_0 - S_f}{\left[1 + \frac{d\left(\frac{V^2}{2g}\right)}{dy} \right]}$$

(5.3.8)

Equations (5.3.7) and (5.3.8) are two expressions of *the differential equation for gradually varied flow*. Equation (5.3.8) can also be expressed in terms of the Froude number. First observe that

$$\frac{d}{dy}\left(\frac{V^2}{2g}\right) = \frac{d}{dy}\left(\frac{Q^2}{2gA^2}\right) = -\left(\frac{Q^2}{gA^3}\right)\frac{dA}{dy}$$

(5.3.9)

By definition, the incremental increase in cross-sectional area of flow dA, due to an incremental increase in the depth dy, is $dA = Tdy$, where T is the top width of flow (see Figure 5.3.1). Also $A/T = D$, which is the hydraulic depth. Equation (5.3.9) can now be expressed as

$$\frac{d}{dy}\left(\frac{V^2}{2g}\right) = -\left(\frac{Q^2}{gA^3}\right)\frac{Tdy}{dy} = -\frac{Q^2}{gA^2}\left(\frac{T}{A}\right) = -\frac{Q^2}{gA^2D}$$

(5.3.10a)

$$= -F_r^2$$

(5.3.10b)

where $F_r = \frac{V}{\sqrt{gD}} = \frac{Q}{A\sqrt{gD}}$. Substituting equation (5.3.10b) into (5.3.8) and simplifying, we find that the gradually varied flow equation in terms of the Froude number is

$$\frac{dy}{dx} = \frac{S_0 - S_f}{1 - F_r^2}$$

(5.3.11)

EXAMPLE 5.3.1

Consider a vertical sluice gate in a wide rectangular channel $(R = A/P = By/(B+2Y) \approx y$ because $B \gg 2y)$. The flow downstream of a sluice gate is basically a jet that possesses a vena contracta (see Figure 5.3.2). The distance from the sluice gate to the vena contracta as a rule is approximated as the same as the sluice gate opening (Chow, 1959). The coefficients of contraction for vertical sluice gates are approximately 0.6, ranging from 0.598 to 0.611 (Henderson, 1966). The objective of this problem is to determine the distance from the vena contracta to a point b downstream where the depth of flow is known to be 0.5 m deep. The depth of flow at the vena contracta is 0.457 m for a flow rate of 4.646 m³/s per meter of width. The channel bed slope is 0.0003 and Manning's roughness factor is $n = 0.020$.

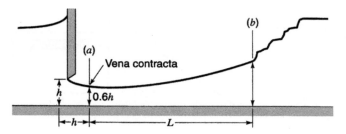

Figure 5.3.2 Flow downstream of a sluice gate in a wide rectangular channel.

SOLUTION

To compute the distance, Δx from y_a to y_b, the gradually varied flow equation (5.3.1) can be used,

$$S_0 \Delta x + \frac{V_1^2}{2g} = \Delta y + \frac{V_2^2}{2g} + S_f \Delta x$$

where $\Delta y = y_2 - y_1$. Solving for Δx, we get

$$\left(S_0 - S_f\right)\Delta x = \Delta y + \left(\frac{V_2^2}{2g} - \frac{V_1^2}{2g}\right)$$

$$\Delta x = \frac{\Delta y + \left(\dfrac{V_2^2}{2g} - \dfrac{V_1^2}{2g}\right)}{S_0 - S_f}$$

The friction slope is computed using Manning's equation (5.1.22) with average values of the hydraulic radius

$$V_{\text{ave}} = \frac{1}{n} R_{\text{ave}}^{2/3} S_f^{1/2}$$

so

$$S_f = \left[\frac{nV_{\text{ave}}}{R_{\text{ave}}^{2/3}}\right]^2$$

Let us use the following values for this example:

Location	y (m)	$R = y$ (m)	V (m/s)	$V^2/2g$ (m)
a	0.457	0.457	10.17	5.27
b	0.500	0.500	9.292	4.40

Now we get

$$V_{\text{ave}} = \frac{10.17 + 9.292}{2} = 9.73 \text{ m/s}$$

$$R_{\text{ave}} = \frac{0.457 + 0.500}{2} = 0.479 \text{ m}$$

$$S_f = \left[\frac{0.020(9.73)}{0.479^{2/3}}\right]^2 = 0.101 \text{ m/m}$$

The distance from a to b, Δx, is

$$\Delta x = \frac{(0.500 - 0.457) + (4.40 - 5.27)}{0.0003 - 0.101} = \frac{-0.827}{-0.101} = 8.21 \text{ m}$$

The distance from the sluice gate to b is $\left(\frac{0.457}{0.6}\right) + 8.21 = 8.97$ m.

5.3.2 Water Surface Profile Classification

Channel bed slopes may be classified as mild (M), steep (S), critical (C), horizontal (H) ($S_0 = 0$), and adverse (A) ($S_0 < 0$). To define the various types of slopes for the mild, steep, and critical slopes, the normal depth y_n and critical depth y_c are used:

$$\text{Mild:} \quad y_n > y_c \quad \text{or} \quad \frac{y_n}{y_c} > 1 \tag{5.3.12a}$$

$$\text{Steep:} \quad y_n < y_c \quad \text{or} \quad \frac{y_n}{y_c} < 1 \tag{5.3.12b}$$

$$\text{Critical:} \quad y_n = y_c \quad \text{or} \quad \frac{y_n}{y_c} = 1 \tag{5.3.12c}$$

The horizontal and adverse slopes are special cases because the normal depth does not exist for them. Table 5.3.1 lists the types and characteristics of the various types of profiles and Figure 5.3.3 shows the classification of gradually varied flow profiles.

Table 5.3.1 Types of Flow Profiles in Prismatic Channels

Channel slope	Designation			Relation of y to y_n and y_c			General type of curve	Type of flow
	Zone 1	Zone 2	Zone 3	Zone 1	Zone 2	Zone 3		
Horizontal $S_o = 0$	None			$y > y_n$	$>$	y_c	None	None
		H2		y_n	$> y$	$> y_c$	Drawdown	Subcritical
			H3	y_n	$>$	$y_c > y$	Backwater	Supercritical
Mild $0 < S_o < S_c$	M1			$y > y_n$	$>$	y_c	Backwater	Subcritical
		M2		y_n	$> y$	$> y_c$	Drawdown	Subcritical
			M3	y_n	$>$	$y_c > y$	Backwater	Supercritical
Critical $S_o = S_c > 0$	C1			$y > y_c$	$=$	y_n	Backwater	Subcritical
		C2		y_n	$= y$	$= y_c$	Parallel to channel bottom	Uniform-critical
			C3	y_c	$=$	$y_n > y$	Backwater	Supercritical
Steep $S_o > S_c > 0$	S1			$y > y_c$	$>$	y_n	Backwater	Subcritical
		S2		y_c	$> y$	$> y_n$	Drawdown	Supercritical
			S3	y_c	$>$	$y_n > y$	Backwater	Supercritical
Adverse $S_o < 0$	None			$y > (y_n)^*$	$>$	y_c	None	None
		A2		$(y_n)^*$	$> y$	$> y_c$	Drawdown	Subcritical
			A3	$(y_n)^*$	$>$	$y_c > y$	Backwater	Supercritical

$^*y_n$ in parentheses is assumed a positive value.

Source: Chow (1959).

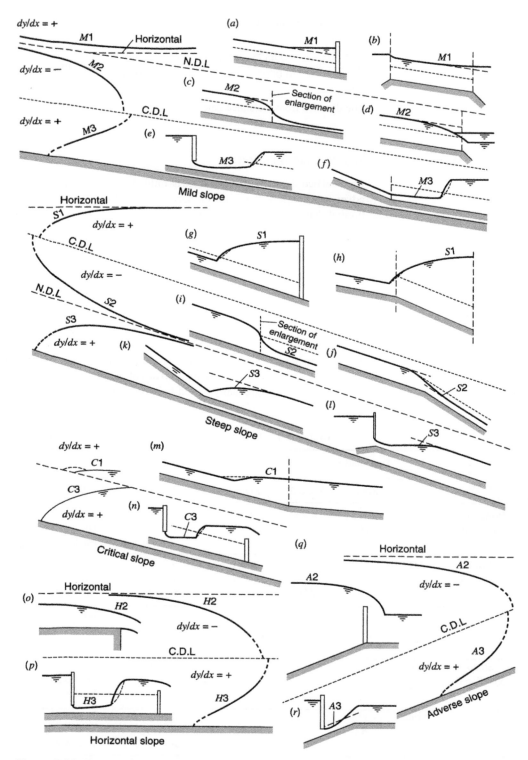

Figure 5.3.3 Flow profiles (from Chow (1959)).

The three zones for mild slopes are defined as

Zone 1: $y > y_n > y_c$

Zone 2: $y_n > y > y_c$

Zone 3: $y_n > y_c > y$

The energy grade line, water surface, and channel bottom are all parallel for uniform flow, i.e., $S_f = S_0 = $ slope of water surface when $y = y_n$. From Manning's equation for a given discharge, $S_f < S_0$ if $y > y_n$.

Now consider the qualitative characteristics using the three zones.

Zone 1 (M1 profile): $y > y_n$; then $S_f < S_0$ or $S_0 - S_f = +$

$F_r < 1$ since $y > y_c$, so $1 - F_r^2 = +$

by equation (5.3.11), $\dfrac{dy}{dx} = \dfrac{S_0 - S_f}{1 - F_r^2} = \dfrac{+}{+} = +$

then y increases with x so that $y \rightarrow y_n$

Zone 2 (M2 profile): $y < y_n$; then $S_f > S_0$ or $S_0 - S_f = -$

$F_r < 1$ since $y > y_c$, so $1 - F_r^2 = +$

by equation (5.3.11), $\dfrac{dy}{dx} = \dfrac{S_0 - S_f}{1 - F_r^2} = \dfrac{-}{+} = -$

then y decreases with x so that $y \rightarrow y_c$

Zone 3 (M3 profile): $y < y_n$; then $S_f > S_0$ or $S_0 - S_f = -$

$F_r > 1$ since $y < y_c$ so $1 - F_r^2 = -$

by equation (5.3.11), $\dfrac{dy}{dx} = \dfrac{S_0 - S_f}{1 - F_r^2} = \dfrac{-}{-} = +$

then y increases with x so that $y \rightarrow y_c$

This analysis can be made of the other profiles. The results are summarized in Table 5.3.1.

EXAMPLE 5.3.2

For the rectangular channel described in examples 5.1.1, 5.1.3, and 5.2.1, classify the type of slope. Determine the types of profiles that exist for depths of 5.0 ft, 2.0 ft, and 1.0 ft with a discharge of 100 ft³/s.

SOLUTION

In example 5.1.3, the normal depth is computed as $y_n = 3.97$ ft, and in example 5.2.1, the critical depth is computed as $y_c = 1.69$ ft. Because $y_n > y_c$, this is a mild channel bed slope.

For a flow depth of 5.0 ft, $5.0 > y_n > y_c$, so that an M1 profile with a backwater curve exists (refer to Table 5.3.1 and Figure 5.3.3). The flow is subcritical.

For a flow depth of 2.0 ft, $y_n > 2.0 > y_c$, so that an M2 profile with a drawdown curve exists. The flow is subcritical.

For a flow depth of 1.0 ft, $y_n > y_c > 1.0$, so that an M3 profile with a backwater curve exists. The flow is supercritical.

Prismatic Channels with Changes in Slope

Consider a channel that changes slope from a mild slope to a steep slope. The critical depth is the same for each slope; however, the normal depth changes, for the upstream mild slope, $y_{n1} > y_c$, and for the downstream steep slope, $y_{n2} < y_c$. The only control is the critical depth at the break in slopes where flow transitions from a subcritical flow to a supercritical flow. The flow profiles are an M2

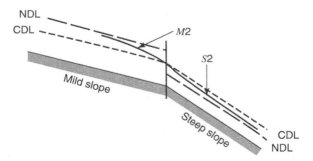

Figure 5.3.4 Water surface profile for prismatic channel with slope change from mild to steep.

profile for the upstream reach and an S2 profile for the downstream reach as shown in Figure 5.3.4. The specific energy for the reaches, E_{n1} and E_{n2}, and the specific energy for critical conditions can be computed. A channel that changes slope from a steep slope to a mild slope is more complicated in that a hydraulic jump forms. This jump could form on the upper steep slope or on the lower mild slope.

5.3.3   Direct Step Method

In example 5.3.1 we computed the location where a specified depth occurred, using the energy equation. This procedure can be extended to compute reach lengths for specified depths at each end of a reach. Computations are performed step by step from one end of the channel reach by reach to the other end. This procedure is called the direct step method and is applicable only to prismatic channels.

The gradually varied flow equation can be expressed in terms of the specific energy. Equation (5.3.1) can be rearranged to

$$S_o \Delta x - \overline{S}_f \Delta x = \left(y_2 + \alpha_2 \frac{V_2^2}{2g} \right) - \left(y_1 + \alpha_1 \frac{V_1^2}{2g} \right) = E_2 - E_1 \tag{5.2.12}$$

where $\overline{S}_f$ is the average friction slope for the channel reach. Solving for Δx,

$$\Delta x = \frac{E_2 - E_1}{S_o - \overline{S}_f} \tag{5.3.13}$$

The direct step method is based on this equation. Manning's equation is used to compute the friction slope at the upstream and downstream ends of each reach for the specified depths using

$$S_f = \frac{n^2 v^2}{2.22 R^{2/3}} \tag{5.3.14}$$

The average friction slope, $\overline{S}_f = \frac{1}{2} \left(S_{f1} + S_{f2} \right)$, is used in equation (5.3.13). The computation procedure must be performed from downstream to upstream if the flow is subcritical and from upstream to downstream if the flow is supercritical.

EXAMPLE 5.3.3

A trapezoidal channel has the flowing characteristics: slope = 0.0016; bottom width = 20 ft; and side slopes = 1 vertical to 2 horizontal ($z = 2$). Water at a downstream location is an embankment where the water depth is 5.0 ft just upstream of the embankment and the discharge is 400 cfs. Determine the distance upstream to where the flow depth is 4.60 ft. (Adapted from Chow, 1959.) (Assume $\alpha = 1.10$.)

SOLUTION

The critical depth is $y_c = 2.22$ ft and the normal depth is $y_n = 3.36$ ft, so the flow is subcritical. Because the depth of 5.0 ft is greater than the normal depth of 3.36 ft, this is an M1 profile. Then the depth will decrease proceeding in the upstream direction because $dy/dx = +$ in the downstream direction. In other words, going upstream the depth will approach normal depth. So we will consider the first reach, going upstream to a depth of 5.0 to 4.8 ft and the second reach upstream is from a depth of 4.8 to 4.6 ft depth.

First reach; At $y = 5.0$ ft then $A = 150$ ft^3, $R = 3.54$ ft, $V = 2.67$ ft/sec, $S_f = 0.000037$, and $E = 5 + (1.1)(2.67)^2/[2(32.2)] = 5.123$ ft, and at $y = 4.80$ ft then $A = 142.1$ ft^2, $R = 3.43$ ft, $V = 2.82$ ft/sec, $S_f = 0.00043$, and $E = 4.936$ ft. Using equation 5.3.14,

where $\bar{S}_f = \dfrac{1}{2}(0.00037 + 0.00043) = 0.00040$,

then $\Delta x = \dfrac{E_2 - E_1}{S_o - \bar{S}_f} = \dfrac{(5.123 - 4.938)}{0.0016 - 0.00040} = 156$ ft.

Second reach: At $y = 4.60$ ft, $A = 134.3$ ft^2, $R = 3.31$ ft, $V = 2.98$ ft/sec, $S_f = 0.00051$, and $E = 4.752$ ft. $\bar{S}_f = \dfrac{1}{2}(0.00043 + 0.00051) = 0.00047$, and $\Delta x = 163$ ft.

So the distance upstream to a depth of 4.60 ft is $156 + 163 = 319$ ft. The procedure can be continued upstream to where the normal depth occurs.

5.4 GRADUALLY VARIED FLOW FOR NATURAL CHANNELS

5.4.1 Development of Equations

As an alternate to the procedure presented above, the gradually varied flow equation can be expressed in terms of the water surface elevation for application to natural channels by considering $w = z + y$ where w is the water surface elevation above a datum such as mean sea level. The total energy H at a section is

$$H = z + y + \alpha \frac{V^2}{2g} = w + \alpha \frac{V^2}{2g} \qquad (5.4.1)$$

including the energy correction factor α. The change in total energy head with respect to location along a channel is

$$\frac{dH}{dx} = \frac{dw}{dx} + \frac{d}{dx}\left(\alpha \frac{V^2}{2g}\right) \qquad (5.4.2)$$

The total energy loss is due to friction losses (S_f) and contraction-expansion losses (S_e):

$$\frac{dH}{dx} = -S_f - S_e \qquad (5.4.3)$$

S_e is the slope term for the contraction-expansion loss. Substituting (5.4.3) into (5.4.2) results in

$$-S_f - S_e = \frac{dw}{dx} + \frac{d}{dx}\left(\alpha \frac{V^2}{2g}\right) \qquad (5.4.4)$$

The friction slope S_f can be expressed using Manning's equation (5.1.23) or (5.1.25):

$$Q = KS_f^{1/2} \qquad (5.4.5)$$

where K is defined as the *conveyance* in SI units

$$K = \frac{1}{n}AR^{2/3} \qquad (5.4.6a)$$

or in U.S. customary units as

$$K = \frac{1.486}{n} AR^{2/3} \tag{5.4.6b}$$

for equations (5.1.23) or (5.1.25), respectively. The friction slope (from equation 5.4.5) is then

$$S_f = \frac{Q^2}{K^2} = \frac{Q^2}{2}\left[\frac{1}{K_1^2} + \frac{1}{K_2^2}\right] \tag{5.4.7}$$

with the conveyance effect $\frac{1}{K^2} = \frac{1}{2}\left[\frac{1}{K_1^2} + \frac{1}{K_2^2}\right]$, where K_1 and K_2 are the conveyances, respectively, at the upstream and the downstream ends of the reach. Alternatively, the friction slope can be determined using an average conveyance, i.e., $\overline{K} = (K_1 + K_2)/2$ and $Q = \overline{K}S_f^{1/2}$; then

$$S_f = \frac{Q^2}{\overline{K}^2} \tag{5.4.8}$$

The *contraction-expansion loss term* S_e can be expressed for a *contraction loss* as

$$S_e = \frac{C_c}{dx}\left|\left[\alpha_2 \frac{V_2^2}{2g} - \alpha_1 \frac{V_1^2}{2g}\right]\right| \quad \text{for} \quad d\left(\alpha \frac{V^2}{2g}\right) = \left(\alpha_2 \frac{V_2^2}{2g} - \alpha_1 \frac{V_1^2}{2g}\right) > 0 \tag{5.4.9}$$

and for an *expansion loss* as

$$S_e = \frac{C_e}{dx}\left|\left[\alpha_2 \frac{V_2^2}{2g} - \alpha_1 \frac{V_1^2}{2g}\right]\right| \quad \text{for} \quad d\left(\alpha \frac{V^2}{2g}\right) < 0 \tag{5.4.10}$$

The *gradually varied flow equation for a natural channel* is defined by substituting equation (5.4.7) into equation (5.4.4):

$$-\frac{Q^2}{2}\left[\frac{1}{K_1^2} + \frac{1}{K_2^2}\right] - S_e = \frac{dw}{dx} + \frac{d}{dx}\left(\alpha \frac{V^2}{2g}\right) \tag{5.4.11a}$$

or

$$-\frac{Q^2}{2}\left[\frac{1}{K_1^2} + \frac{1}{K_2^2}\right] - S_e = \frac{w_2 - w_1}{\Delta x} + \frac{1}{\Delta x}\left[\alpha_2 \frac{V_2^2}{2g} - \alpha_1 \frac{V_1^2}{2g}\right] \tag{5.4.11b}$$

Rearranging yields

$$w_1 + \alpha_1 \frac{V_1^2}{2g} = w_2 + \alpha_2 \frac{V_2^2}{2g} + \frac{Q^2}{2}\left[\frac{1}{K_1^2} + \frac{1}{K_2^2}\right]\Delta x + S_e \Delta x \tag{5.4.12}$$

EXAMPLE 5.4.1

Derive an expression for the change in water depth as a function of distance along a prismatic channel (i.e., constant alignment and slope) for a gradually varied flow.

SOLUTION

We start with equation (5.3.11):

$$\frac{dy}{dx} = \frac{S_o - S_f}{1 - F_r^2}$$

with

$$F_r^2 = \left(V / \sqrt{gD} \right)^2 = \left(Q/A \sqrt{gD} \right)^2 = \left(Q/A \sqrt{gA/T} \right)^2 = \frac{Q^2 T}{gA^3}.$$

Then

$$\frac{dy}{dx} = \frac{S_0 - S_f}{\left(1 - \dfrac{Q^2 T}{gA^3} \right)}$$

To determine a gradually varied flow profile, this equation is integrated.

EXAMPLE 5.4.2

For a river section with a subcritical discharge of 6500 ft^3/s, the water surface elevation at the downstream section is 5710.5 ft with a velocity head of 3.72 ft. The next section is 500 ft upstream with a velocity head of 1.95 ft. The conveyances for the downstream and upstream sections are 76,140 and 104,300, respectively. Using expansion and contraction coefficients of 0.3 and 0.1, respectively, determine the water surface elevation at the upstream section.

SOLUTION

Using equation (5.4.12), the objective is to solve for the upstream water surface elevation w_1:

$$w_1 = w_2 + \alpha_2 \frac{V_2^2}{2g} - \alpha_1 \frac{V_1^2}{2g} + \frac{Q^2}{2} \left[\frac{1}{K_1^2} + \frac{1}{K_2^2} \right] \Delta x + S_e \Delta x$$

$$= w_2 + \Delta \left(\alpha \frac{V^2}{2g} \right) + \frac{Q^2}{2} \left[\frac{1}{K_1^2} + \frac{1}{K_2^2} \right] \Delta x + S_e \Delta x$$

The friction slope term is

$$\frac{Q^2}{2} \left[\frac{1}{K_1^2} + \frac{1}{K_2^2} \right] \Delta x = \frac{6500^2}{2} \left[\frac{1}{76,140^2} + \frac{1}{104,300^2} \right] 500 = 2.79 \text{ ft}$$

$$\Delta \left(\alpha \frac{V^2}{2g} \right) = 3.72 - 1.95 = +1.77 \text{ ft}$$

Because $\Delta \left[\alpha \left(V^2/2g \right) \right] > 0$, a contraction exists for flow from cross-section 1 to cross-section 2. Then

$$S_e \Delta x = C_c \left[\alpha_2 \frac{V_2^2}{2g} - \alpha_1 \frac{V_1^2}{2g} \right] = 0.1(1.77) = 0.177$$

The water surface elevation at the upstream cross-section is then

$$w_1 = 5710.5 + 1.77 + 2.79 + 0.177 = 5715.2 \text{ ft}$$

5.4.2 Energy Correction Factor

In section 3.7, the formula for the kinetic energy correction factor is derived as

$$\alpha = \frac{1}{AV^3} \int_A v^3 dA \tag{3.7.4}$$

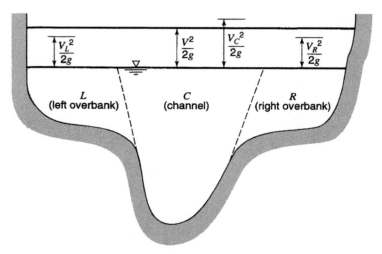

Figure 5.4.1 Compound channel section.

which can be approximated as

$$\alpha = \frac{\sum v^3 \Delta A}{V^3 A} \tag{5.4.13}$$

where V is the mean velocity.

Consider a compound channel section as shown in Figure 5.4.1 that has three flow sections. The objective is to derive an expression for the energy coefficient in terms of the conveyance for a compound channel, so that the velocity head for the entire channel is $\alpha \left(V^2/2g \right)$ where V is the mean velocity in the compound channel. Equation (5.4.13) can be expressed as

$$\alpha \approx \frac{\displaystyle\sum_{i=1}^{N} V_i^3 A_i}{V^3 \displaystyle\sum_{i=1}^{N} A_i} \tag{5.4.14}$$

where N is the number of sections (subareas) of the channel (e.g., in Figure 5.4.1, $N = 3$), V is the mean velocity in each section (subarea), and A_i is the cross-sectional area of flow in each section (subarea).

The mean velocity can be expressed as

$$V = \frac{\displaystyle\sum_{i=1}^{N} V_i A_i}{\displaystyle\sum_{i=1}^{N} A_i} \tag{5.4.15}$$

Substituting $V_i = Q_i/A_i$ and equation (5.4.15) for V into (5.4.14) and simplifying yields

$$\alpha = \frac{\displaystyle\sum_{i=1}^{N} \left(\frac{Q_i}{A_i} \right)^3 A_i}{\left[\dfrac{\displaystyle\sum_{i=1}^{N} \left(\frac{Q_i}{A_i} \right) A_i}{\displaystyle\sum_{i=1}^{N} A_i} \right]^3 \left(\displaystyle\sum_{i=1}^{N} A_i \right)} = \frac{\displaystyle\sum_{i=1}^{N} \left(\frac{Q_i^3}{A_i^2} \right) \left(\displaystyle\sum_{i=1}^{N} A_i \right)^2}{\left(\displaystyle\sum_{i=1}^{N} Q_i \right)^3} \tag{5.4.16}$$

Now using equation (5.4.5) for each section, we get

$$Q_i = K_i S_{f_i}^{1/2} \tag{5.4.17}$$

and solving for $S_{f_i}^{1/2}$ yields

$$S_{f_i}^{1/2} = \frac{Q_i}{K_i} \tag{5.4.18}$$

Assuming that the friction slope is the same for all sections, $S_{f_i} = S_f (i = 1, \ldots, N)$, then according to equation (5.4.18)

$$\frac{Q_1}{K_1} = \frac{Q_2}{K_2} = \frac{Q_3}{K_3} = \cdots = \frac{Q_N}{K_N} \tag{5.4.19}$$

This leads to

$$Q_1 = K_1 \frac{Q_N}{K_N} \quad Q_2 = K_2 \frac{Q_N}{K_N} \cdots Q_N = K_N \frac{Q_N}{K_N} \tag{5.4.20}$$

and the total discharge is

$$Q = \sum Q_i = \frac{Q_N}{K_N} \sum K_i \tag{5.4.21}$$

Substituting the above expression for $\sum Q_i$ and

$$Q_i = K_i \frac{Q_N}{K_N} \tag{5.4.22}$$

into equation (5.4.16) and simplifying, we get

$$\alpha = \frac{\sum\limits_{i=1}^{N} \left(\dfrac{K_i^3}{A_i^2} \right) \left(\sum\limits_{i=1}^{N} A_i \right)^2}{\left(\sum\limits_{i=1}^{N} K_i \right)^3} \tag{5.4.23}$$

or

$$\alpha = \frac{\sum\limits_{i=1}^{N} \left(\dfrac{K_i^3}{A_i^2} \right) (A_t)^2}{(K_t)^3} \tag{5.4.24}$$

where A_t and K_t are the totals.

The friction slope for the reach is

$$S_{f_i} = \left(\frac{\sum Q_i}{\sum K_i} \right)^2 = \frac{Q^2}{(\sum K_i)^2} \tag{5.4.25}$$

by eliminating Q_N / K_N from equations (5.4.19) through (5.4.21).

EXAMPLE 5.4.3   For the compound cross-section at river mile 1.0 shown in Figure 5.4.2, determine the energy correction factor α. The discharge is $Q = 11,000$ cfs and the water surface elevation is 125 ft.

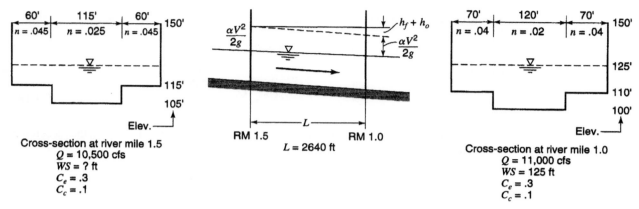

Figure 5.4.2 Cross-section and reach length data for example 5.4.3 (from Hoggan (1997)).

SOLUTION

Step 1   Compute the cross-sectional areas of flow for the left overbank (L), channel (C), and right overbank (R):

$A_L = 1050 \text{ ft}^2, A_c = 3000 \text{ ft}^2, A_R = 1050 \text{ ft}^2$

Step 2   Compute the hydraulic radius for L, C, and R:

$R_L = 1050/85 = 12.35 \text{ ft}; \; R_c = 3000/140 = 21.40 \text{ ft}; \; R_R = 1050/85 = 12.35 \text{ ft}$

Step 3   Compute the conveyance factor for L, C, and R:

$$K_L = \frac{1.486 A_L R_L^{2/3}}{n} = \frac{1.486(1050)(12.35)^{2/3}}{0.04} = 208,700$$

$$K_C = \frac{1.486(3000)(21.4)^{2/3}}{0.02} = 1,716,300$$

$$K_R = 208,700 (K_R = K_L)$$

Step 4   Compute totals A_t and K_t:

$$K_t = K_L + K_C + K_R = 2,133,700$$
$$A_t = A_L + A_C + A_R = 5100 \text{ ft}^2$$

Step 5   Compute K^3/A^2 and $\sum K^3/A^2$:

$$K_L^3/A_L^2 = 8.25 \times 10^9$$

$$K_C^3/A_C^2 = 561.8 \times 10^9$$

$$K_R^3/A_R^2 = 8.25 \times 10^9$$

$$\sum K^3/A^2 = 578.3 \times 10^9$$

Step 6   Use equation (5.4.24) to compute α:

$$\alpha = \frac{\sum_{i=1}^{N} \left(\frac{K_i^3}{A_i^2}\right)(A_t)^2}{(K_t)^3} = \frac{(578.3 \times 10^9)(5100)^2}{(2,133,700)^3} = 1.55$$

EXAMPLE 5.4.4

For the data in Figure 5.4.2, start with the known water surface elevation at river mile 1.0 and determine the water surface at river mile 1.5 (adapted from Hoggan, 1997).

SOLUTION   Computations are presented in Table 5.4.1.

Table 5.4.1   Standard Step Backwater Computation

(1) River mile	(2) Water surface elevation W_k	(3) Area A (ft^2)	(4) Hydraulic radius R (ft)	(5) Manning roughness n	(6) Conveyance K	(7) Average conveyance $\overline{K}$	(8) Average friction slope $\overline{S}_f$	(9) Friction loss h_L (ft)	(10) K^3/A^2 (10^9)	(11) Energy correction factor α	(12) Velocity V (ft/sec)	(13) $\alpha\frac{V^2}{2g}$ (ft/sec)	(14) $\Delta\left(\frac{V^2}{2g}\right)$ (ft/sec)	(15) $h_o{}^{**}$ (ft)	(16) $W_k{}^{***}$ (ft)
1.0	125.0	1050	12.35	0.040	208,700				8.25						
		3000	21.43	0.020	1,716,300				561.80						
		1050	12.35	0.040	208.700				8.25						
		5100			2,133,700				578.30	1.55	2.16	0.11			
1.5	126.1	666.0	9.37	0.045	97,650				2.10						
		2426.5	17.97	0.025	989,400				164.50						
		666.0	9.37	0.045	97,650				2.10						
		3758.5			1,184,700	1,659,200	0.000042	0.111	168.70	1.43	2.79	0.17	−0.06	0.02	125.07
1.5	125.0	600	8.57	0.045	83,000				1.59						
		2300	17.04	0.025	905,300				140.22						
		600	8.57	0.045	83,000				1.59						
		3500			1,071,300	1,602,350	0.000043	0.113	143.30	1.43	3.00	0.20	−0.09	0.03	125.05

$$^*\alpha = \frac{(A_t)^2 \sum K_i^3/A_i^2}{(K_t)^3}$$

$^{**}h_o = C_e|\Delta(\alpha V^2/2g)|$ for $\Delta(\alpha V^2/2g) < 0$ (loss due to channel expansion); $h_o = C_c|\Delta(\alpha V^2/2g)|$ for $\Delta(\alpha V^2/2g) > 0$ (loss due to channel contraction).

$^{***}W_2 = W_1 + \Delta(\alpha V^2/2g) + h_L + h_o = 125.0 + (-0.06) + 0.111 + 0.02 = 125.066 \cong 125.07$.

Source: Hoggan (1997).

5.4.3   Application for Water Surface Profile

The change in head with respect to distance x along the channel has been expressed in equation (5.4.2) as

$$\frac{dH}{dx} = \frac{dw}{dx} + \frac{d}{dx}\left(\alpha\frac{V^2}{2g}\right) \tag{5.4.2}$$

The total energy loss term is $dH/dx = -S_f - S_e$, where S_f is the friction slope defined by equation (5.4.7) or equation (5.4.8) and S_e is the slope of the contraction or expansion loss. The differentials dw and $d[\alpha(V^2/2g)]$ are defined over the channel reach as $dw = w_k - w_{k+1}$ and $d[\alpha(V^2/2g)] = \alpha_{k+1}\frac{V_{k+1}^2}{2g} - \alpha_k\frac{V_k^2}{2g}$, where we define $k+1$ at the downstream and k at the upstream. River cross-sections are normally defined from downstream to upstream for gradually varied flow. Equation (5.4.2) is now expressed as

$$w_k + \alpha_k\frac{V_k^2}{2g} = w_{k+1} + \alpha_{k+1}\frac{V_{k+1}^2}{2g} + S_f dx + S_e dx \tag{5.4.26}$$

The *standard step procedure* for water surface computations is described in the following steps:

a. Start at a point in the channel where the water surface is known or can be approximated. This is the *downstream boundary condition* for subcritical flow and the *upstream boundary condition* for supercritical flow. Computation proceeds upstream for subcritical flow and downstream for supercritical flow. Why?

b. Choose a water surface elevation w_k at the upstream end of the reach for subcritical flow or w_{k+1} at the downstream end of the reach for supercritical flow. This water surface elevation will be slightly lower or higher depending upon the type of profile (see Chow (1959); Henderson (1966); French (1985); or Chaudhry (1993)).

c. Next compute the conveyance, corresponding friction slope, and expansion and contraction loss terms in equation (5.4.26) using the assumed water surface elevation.

d. Solve equation (5.4.26) for w_{k+1} (supercritical flow) or w_k (subcritical flow).

e. Compare the calculated water surface elevation w with the assumed water surface elevation w'. If the calculated and assumed elevations do not agree within an acceptable tolerance (e.g., 0.01 ft), then set $w'_{k+1} = w_{k+1}$ (for supercritical flow) and $w'_k = w_k$ (for subcritical flow) and return to step (c).

Computer models for determining water surface profiles using the standard step procedure include the HEC-2 model and the newer HEC-RAS model. HEC-RAS River Analysis System (developed by the U.S. Army Corps of Engineers (USACE) Hydrologic Engineering Center) computes water surface profiles for one-dimensional steady, gradually varied flow in rivers of any cross-section (HEC, 1997a–c). HEC-RAS can simulate flow through a single channel, a dendritic system of channels, or a full network of open channels (sometimes called a fully looped system).

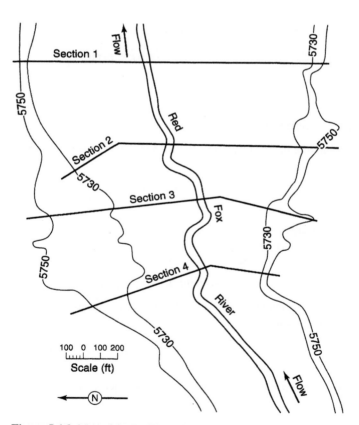

Figure 5.4.3 Map of the Red Fox River indicating cross-sections for water surface profile analysis (from U.S. Bureau of Reclamation (1957)).

HEC-RAS can model sub- or supercritical flow, or a mixture of each within the same analysis. A graphical user interface provides input data entry, data modifications, and plots of stream cross-sections, profiles, and other data. Program options include inserting trapezoidal excavations on cross-sections, and analyzing the potential for bridge scour. The water surface profile through structures such as bridges, culverts, weirs, and gates can be computed. The World Wide Web address to obtain the HEC-RAS model is www.hec.usace.army.mil.

EXAMPLE 5.4.5

A plan view of the Red Fox River in California is shown in Figure 5.4.3, along with the location of four cross-sections. Perform the standard step calculations to determine the water surface elevation at cross-section 3 for a discharge of 6500 ft^3/s. Figures 5.4.4a, b, and c are plots of cross-sections at 1, 2, and 3, respectively. Figures 5.4.5a, b, and c are the area and hydraulic radius curves for cross-sections 1, 2, and 3, respectively. Use expansion and contraction coefficients of 0.3 and 0.1, respectively. Manning's roughness factors are presented in Figure 5.4.4. The downstream starting water surface elevation at cross-section 1 is 5710.5 ft above mean sea level. This example was originally adapted by the U.S. Army Corps of Engineers from material developed by the U.S. Bureau of Reclamation (1957). Distance between cross-sections 1 and 2 is 500 ft, between cross-sections 2 and 3 is 400 ft, and between cross-sections 3 and 4 is 400 ft.

SOLUTION

The computations for this example are illustrated in Table 5.4.2.

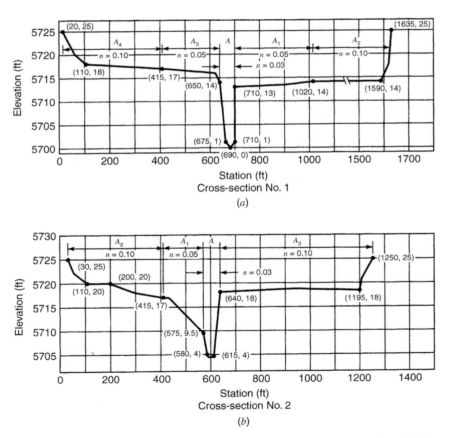

Figure 5.4.4 Cross-sections of the Red Fox River (from U.S. Bureau of Reclamation (1957)).

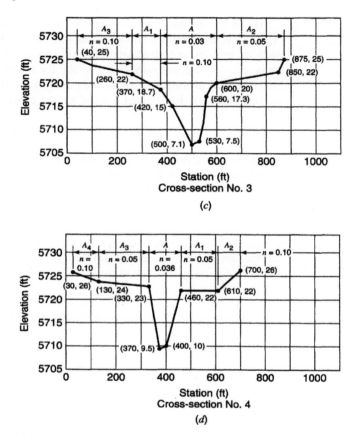

Figure 5.4.4 (*Continued*)

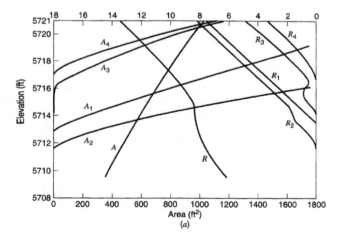

Figure 5.4.5 Area elevation and hydraulic radius—elevation curves for cross-sections 1 to 4. (*a*) Cross-section 1; (*b*) Cross-section 2; (*c*) Cross-section 3; (*d*) Cross-section 4 (from U.S. Bureau of Reclamation (1957)).

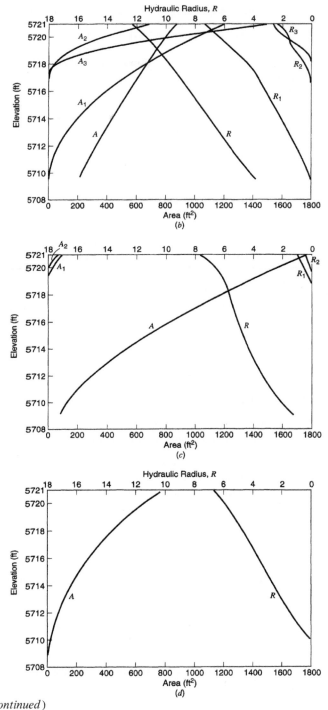

Figure 5.4.5 (*Continued*)

151

Table 5.4.2   Standard Step Backwater Computation for Red Fox River

(1)	(2) Water surface elevation W_{k+1} (ft²)...	(3) Area A (ft²)	(4) Hydraulic radius R (ft)	(5) Manning roughness n	(6) Conveyance K	(7) Average conveyance $\bar{K}$	(8) Average friction slope $S_f(10^{-3})$ (ft/ft)	(9) Friction loss h_L (ft)	(10) $\frac{K^3}{A^2}$ (10^6)	(11) Energy correction factor α	(12) Velocity V (ft/sec)	(13) $\alpha\frac{V^2}{2g}$ (ft)	(14) $\Delta\left(\alpha\frac{V^2}{2g}\right)$ (ft)	(15) h_o (ft)	(16) W_k (ft)
1	5710.5*	420	7.0	0.03	76,100					1.0	15.5	3.72			
2	5714.7	470	7.6	0.03	90,100				3311.1						
		260	2.5	0.05	14,200				42.0						
		730			104,300	90,200	5.19	2.60	3353.5	1.58	8.90	1.95	+1.77	0.18	5715.07**
	5715.0														
		500	7.85	0.03	97,800				3741.8						
		300	2.7	0.05	17,300				57.5						
		800			115,100	95,600	4.62	2.31	3799.3	1.59	8.13	1.63	+2.09	0.21	5715.1
3	5718.0	1145	5.85	0.03	184,100	149,600	1.89	0.76		1.0	5.68	0.50	+1.13	0.11	5717.1
	5717.1	970	5.6	0.03	151,500	133,300	2.38	0.95		1.0	6.70	0.70	+.93	0.09	5717.1

*Known starting water surface elevation.

**$W_{k+1} = 5710.7 + 1.77 + 2.60 + 0.18 = 5715.07 = 5715.1$; $\alpha = (A_t)^2 \sum K_i^3/A_i^2/(K_t)^3$; $h_o = C_e|(\alpha V^2/2g)|$ for $\Delta(\alpha V^2 2g) < 0$ (loss due to channel expansion); $h_o = C_c|\Delta(\alpha V^2/2g)|$ for $\Delta(\alpha V^2/2g) > 0$ (loss due to channel contraction); $W_{k+1} = W_k + \Delta(\alpha V^2/2g) + h_L + h_o$.

Source: Hoggan (1997).

5.5   RAPIDLY VARIED FLOW

Rapidly varied flow occurs when a water flow depth changes abruptly over a very short distance. The following are characteristic features of rapidly varied flow (Chow, 1959):

- Curvature of the flow is pronounced, so that pressure distribution cannot be assumed to be hydrostatic.
- The rapid variation occurs over a relatively short distance so that boundary friction is comparatively small and usually insignificant.
- Rapid changes of water area occur in rapidly varied flow, causing the velocity distribution coefficients α and β to be much greater than 1.0.

Examples of rapidly varied flow are hydraulic jumps, transitions in channels, flow over spillways, flow in channels of nonlinear alignment, and flow through nonprismatic channel sections such as flow in channel junctions, flow through trash racks, and flow between bridge piers.

The discussion presented in this chapter is limited to the hydraulic jump. The *hydraulic jump* occurs when a rapid change in flow depth occurs from a small depth to a large depth such that there is an abrupt rise in water surface. A hydraulic jump occurs wherever super-critical flow changes to subcritical flow. Hydraulic jumps can occur in canals downstream of regulating sluices, at the foot of spillways, or where a steep channel slope suddenly becomes flat.

Figure 5.5.1 illustrates a hydraulic jump along with the specific energy and specific force curves. The depths of flow upstream and downstream of the jump are called *sequent depths* or *conjugate depths*. Because hydraulic jumps are typically short in length, the losses due to shear along the wetted perimeter are small compared to the pressure forces. Neglecting these forces and assuming a horizontal channel ($F_g = 0$), the momentum principle can be applied as in section 5.2.3 to derive

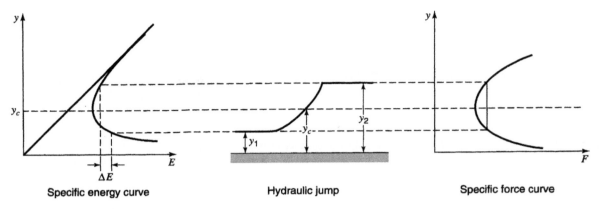

Figure 5.5.1 Hydraulic jump.

equation (5.2.24):

$$\frac{Q^2}{gA_1} + A_1\bar{y}_1 = \frac{Q^2}{gA_2} + A_2\bar{y}_2 \tag{5.2.24}$$

Consider a rectangular channel of width $B > 0$, so $Q = A_1V_1 = A_2V_2, A_1 = By_1, A_2 = By_2$, $\bar{y}_1 = y_1/2$, and $\bar{y}_2 = y_2/2$:

$$\frac{Q^2}{gBy_1} + By_1\left(\frac{y_1}{2}\right) = \frac{Q^2}{gBy_2} + By_2\left(\frac{y_2}{2}\right) \tag{5.5.1}$$

Simplifying yields

$$\frac{Q^2}{g}\left(\frac{1}{y_1} - \frac{1}{y_2}\right) = \frac{1}{2}B^2\left(y_2^2 - y_1^2\right)$$

$$\frac{Q^2}{g}(y_2 - y_1) = \frac{1}{2}B^2 y_1 y_2\left(y_2^2 - y_1^2\right)$$

$$\frac{B^2 y_1^2 V_1^2}{g}(y_2 - y_1) = \frac{1}{2}B^2 y_1 y_2(y_2 + y_1)(y_2 - y_1)$$

$$\frac{y_1 V_1^2}{g} = \frac{1}{2}y_2(y_2 + y_1)$$

Dividing by y_1^2, we get

$$\frac{2V_1^2}{gy_1} = \frac{y_2}{y_1}\left(\frac{y_2}{y_1} + 1\right) \tag{5.5.2}$$

The Froude number for a rectangular channel is $F_{r_1} = V_1/\sqrt{gD_1} = V_1/\sqrt{gy_1}$; therefore equation (5.5.2) reduces to

$$\left(\frac{y_2}{y_1}\right)^2 + \frac{y_2}{y_1} - 2F_{r_1}^2 = 0 \tag{5.5.3}$$

or

$$\frac{y_2}{y_1} = \frac{1}{2}\left(-1 + \sqrt{1 + 8F_{r_1}^2}\right) \tag{5.5.4}$$

using the quadratic formula and discarding the negative roots.

Alternatively, $Q = By_2V_2$ and $F_{r_2} = V_2/\sqrt{gy_2}$ could have been used to derive

$$\frac{y_1}{y_2} = \frac{1}{2}\left(-1 + \sqrt{1 + 8F_{r_2}^2}\right) \tag{5.5.5}$$

Equations (5.5.4) and (5.5.5) can be used to find the sequent depths of a hydraulic jump. The use of hydraulic jumps as energy dissipaters is further discussed in Chapters 15 and 17.

EXAMPLE 5.5.1

Consider the 8-ft wide rectangular channel used in examples 5.1.1, 5.1.3, and 5.3.1 with a discharge of 100 cfs. If a weir were placed in the channel and the depth upstream of the weir were 5 ft, would a hydraulic jump form upstream of the weir?

SOLUTION

For the discharge of 100 cfs, the normal depth is $y_n = 3.97$ ft from example 5.1.3, and the critical depth is $y_c = 1.69$. Because $y_c < y_n < 5$ ft, a hydraulic jump would not form. As a result of $y_n > y_c$, a mild slope exists. For a jump to form, $y_n < y_c$, which is a steep slope.

EXAMPLE 5.5.2

For example 5.3.1, determine whether a hydraulic jump will occur.

SOLUTION

The normal depth and critical depth must be computed and compared. Using equation (5.2.14) with $q = 4.646$ m³/s per meter of width, we get

$$y_c = \left(\frac{q^2}{g}\right)^{1/3} = \left(\frac{4.646^2}{9.81}\right)^{1/3} = 1.30 \text{ m}.$$

The depths of flow at $y_a = 0.457$ m and at $y_b = 0.5$ m, so this flow is supercritical flow.

Next, the normal depth is computed using Manning's equation (5.1.23):

$$Q = \frac{1}{n}AR^{2/3}S_0^{1/2}$$

or

$$q = \frac{1}{n}(y_n)(y_n)^{2/3}S_0^{1/2} = \frac{1}{n}y_n^{5/3}S_0^{1/2}$$

$$4.646 = \frac{1}{0.020}y_n^{5/3}(0.0003)^{1/2}$$

Thus $y_n = (5.365)^{3/5} = 2.74$ m. Under these conditions $y_n > y_c > 0.5$ m, an M3 water surface profile exists and a hydraulic jump occurs. If normal depth occurs downstream of the jump, what is the depth before the jump? Using equation (5.5.5) with $y_2 = 2.74$ m, we find that y_1 is 0.5 m.

EXAMPLE 5.5.3

A rectangular channel is 10.0 ft wide and carries a flow of 400 cfs at a normal depth of 3.00. Manning's $n = 0.017$. An obstruction causes the depth just upstream of the obstruction to be 8.00 ft deep. Will a jump form upstream from the obstruction? If so, how far upstream? What type of curve will be present?

SOLUTION

First a determination must be made whether a jump will form by comparing the normal depth and critical depth. Using equation (5.2.14), we find

$$y_c = \left(\frac{q^2}{g}\right)^{1/3} = \left(\frac{40^2}{32.2}\right)^{1/3} = 3.68 \text{ ft}$$

and $y_c > y_n = 3.0$ ft, therefore the channel is steep. Because $y_n < y_c < 8$ ft, a subcritical flow exists on a steep channel, a hydraulic jump forms upstream of the obstruction. If the depth before the jump is considered to be normal depth, $y_n = y_1 = 3$, then the conjugate depth y_2 can be computed using equation (5.5.4), where the Froude number is

$$F_{r_1}^2 = \left(\frac{V_1}{\sqrt{gy_1}}\right)^2 = \frac{V_1^2}{gy_1} = \frac{q^2}{gy_1^3}$$

so

$$y_2 = \frac{y_1}{2}\left(-1 + \sqrt{1 + \frac{8q^2}{gy_1^3}}\right)$$

$$= \frac{3}{2}\left(-1 + \sqrt{1 + \frac{8(40)^2}{g(3)^3}}\right)$$

$$= 4.45 \text{ ft}$$

Next the distance Δx from the depth of 4.45 ft to the depth of 8 ft is determined using equation (5.3.1):

$$S_0\Delta x + \frac{V_2^2}{2g} = \Delta y + \frac{V_3^2}{2g} + S_f\Delta x$$

which can be rearranged to yield

$$(S_0 - S_f)\Delta x = y_3 + \frac{V_3^2}{2g} - y_2 - \frac{V_2^2}{2g}$$

$$= E_3 - E_2$$

so

$$\Delta x = \frac{E_3 - E_2}{S_0 - S_f} = \frac{\Delta E}{S_0 - S_f}$$

To solve for Δx, first compute E_2 and E_3:

Depth (ft)	A (ft^2)	R (ft)	V (ft/s)	$V^2/2g$ (ft)	E (ft)	R_{ave} (ft)	V_{ave} (ft/s)
8	80	3.08	5.00	0.388	8.388		
4.45	44.5	2.35	8.99	1.25	5.70	2.72	7.00

Now compute S_f from Manning's equation using equation (5.1.24) with V_{ave} and R_{ave}, and rearrange to yield

$$S_f = \frac{n^2 V_{ave}^2}{2.22 R_{ave}^{4/3}} = \frac{0.017^2 \times 7^2}{2.22 \times 2.72^{4/3}} = 0.00168$$

Compute S_0 using Manning's equation with the normal depth:

$$Q = \frac{1.49}{n} AR^{2/3} S_0^{1/2}$$

$$400 = \frac{1.49}{0.017} \times 3 \times 10 \times \left(\frac{30}{16}\right)^{2/3} \sqrt{S_0}$$

Thus, $S_0 = 0.0100$. Now, using $\Delta x = \dfrac{\Delta E}{S_0 - S_f} = \dfrac{8.388 - 5.71}{0.0100 - 0.00168} = 322$ ft, the distance from the conjugate depth of the jump $y_2 = 4.45$ ft downstream to the depth y_3 of 8 ft (location of the obstruction) is 322 ft. In other words, the hydraulic jump occurs approximately 322 ft upstream of the obstruction. The type of water surface profile after the jump is an S1 profile.

EXAMPLE 5.5.4

A hydraulic jump occurs in a rectangular channel 3.0 m wide. The water depth before the jump is 0.6 m, and after the jump is 1.6 m. Compute (a) the flow rate in the channel, (b) the critical depth, and (c) the head loss in the jump.

SOLUTION

(a) To compute the flow rate knowing $y_1 = 0.6$ m and $y_2 = 1.6$ m, equation (5.5.4) can be used:

$$y_2 = \frac{y_1}{2}\left[-1 + \sqrt{1 + \frac{8q^2}{gy_1^3}}\right]$$

in which $F_{r_1} = q / \sqrt{gy_1^3}$ has been substituted:

$$1.6 = \frac{0.6}{2}\left[-1 + \sqrt{1 + \frac{8q^2}{9.81(0.6)^3}}\right]$$

$$6.33 = \sqrt{1 + 3.775q^2}$$

$$40.07 = 3.775q^2$$

$q = 3.26$ m³/s per meter width of channel

and

$Q = 3q = 9.78$ m³/s.

(b) Critical depth is computed using equation (5.2.14):

$$y_c = \left(\frac{q^2}{g}\right)^{1/3} = \left(\frac{3.26^2}{9.81}\right)^{1/3} = 1.03 \text{ m}$$

(c) The headloss in the jump is the change in specific energy before and after the jump

$$h_L = \Delta E = E_1 - E_2$$

so

$$h_L = y_1 + \frac{V_1^2}{2g} - y_2 - \frac{V_2^2}{2g}$$

so

$$V_1 = Q/A_1 = 9.78/(3 \times 0.6) = 5.43 \text{ m/s}$$
$$V_2 = Q/A_2 = 9.78/(3 \times 1.6) = 2.04 \text{ m/s}$$
$$h_L = 0.6 + \frac{5.43^2}{2(9.81)} - 1.6 - \frac{2.04^2}{2(9.81)}$$

$$= 0.6 + 1.5 - 1.6 - 0.21$$

$$h_L = 0.29 \text{ m}.$$

EXAMPLE 5.5.5

Derive an equation to approximate the headloss (energy loss) of a hydraulic jump in a horizontal rectangular channel in terms of the depths before and after the jump, y_1 and y_2, respectively.

SOLUTION

The energy loss can be approximated by

$$h_L = E_1 - E_2$$

$$= \left(y_1 + \frac{V_1^2}{2g}\right) - \left(y_2 + \frac{V_2^2}{2g}\right)$$

The velocities can be expressed as $V_1 = Q/A_1 = q/y_1$ and $V_2 = q/y_2$, so

$$h_L = y_1 + \frac{1}{2g}\frac{q^2}{y_1^2} - y_2 - \frac{1}{2g}\frac{q^2}{y_2^2}$$

$$= \frac{q^2}{2g}\left(\frac{1}{y_1^2} - \frac{1}{y_2^2}\right) + (y_1 - y_2)$$

The balance between hydrostatic forces and the momentum flux per unit width of the channel can be expressed using equation (5.2.23):

$$\sum F = F_1 - F_2 = \rho q(V_2 - V_1) = \gamma A_1 \bar{y}_1 - \gamma A_2 \bar{y}_2 = \rho q V_2 - \rho q V_1$$

where $A = (1)(y_1)$ and $\bar{y}_1 = y_1/2$, so

$$\frac{\gamma}{2}y_1^2 - \frac{\gamma}{2}y_2^2 = \rho q\left(\frac{q}{y_2} - \frac{q}{y_1}\right)$$

Solving, we get

$$\frac{q^2}{g} = y_1 y_2 \left(\frac{y_1 + y_2}{2}\right)$$

Substituting this equation into the above equation for the headloss and simplifying gives

$$\Delta E = \frac{(y_2 - y_1)^3}{4 y_1 y_2}$$

Prismatic Channels with Change in Slope

Consider the channel that changes slope from a steep slope to a mild slope shown in Figure 5.5.2. In this case $y_{n_2} > y_{n_1}$. The conjugate depth of y_{n_1} is y' computed using equation (5.5.4). If $y' < y_{n_2}$, the hydraulic jump occurs upstream of the slope break with control at A, and if $y' > y_{n_2}$, the jump occurs downstream of the break in slope with control at B. If $\left(E_{n_1} - \Delta E_{j_1}\right) > E_{n_2}$, then a hydraulic jump occurs on the downstream mild slope. If $E_{n_2} > \left(E_{n_1} - \Delta E_{j_1}\right)$, then the hydraulic jump occurs on the upstream steep slope. ΔE_{j_1} is the energy loss in the jump in reach 1. Another way to look at this is if $E_{y'} > E_{n_2}$ control is at B because the energy loss ΔE_{j_1} in the hydraulic jump in reach 1 is not large enough to decrease the energy from E_{n_1} to E_{n_2}.

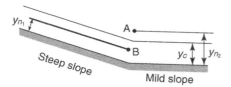

Figure 5.5.2 Rapidly varied flow caused by slope change from steep to mild.

5.6 DISCHARGE MEASUREMENT

5.6.1 Weir

A *weir* is a device (or overflow structure) that is placed normal to the direction of flow. The weir essentially backs up water so that in flowing over the weir, the water goes through critical depth. Weirs have been used for the measurement of water flow in open channels for many years. Weirs can generally be classified as *sharp-crested weirs* and *broad-crested weirs*. Weirs are discussed in detail in Bos et al. (1984), Brater et al. (1996), and Replogle et al. (1999).

A *sharp-crested weir* is basically a thin plate mounted perpendicular to the flow with the top of the plate having a beveled, sharp edge, which makes the nappe spring clear from the plate (see Figure 5.6.1). The rate of flow is determined by measuring the head, typically in a stilling well (see Figure 5.6.2) at a distance upstream from the crest. The head H is measured using a gauge.

Suppressed Rectangular Weir

These sharp-crested weirs are as wide as the channel, and the width of the nappe is the same length as the crest. Referring to Figure 5.6.1, consider an elemental area $dA = Bdh$ and assume the velocity is $\sqrt{2gh}$; then the elemental flow is

$$dQ = Bdh\sqrt{2gh} = B\sqrt{2g}\,h^{1/2}dh \tag{5.6.1}$$

The discharge is expressed by integrating equation (5.6.1) over the area above the top of the weir crest:

$$Q = \int_0^H dQ = \sqrt{2g}\,B\int_0^H h^{1/2}dh = \frac{2}{3}\sqrt{2g}\,BH^{3/2} \tag{5.6.2}$$

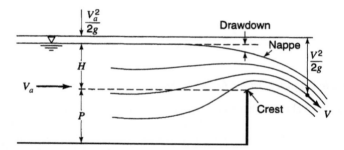

Figure 5.6.1 Flow over sharp-crested weir.

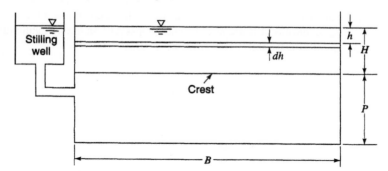

Figure 5.6.2 Rectangular sharp-crested weir without end contraction.

Friction effects have been neglected in derivation of equation (5.6.2). The drawdown effect shown in Figure 5.6.1 and the crest contraction indicate that the streamlines are not parallel or normal to the area in the plane. To account for these effects a coefficient of discharge C_d is used, so that

$$Q = C_d \frac{2}{3} \sqrt{2g}\, BH^{3/2} \tag{5.6.3}$$

where C_d is approximately 0.62. This is the basic equation for a suppressed rectangular weir, which can be expressed more generally as

$$Q = C_w BH^{3/2} \tag{5.6.4}$$

where C_w is the weir coefficient, $C_w = C_d \frac{2}{3} \sqrt{2g}$. For U.S. customary units, $C_w \approx 3.33$, and for SI units $C_w \approx 1.84$.

If the velocity of approach V_a where H is measured is appreciable, then the integration limits are

$$Q = \sqrt{2g}B \int_{\frac{v_a^2}{2g}}^{H+\frac{v_a^2}{2g}} h^{1/2}\, dh = C_w B \left[\left(H + \frac{V_a^2}{2g} \right)^{3/2} - \left(\frac{V_a^2}{2g} \right)^{3/2} \right] \tag{5.6.5a}$$

When $\left(\dfrac{V_a^2}{2g} \right)^{3/2} \approx 0$, equation (5.6.5a) can be simplified to

$$Q = C_w B \left(H + \frac{V_a^2}{2g} \right)^{3/2} \tag{5.6.5b}$$

Contracted Rectangular Weirs

A *contracted rectangular weir* is another sharp-crested weir with a crest that is shorter than the width of the channel and one or two beveled end sections so that water contracts both horizontally and vertically. This forces the nappe width to be less than B. The effective crest length is

$$B' = B - 0.1\, nH \tag{5.6.6}$$

where $n = 1$ if the weir is placed against one side wall of the channel so that the contraction on one side is suppressed and $n = 2$ if the weir is positioned so that it is not placed against a side wall.

Triangular Weir

Triangular or *V-notch weirs* are sharp-crested weirs that are used for relatively small flows, but have the advantage that they can also function for reasonably large flows as well. Referring to Figure 5.6.3,

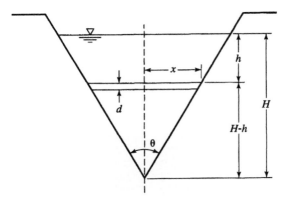

Figure 5.6.3 Triangular sharp-crested weir.

the rate of discharge through an elemental area, dA, is

$$dQ = C_d\sqrt{2gh}\,dA \tag{5.6.7}$$

where $dA = 2xdh$, and, $x = (H-h)\tan\dfrac{\theta}{2}$ so $dA = 2(H-h)\tan\left(\dfrac{\theta}{2}\right)dh$. Then

$$dQ = C_d\sqrt{2gh}\left[2(H-h)\tan\left(\frac{\theta}{2}\right)dh\right] \tag{5.6.8}$$

and

$$
\begin{aligned}
Q &= C_d 2\sqrt{2g}\ \tan\left(\frac{\theta}{2}\right)\int_0^H (H-h)h^{1/2}dh \\
&= C_d\left(\frac{8}{15}\right)\sqrt{2g}\ \tan\left(\frac{\theta}{2}\right)H^{5/2} \\
&= C_w H^{5/2}
\end{aligned}
\tag{5.6.9}
$$

The value of C_w for a value of $\theta = 90°$ (the most common) is $C_w = 2.50$ for U.S. customary units and $C_w = 1.38$ for SI units.

Broad-Crested Weir

Broad-crested weirs (refer to Figure 5.6.4) are essentially critical-depth weirs in that if the weirs are high enough, critical depth occurs on the crest of the weir. For critical flow conditions $y_c = \left(q^2/g\right)^{1/3}$ and $E = \dfrac{3}{2}y_c$ for rectangular channels:

$$Q = B\cdot q = B\sqrt{gy_c^3} = B\sqrt{g\left(\frac{2}{3}E\right)^3} = B\left(\frac{2}{3}\right)^{3/2}\sqrt{g}E^{3/2}$$

or, assuming the approach velocity is negligible:

$$Q = B\left(\frac{2}{3}\right)^{3/2}\sqrt{g}\,H^{3/2}$$

$$Q = C_w B H^{3/2} \tag{5.6.10}$$

Figure 5.6.5 illustrates a broad-crested weir installation in a concrete-lined canal.

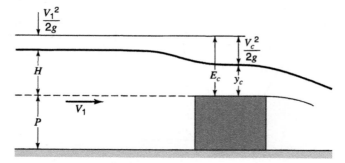

Figure 5.6.4 Broad-crested weir.

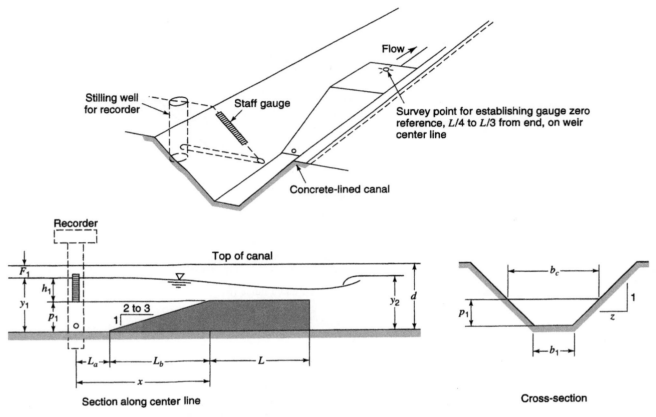

Figure 5.6.5 Broad-crested weir in concrete-lined canal (from Bos et al. (1984)).

EXAMPLE 5.6.1

A rectangular, sharp-crested suppressed weir 3 m long is 1.0 m high. Determine the discharge when the head is 150 mm.

SOLUTION

Using equation (5.6.4), $Q = 1.84\, BH^{1.5}$, the discharge is

$$Q = 1.84(3)\left(\frac{150}{1000}\right)^{1.5} = 0.321 \text{ m}^3/\text{s}$$

5.6.2 Flumes

Bos et al. (1984) provide an excellent discussion of flumes. A weir is a control section that is formed by raising the channel bottom, whereas a *flume* is formed by narrowing a channel. When a control section is formed by raising both the channel bottom and narrowing it, the structure is usually called a *flume*. Figure 5.6.6 shows a distinction between weirs and flumes.

Weirs can result in relatively large headlosses and, if the water has suspended sediment, can cause deposition upstream of the weir, resulting in a gradual change of the weir coefficient. These disadvantages can be overcome in many situations by the use of flumes.

Figure 5.6.7 illustrates the general layout of a flow measuring structure. Most flow measurement and flow regulating structures consist of (a) a *converging transition*, where subcritical flow is accelerated and guided into the throat without flow separation; (b) a *throat* where the water

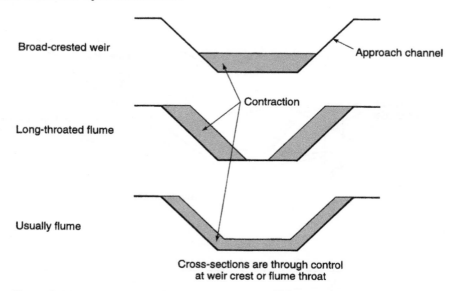

Figure 5.6.6 Distinction between a weir and a flume (from Bos et al. (1984)).

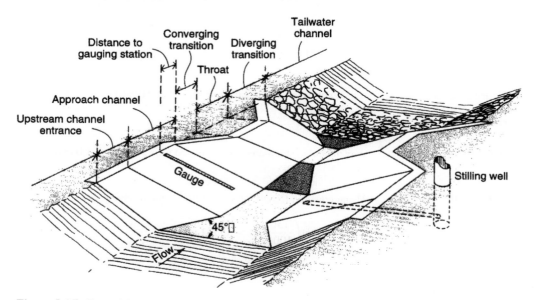

Figure 5.6.7 General layout of a flow-measuring structure (from Bos et al. (1984)).

accelerates to supercritical flow so that the discharge is controlled; and (c) a *diverging transition* where flow velocity is gradually reduced to subcritical flow and the potential energy is recovered.

One of the more widely used flumes is the *Parshall flume* (U.S. Bureau of Reclamation, 1981), which is a Venturi-type flume illustrated in Figure 5.6.8. The discharge equation for these flumes with widths of 1 ft (0.31 m) to 8 ft (2.4 m) is

$$Q = 4WH_a^{1.522W^{0.026}} \qquad (5.6.11)$$

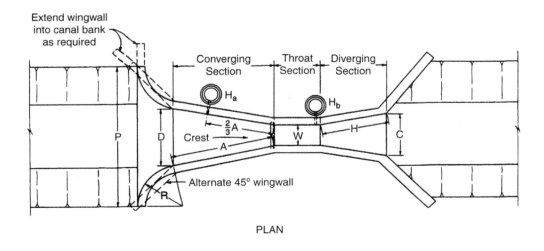

PLAN

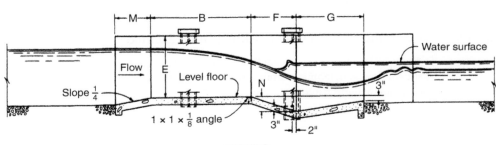

PROFILE

W		A		⅔A		B		C		D		E		F		G		M		N		P		R		FREE-FLOW CAPACITY	
																										MINIMUM	MAXIMUM
FT.	IN.	FT.	IN.	FT.	IN.	FT.	IN.	FT.	IN.	FT.	IN.	FT.	IN.	FT.	IN.	FT.	IN.	FT.	IN.	FT.	IN.	FT.	IN.	FT.	IN.	CFS	CFS
0	6	2	$\frac{7}{16}$	1	$4\frac{5}{16}$	2	0	1	$3\frac{1}{2}$	1	$3\frac{5}{8}$	2	0	1	0	2	0	1	0	0	$4\frac{1}{2}$	2	$11\frac{1}{2}$	1	4	.05	3.9
	9	2	$10\frac{5}{8}$	1	$11\frac{1}{8}$	2	10	1	3	1	$10\frac{5}{8}$	2	6	1	0	1	6	1	0		$4\frac{1}{2}$	3	$6\frac{1}{2}$	1	4	.09	8.9
1	0	4	6	3	0	4	$4\frac{7}{8}$	2	0	2	$9\frac{1}{4}$	3	0	2	0	3	0	1	3		9	4	$10\frac{3}{4}$	1	8	.11	16.1
1	6	4	9	3	2	4	$7\frac{7}{8}$	2	6	3	$4\frac{3}{8}$	3	0	2	0	3	0	1	3		9	5	6	1	8	.15	24.6
2	0	5	0	3	4	4	$10\frac{7}{8}$	3	0	3	$11\frac{1}{2}$	3	0	2	0	3	0	1	3		9	6	1	1	8	.42	33.1
3	0	5	6	3	8	5	$4\frac{3}{4}$	4	0	5	$1\frac{7}{8}$	3	0	2	0	3	0	1	3		9	7	$3\frac{1}{2}$	1	8	.61	50.4
4	0	6	0	4	0	5	$10\frac{5}{8}$	5	0	6	$4\frac{1}{4}$	3	0	2	0	3	0	1	6		9	8	$10\frac{3}{4}$	2	0	1.3	67.9
5	0	6	6	4	4	6	$4\frac{1}{2}$	6	0	7	$6\frac{5}{8}$	3	0	2	0	3	0	1	6		9	10	$1\frac{1}{4}$	2	0	1.6	85.6
6	0	7	0	4	8	6	$10\frac{3}{4}$	7	0	8	9	3	0	2	0	3	0	1	6		9	11	$3\frac{1}{2}$	2	0	2.6	103.5
7	0	7	6	5	0	7	$4\frac{1}{4}$	8	0	9	$11\frac{3}{8}$	3	0	2	0	3	0	1	6		9	12	6	2	0	3.0	121.4
8	0	8	0	5	4	7	$10\frac{1}{8}$	9	0	11	$1\frac{3}{4}$	3	0	2	0	3	0	1	6		9	13	$8\frac{1}{4}$	2	0	3.5	139.5

Figure 5.6.8 Standard Parshall flume dimensions. 103-D-1225 (from U.S. Bureau of Reclamation (1978)).

where Q is the discharge in ft^3/s, W is the width of the flume throat, and H_a is the upstream head in ft. For smaller flumes, e.g., 6-inch flumes,

$$Q = 2.06H_a^{1.58} \qquad (5.6.12)$$

and for 9-inch flumes

$$Q = 3.07H_a^{1.53} \qquad (5.6.13)$$

EXAMPLE 5.6.2   Water flows through a Parshall flume with a throat width of 4.0 ft at a depth of 2.0 ft. What is the flow rate?

SOLUTION   Using equation (5.6.11).

$$Q = 4.0(4.0)(2.0)^{(1.522)(4.0)^{0.026}} = 47.8 \text{ ft}^3/\text{s}$$

Figure 5.6.9 illustrates a flow-measuring structure for unlined (earthen) channels that are longer, and consequently more expensive, than structures for concrete-lined channels. For concrete-lined channels, the approach channel and sides of the control section are already available.

5.6.3   Stream Flow Measurement: Velocity-Area-Integration Method

As for weirs and flumes, stream flow is not directly measured. Instead, water level is measured and stream flow is determined from a *rating curve*, which is the relationship between water surface elevation and discharge.

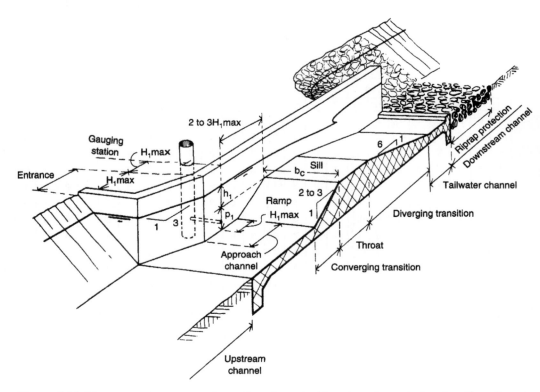

Figure 5.6.9 Flow-measuring structure for earthen channel with rectangular control section (from Bos et al. (1984)).

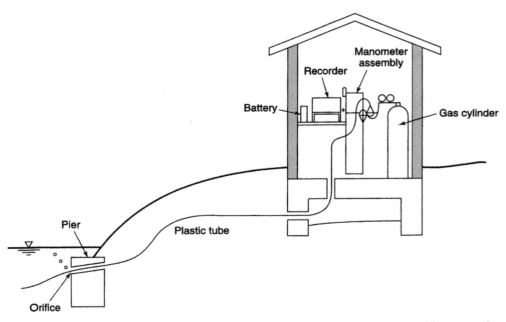

Figure 5.6.10 Water level measurement using a bubble gauge recorder. The water level is measured as the back pressure on the bubbling stream of gas by using a mercury manometer (from Rantz et al. (1982)).

Water surface level (elevation) can be measured manually or automatically. *Crest stage gauges* are used to measure flood crests. They consist of a wooden staff gauge placed inside a pipe with small holes for water entry. Cork in the pipe floats as the water rises and adheres to the staff (scale) at the highest water level. *Bubble gauges* (shown in Figure 5.6.10) sense the water surface level by bubbling a continuous stream of gas (usually carbon dioxide) into the water. The pressure required to continuously force the gas stream out beneath the water surface is a measure of the depth of water over the nozzle of the bubble stream. The pressure is measured with a manometer assembly to provide a continuous record of water level in the stream (gauge height).

Rating curves are developed using a set of measurements of discharge and gauge height in the stream. The discharge is $Q = AV$ where V is the mean velocity normal to the cross-sectional area of flow A, which is a function of the gauge height. So in order to measure discharge the velocity and the gauge height must be determined. In a stream or river, the velocity varies with depth, as discussed in Chapter 3 (Figure 3.7.1). Therefore, the velocity must be recorded at various locations and depths across the stream.

Referring to Figure 5.6.11, the total discharge is computed by summing the incremental discharge calculated from each measurement i, $i = 1, 2, \ldots, n$, of velocity V_i, and depth y_i. These

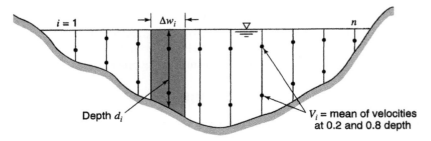

Figure 5.6.11 Computation of discharge from stream gauging data.

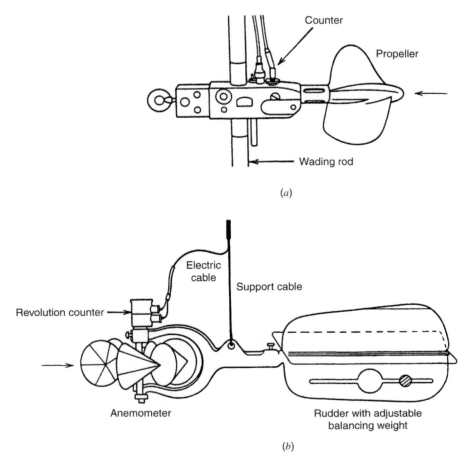

Figure 5.6.12 (*a*) Propeller- and (*b*) Price-type current meters (from James (1988)).

measurements represent average values over the width Δw_i of the stream. The total discharge is computed using

$$Q = \sum_{i=1}^{n} V_i y_i \Delta w_i \qquad (5.6.14)$$

Both theory and experimental evidence indicate that the mean velocity in a vertical section can be closely approximated by the average of the velocities at 0.2 depth and 0.8 depth below the water surface, as shown in Figure 5.6.11. If the stream is shallow, it may be possible to take a single measurement of velocity at a 0.6 depth.

To measure the velocity in a stream, a *current meter*, which is an impellor device, can be used. The speed at which the impellor rotates is proportional to the flow velocity. Figure 5.6.12*a* shows a propeller type current meter on a wading rod and Figure 5.6.12*b* shows a Price current meter, which is the most commonly used velocity meter in the United States. Refer to Wahl et al. (1995) for detailed descriptions on the stream-gauging program of the U.S. Geological Survey.

PROBLEMS

5.1.1 Compute the hydraulic radius and hydraulic depth for a trapezoidal flood control channel with a bottom width of 20 ft, side slopes 2:1 (h:v), and a top width of 40 ft.

5.1.2 Compute the hydraulic radius and hydraulic depth for a trapezoidal flood control channel with a bottom width of 4 m, side slopes 2:1 (h:v), and a top width of 8 m.

5.1.3 Compute the hydraulic radius and hydraulic depth for a 36-inch diameter culvert with a depth of flow of 24 in.

5.1.4 Compute the hydraulic radius and hydraulic depth for a 1.5-m diameter culvert with a depth of flow of 1.24 m.

5.1.5 A 2-m wide rectangular channel with a bed slope of 0.0005 has a depth of flow of 1.5 m. Manning's roughness coefficient is 0.015. Determine the steady uniform discharge in the channel.

5.1.6 Determine the uniform flow depth in a rectangular channel 2.5 m wide with a discharge of 3 m^3/s. The slope is 0.0004 and Manning's roughness factor is 0.015.

5.1.7 Determine the uniform flow depth in a trapezoidal channel with a bottom width of 8 ft and side slopes of 1 vertical to 2 horizontal. The discharge is 100 ft^3/s. Manning's roughness factor is 0.015 and the channel bottom slope is 0.0004.

5.1.8 Determine the uniform flow depth in a trapezoidal channel with a bottom width of 2.5 m and side slopes of 1 vertical to 2 horizontal with a discharge of 3 m^3/s. The slope is 0.0004 and Manning's roughness factor is 0.015.

5.1.9 Determine the cross-section of the greatest hydraulic efficiency for a trapezoidal channel with side slope of 1 vertical to 2 horizontal if the design discharge is 10 m^3/s. The channel slope is 0.001 and Manning's roughness factor is 0.020.

5.1.10 For a trapezoidal-shaped channel ($n = 0.014$ and slope S_o of 0.0002 with a 20-ft bottom width and side slopes of 1 vertical to 1.5 horizontal), determine the normal depth for a discharge of 1000 cfs.

5.1.11 Show that the best hydraulic trapezoidal section is one-half of a hexagon.

5.1.12 A trapezoidal channel has a bottom width of 10 ft and side slopes of 2:1 (h:v). The channel has a slope of 0.0001 and a Manning's roughness of 0.018. If the uniform flow depth is 4 ft, what is the discharge in the channel?

5.1.13 Compute the normal depth of flow in a 36-in diameter culvert with a slope of 0.0016 and Manning's n of 0.015 for a discharge of 20 cfs.

5.1.14 A 6-ft diameter concrete-lined sewer has a bottom slope of 1.5 ft/mi Find the depth of flow for a discharge of 20 cfs.

5.1.15 Compute the uniform flow depth in a trapezoidal channel with a bottom width of 20 ft, slope of 0.0016, Manning's n of 0.025, and side slopes of 2:1 (h:v) for a discharge of 500 cfs. What is the velocity of flow?

5.1.16 Design a trapezoidal concrete-lined channel ($n = 0.015$) to convey 100 cfs on a slope of 0.001. Assume the use of a best hydraulic section for the design.

5.1.17 Design a trapezoidal concrete-lined channel ($n = 0.015$) to convey 20 m^3/s on a slope of 0.0001. Assume the use of a best hydraulic section for the design.

5.1.18 Phillips and Ingersoll (1998) presented equations for determining the Manning's roughness factor for gravel-bed

streams using the relative roughness defined as (R/d_{50}) where R is the hydraulic radius and d_{50} is the median grain size. The equation was verified for Arizona (for the range in d_{50} 0.28 to 0.36 ft) is $n = (0.0926R^{1/6})/[1.46 + 2.23\log(R/d_{50})]$. Use this equation to develop a graph illustrating the relation of n, R, and d_{50}.

5.1.19 Using the slope-area method, compute the flood discharge through a river reach of 800 ft having known values of water areas, conveyances, and energy coefficients of the upstream and downstream end sections. The fall of the water surface is 1.0 ft.

$$A_u = 11,070 \text{ ft}^2$$
$$A_d = 10,990 \text{ ft}^2$$
$$K_u = 3.034 \times 10^6$$
$$K_d = 3.103 \times 10^6$$
$$\alpha_u = 1.134$$
$$\alpha_d = 1.177$$

5.2.1 Solve example 5.2.2 for discharges of 0, 25, 75, 125, and 200 ft^3/s.

5.2.2 Rework example 5.2.3 for a 30-cm high hump and a side wall constriction that reduces the channel width to 1.6 m.

5.2.3 Compute the critical depth for the channel in problem 5.1.5.

5.2.4 Compute the critical depth for the channel in problem 5.1.6.

5.2.5 Rework example 5.2.4 with discharges of 0, 25, 75, 125, and 200 cfs.

5.2.6 Compute the critical depth in a 36-in diameter culvert with a slope of 0.0016 for a discharge of 20 cfs.

5.2.7 Compute the critical flow depth in a trapezoidal channel with a bottom width of 20 ft a slope of 0.0016, Manning's n of 0.025, and side slopes of 2:1 (h:v) for a discharge of 500 cfs.

5.2.8 A trapezoidal channel has a bottom width of 10 ft and side slopes of 2:1 (h:v). Manning's roughness factor is 0.018. For a uniform flow depth of 2.9 ft, what is the normal slope (corresponding to uniform flow depth) and the critical slope of the channel for a discharge of 200 cfs?

5.3.1 Resolve example 5.3.1 for a channel bed slope of 0.003.

5.3.2 A 2.45-m wide rectangular channel has a bed slope of 0.0004 and Manning's roughness factor of 0.015. For a discharge of 2.83 m^3/sec, determine the type of water surface profile for depths of 1.52 m, 0.61 m, and 0.30 m.

5.3.3 Rework problem 5.3.2 with a bed slope of 0.004.

5.3.4 If the channel of problem 5.1.10 is preceded by a steep slope and followed by a mild slope and a sluice gate as shown in Figure P5.3.4, sketch a possible water surface profile with the elevations to a scale of 1 in to 10 ft. Consider a discharge of 1500 cfs. For this discharge, the normal depth for a slope of 0.0003 is 8.18 ft and for a slope of 0.0002 is 9.13 ft.

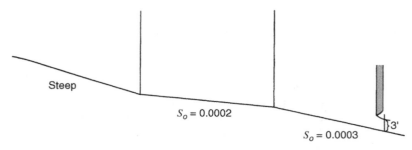

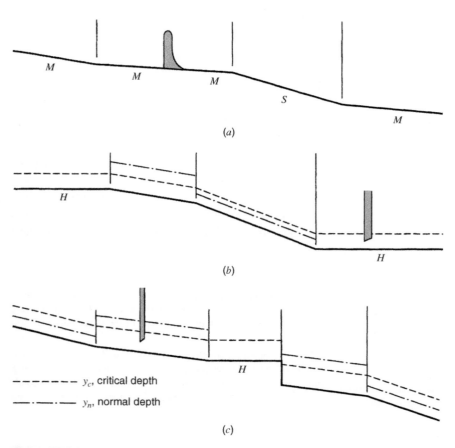

Figure P5.3.4

Figure P5.3.5

5.3.5 Sketch possible water surface profiles for the channel in Figure P5.3.5. First locate and mark the control points, then sketch the profiles, marking each profile with the appropriate designation. Show any hydraulic jumps that occur.

5.3.6 Show that for depths less than the normal depth ($y < y_n$) that $S_f > S_o$ and that for $y > y_n$ then $S_f < S_o$.

5.3.7 Show that $dy/dx = +$ for the S1 profile; $dy/dx = -$ for the S2 profile; and that $dy/dx = +$ for the S3 profile.

5.3.8 Using the gradually varied flow equation $dy/dx = (S_o - S_f)/(1 - F_r^2)$ define dy/dx for critical flow and for uniform flow.

5.3.9 A trapezoidal channel with a bottom width of 20 ft, a slope of 0.0016, Manning's n of 0.025, and side slopes of 2:1 (h:v) has a discharge of 500 cfs. An obstruction in this channel causes a backwater profile with a depth of 6.5 ft just upstream of the obstruction. What would be the depth of flow 200 ft and 400 ft upstream of the obstruction? How far upstream does the normal depth occur? Assume an energy coefficient of 1.1.

5.3.10 Consider a concrete ($n = 0.013$) wide rectangular channel that discharges 2.0 m³/s per unit width of flow. The channel bottom slope is 0.0001. There is a step rise of 0.2 m. Determine the flow depth downstream of the step assuming no transition loses. Does the water rise or fall at the step?

5.3.11 Consider a 5-m wide rectangular channel with a discharge of 12.5 m³/s. The depth of flow upstream of the step is 2.5 m. There is a step rise of 0.25 m in the bottom of the channel. Determine the flow depth downstream of the step assuming no transition losses. Does the water rise or fall at the step?

5.3.12 Consider the trapezoidal channel with a bottom width of 20 ft, a slope of 0.0016, Manning's n of 0.025, and side slopes of 2:1 (h:v) having a discharge of 500 cfs. Now a step rise of 1.0 ft is placed in the channel bottom. Determine the flow depth downstream of the bottom step assuming no transition losses. Does the water surface rise or fall at the step?

5.3.13 A 5-m wide rectangular channel with two reaches, each with a different slope, conveys 50 m³/s of water. The channel slope for the first reach is 0.001 and then a sudden change to a slope of 0.010. The Manning's n for the channel is 0.015. Perform the necessary computations to sketch the water surface profile and define the type of profiles.

5.3.14 Consider a concrete ($n = 0.013$) wide rectangular channel ($R = y$) that discharges 2.0 m³/sec per unit width of flow. The slope of the channel is 0.001. A low dam causes a backwater depth of 2.0 m immediately behind (upstream of) the dam. Compute the distance upstream of the dam to where the normal depth occurs.

5.3.15 Consider a concrete ($n = 0.013$) wide rectangular channel ($R = y$) that discharges 2.0 m³/sec per unit width of flow. The slope of the channel is 0.0001. A low dam causes a backwater depth of 2.0 m immediately behind (upstream of) the dam. Compute the distance upstream of the dam to where the normal depth occurs.

5.3.16 Consider a concrete trapezoidal channel with a 4-m bottom width, side slopes of 2:1 (h:v), and a bottom slope of 0.005. Determine the depth 150 m upstream from a section that has a measured depth of 2.0 m.

5.3.17 A wide rectangular channel changes in slope from 0.002 to 0.025. Sketch the water surface profile for a discharge of 1.7 m³/s/m and Manning's $n = 0.025$.

5.3.18 A 5-m wide rectangular channel with two reaches, each with a different slope, conveys 40 m³/s of water. The channel slope for the first reach is 0.0005 and then a sudden change to a slope of 0.015 so that critical flow occurs at the transition. Determine the depths of flow at locations 10 m, 20 m and 30 m upstream of the critical depth. The Manning's n for the channel is 0.015.

5.3.19 A 500-ft, 6-ft diameter reinforced concrete pipe culvert ($n = 0.012$) is used to convey stormwater from a detention reservoir to a downstream flood control channel. The slope of the

culvert is 0.02. The outlet of the culvert is placed at an elevation so that it will not be submerged. For a discharge of 230 cfs, compute the water surface profile. Develop a spreadsheet to perform the computations.

5.3.20 A 500-ft, 6-ft diameter reinforced concrete pipe culvert ($n = 0.012$) is used to convey stormwater from a detention reservoir to a downstream flood control channel. The slope of the culvert is 0.001. The outlet of the culvert is placed at an elevation so that it will not be submerged, and the flow falls freely into the flood control channel. For a discharge of 75 cfs, compute the water surface profile. Develop a spreadsheet to perform the computations.

5.4.1 Rework example 5.4.2 using equation (5.4.8), $S_f = Q^2/\overline{K}^2$.

5.4.2 Resolve example 5.4.3 with a discharge of 10,000 cfs and a downstream water surface elevation of 123.5 ft.

5.4.3 Rework example 5.4.4 using a discharge at river mile 1.0 of 8000 cfs and a discharge of 7500 cfs at river mile 1.5. The water surface elevation at river mile 1.0 is 123.5 ft. All other data are the same.

5.4.4 Consider a starting (assumed) water surface elevation of 5719.5 ft at cross-section 1 for example 5.4.5 and determine the water surface elevation at cross-section number 4.

5.4.5 Consider a starting (assumed) water surface elevation of 5717.6 ft at cross-section 1 for example 5.4.5 and determine the computed water surface elevation at cross-section 4.

5.4.6 Perform the backwater computations at mile 2.0 for the situation in example 5.4.4. The Manning's n values at mile 2 are 0.02 for the main channel and 0.04 for the overbanks. Use an assumed trial water surface elevation of 125.5 ft. Cross-sections at miles 2.0 and 1.5 are the same.

5.4.7 For most natural channels and many designed channels, the roughness varies along the wetted perimeter of the channel. In order to perform normal flow computations for these composite channels it is necessary to compute the composite (equivalent or effective) roughness factor. For the composite channel in Figure 5.4.2, compute the effective roughness factor (n_e) at river mile 1.0 for a water surface elevation of 125 ft using the following equations:

$$n_e = \left\{ \left[\sum_{i=1}^{N} P_i n_i^{3/2} \right] \Big/ P \right\}^{2/3}$$

and

$$n_e = \left\{ \left[\sum_{i=1}^{N} P_i n_i^2 \right] \Big/ P \right\}^{1/2}$$

where P_i and n_i are, respectively, the wetted perimeter and Manning's n for each subsection of the channel; $P = \sum_{i=1}^{N} P_i$ is the wetted perimeter of the complete channel section, and N is the number of subsections of the channel.

5.4.8 Use the U.S. Army Corps of Engineers HEC-RAS computer code to solve Example 5.4.4.

5.4.9 Use the U.S. Army Corps of Engineers HEC-RAS computer code to solve example 5.4.5.

5.5.1 Consider a 2.45-m wide rectangular channel with a bed slope of 0.0004 and a Manning's roughness factor of 0.015. A weir is placed in the channel and the depth upstream of the weir is 1.52 m for a discharge of 5.66 m³/s. Determine whether a hydraulic jump forms upstream of the weir.

5.5.2 A hydraulic jump occurs in a rectangular channel 4.0 m wide. The water depth before the jump is 0.4 m and after the jump is 1.7 m. Compute the flow rate in the channel, the critical depth, and the headloss in the jump.

5.5.3 Rework example 5.5.3 with a flow rate of 450 cfs at a normal depth of 3.2 ft. All other data remain the same.

5.5.4 Rework example 5.5.4 if the depth before the jump is 0.8 m and all other data remain the same.

5.5.5 A rectangular channel is 3.0 m wide ($n=0.018$) with a discharge of 14 m³/s at a normal depth of 1.0 m. An obstruction causes the depth just upstream of the obstruction to be 2.7 m deep. Will a jump form upstream of the obstruction? If the jump does form, how far upstream is it located?

5.5.6 Rework example 5.5.4 if the depth after the jump is 1.8 m and all other data remain the same.

5.5.7 A 10-ft wide rectangular channel ($n=0.015$) has a discharge of 251.5 cfs at a uniform flow (normal) depth of 2.5 ft. A sluice gate at the downstream end of the channel controls the flow depth just upstream of the gate to a depth z. Determine the depth z so that a hydraulic jump is formed just upstream of the gate. What is the channel bottom slope? What is the headloss (energy loss) in the hydraulic jump?

5.5.8 A 3-m wide rectangular channel ($n=0.02$) has a discharge of 10 m³/s at a uniform flow (normal) depth of 0.8 m. A sluice gate at the downstream end of the channel controls the flow depth just upstream of the gate to a depth z. Determine the depth z so that a hydraulic depth is formed just upstream of the gate. What is the channel bottom slope? What is the headloss (energy loss) in the hydraulic jump?

5.5.9 A 5-m wide rectangular channel with two reaches, each with a different slope, conveys 40 m³/s of water. The channel slope for the first reach is 0.015 and then a sudden change to a slope of 0.0005. The Manning's n for the channel is 0.015. Does a hydraulic jump occur in the channel? If there is a hydraulic jump, where does it occur: on the first reach or the second reach?

5.5.10 A 5-m wide rectangular channel with two reaches, each with a different slope, conveys 80 m³/s of water. The channel slope for the first reach is 0.01 and then a sudden change to a slope of 0.001. The Manning's n for the channel is 0.015. Does a hydraulic jump occur in the channel? If there is a hydraulic jump, where does it occur: on the first reach or the second reach?

5.5.11 A hydraulic jump is formed in a 10-ft wide channel just downstream of a sluice gate for a discharge of 450 cfs. If the depths of flow are 30 ft and 2 ft just upstream and downstream of the gate, respectively, determine the depth of flow downstream of the jump. What is the energy loss in the jump? What is the thrust $[F_{gate} = \gamma(\Delta F)]$ on the gate? Illustrate the thrust on the specific force and specific energy diagrams for this problem.

5.5.12 Consider a 40-ft wide horizontal rectangular channel with a discharge of 400 cfs. Determine the initial and sequent depths of a hydraulic jump, if the energy loss is 5 ft.

5.6.1 A rectangular, sharp-crested weir with end contraction is 1.6 m long. How high should it be placed in a channel to maintain an upstream depth of 2.5 m for 0.5 m³/s flow rate?

5.6.2 For a sharp-crested suppressed weir ($C_w = 3.33$) of length $B = 8.0$ ft, $P = 2.0$ ft, and $H = 1.0$ ft, determine the discharge over the weir. Neglect the velocity of approach head.

5.6.3 Rework problem 5.6.2 incorporating the velocity of approach head (equation (5.6.5a)).

5.6.4 Rework example 5.6.2 using equation (5.6.5b).

5.6.5 A rectangular sharp-crested weir with end contractions is 1.5 m long. How high should the weir crest be placed in a channel to maintain an upstream depth of 2.5 m for 0.5 m³/s flow rate?

5.6.6 Determine the head on a 60° V-notch weir for a discharge of 150 l/s. Take $C_d = 0.58$.

5.6.7 The head on a 90° V-notch weir is 1.5 ft. Determine the discharge.

5.6.8 Determine the weir coefficient of a 90° V-notch weir for a head of 180 mm for a flow rate of 20 l/s.

5.6.9 Determine the required head for a flow of 3.0 m³/s over a broad-crested weir 1.5 m high and 3 m long with a well-rounded upstream corner ($C_w = 1.67$).

5.6.10 Water flows through a Parshall flume with a throat width of 4.0 ft at a depth of 7.5 ft. Determine the flow rate.

5.6.11 Water flows through a Parshall flume with a throat width of 5.0 ft at a depth of 3.4 ft. Determine the flow rate.

5.6.12 The following information was obtained from a discharge measurement on a stream. Determine the discharge.

Distance from bank (ft)	Depth (ft)	Mean velocity (ft)
0	0.0	0.00
12	0.1	0.37
32	4.4	0.87
52	4.6	1.09
72	5.7	1.34
92	4.3	0.71
100	0.0	0.00

REFERENCES

Barnes, H. H., Jr., *Roughness Characteristics of Natural Channels*, U.S. Geological Survey, Water Supply Paper 1849, U.S. Government Printing Office, Washington, DC, 1962.

Bos, M. G., J. A. Replogle, and A. J. Clemmens, *Flow Measuring Flumes for Open Channel System*, John Wiley & Sons, New York, 1984.

Brater, E. F., H. W. King, J. E. Lindell, and C. Y. Wei, *Handbook of Hydraulics*, 7th edition, McGraw-Hill, New York, 1996.

Chaudhry, M. H., *Open-Channel Flow*, Prentice Hall, Englewood Cliffs, NJ, 1993.

Chow, V. T., *Open-Channel Hydraulics*, McGraw-Hill, New York, 1959.

French, R. H., *Open-Channel Hydraulics*, McGraw-Hill, New York, 1985.

Henderson, F. M., *Open-Channel Flow*, Macmillan, New York, 1966.

Hoggan, D. H., *Computer-Assisted Floodplain Hydrology and Hydraulics*, 2nd edition, McGraw-Hill, New York, 1997.

Hydrologic Engineering Center (HEC), *HEC-RAS River System Analysis System, User's Manual, Version 2.2*, U. S. Army Corps of Engineers Water Resources Support Center, Davis, CA, 1997a.

Hydrologic Engineering Center (HEC), *HEC-RAS River Analysis System, Hydraulic Reference Manual, Version 2.0*, U.S. Army Corps of Engineers Water Resources Support Center, Davis, CA, 1997b.

Hydrologic Engineering Center (HEC), *HEC-RAS River Analysis System, Applications Guide, Version 2.0*, U. S. Army Corps of Engineers Water Resources Support Center, Davis, CA, 1997c.

Jain, S. C., *Open-Channel Flow*, John Wiley & Sons, Inc., New York, 2001.

James, L. G., *Principles of Farm Irrigation System Design*, John Wiley and Sons, Inc., New York, 1988.

Manning, R., "On the Flow of Water in Open Channels and Pipes," *Transactions Institute of Civil Engineers of Ireland*, vol. 20, pp. 161–209, Dublin, 1891; Supplement, vol. 24, pp. 179–207, 1895.

Phillips, J. V., and T. L. Ingersoll, "*Verification of Roughness Coefficients for Selected Natural and Constructed Stream Channels in Arizona*," U.S. Geological Survey Prof. Paper 1584, 1998.

Rantz, S. E. et al., *Measurement and Computation of Stream Flow*, vol. 1, *Measurement of Stage and Discharge*, Water Supply Paper 2175, U.S. Geological Survey, 1982.

Replogle, J. A., A. J. Clemmens, and C. A. Pugh, "Hydraulic Design of Flow Measuring Structures," in *Hydraulic Design Handbook*, edited by L. W. Mays, McGraw-Hill, New York, 1999.

Sturm, T., *Open Channel Hydraulics*, McGraw-Hill, New York, 2001.

Townson, J. M., *Free-Surface Hydraulics*, Unwin Hyman, London, 1991.

U.S. Bureau of Reclamation, *Guide for Computing Water Surface Profiles*, 1957.

U.S. Bureau of Reclamation, *Design of Small Canal Structures*, U.S. Government Printing Office, Denver, CO, 1978.

U.S. Bureau of Reclamation, *Water Measurement Manual*, U.S. Government Printing Office, Denver, CO, 1981.

Wahl, K. L., W. O. Thomas, Jr., and R. M. Hirsch, "Stream-Gauging Program of the U.S. Geological Survey," *Circular 11123*, U.S. Geological Survey, Reston, VA, 1995.

Chapter 6

Hydraulic Processes: Groundwater Flow

Groundwater hydrology is the science that considers the occurrence, distribution, and movement of water below the surface of the earth (Todd and Mays, 2005). It is concerned with both the quantity and quality aspects of this water (Charbeneau, 2000). This chapter only considers the quantity aspects, in particular the hydraulic flow processes, emphasizing *aquifer hydraulics* (also see Batu, 1998; Charbeneau, 2000; and Delleur, 1999).

6.1 GROUNDWATER CONCEPTS

Groundwater flows through *porous media, fractured media*, and *large passages (karst)*. Porous media consist of solid material and *voids* or *pore space*. Porous media contain relatively small openings in the solid and are permeable media allowing the flow of water. The porous media that we typically are interested in for groundwater flow include natural soils, unconsolidated sediments, and sedimentary rocks. The size range of particles in a soil is referred to as the *soil texture*. Grain size determines the particle size classification. The fraction of clay, silt, and sand in a soil texture is described by the *soil texture triangle*, shown in Figure 6.1.1. Each point on the triangle corresponds to different percentages by mass (weight) of clay, sand, and silt.

The subsurface occurrence of groundwater, as shown in Figure 6.1.2, can be divided into the *vadose zone (zone operation)* and the *zone of saturation*. The vadose zone, also called the *unsaturated* or *partially saturated zone*, is the subsurface media above the water table. The term vadose is derived from the Latin *vadosus*, meaning shallow. Flow in the unsaturated, or vadose, zone is discussed further in Section 7.4. This chapter is focused on saturated flow, which is referred to as *groundwater flow*.

Groundwater originates through infiltration, influent streams, seepage from reservoirs, artificial recharge, condensation, seepage from oceans, water trapped in the sedimentary rock (*connate water*), and *juvenile* water (volcanic, magmatic, and cosmic). Any significant quantity of subsurface water is stored in subsurface formations defined as *aquifers*. An aquifer may be defined as a formation that contains sufficient saturated permeable material to yield significant quantities of water to wells (Lohman et al., 1972). Aquifers are usually of large areal extent and are essentially underground storage reservoirs. They may be overlain or underlain by a *confining bed*, which is a relatively impermeable material adjacent to the aquifer.

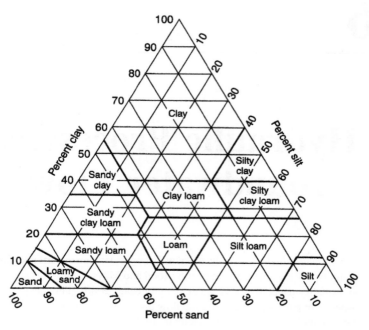

Figure 6.1.1 Triangle of soil textures for describing various combinations of sand, silt, and clay (from U.S. Soil Conservation Service (1951)).

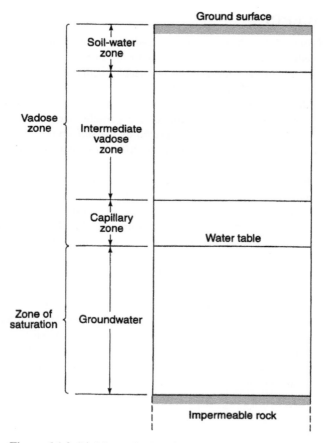

Figure 6.1.2 Divisions of subsurface water (Todd and Mays, 2005).

The following are different types of confining beds (Todd and Mays, 2005):

- *Aquiclude*—A saturated but relatively impermeable material that does not yield appreciable quantities of water to wells; clay is an example.
- *Aquifuge*—A relatively impermeable formation neither containing nor transmitting water; solid granite belongs in this category.
- *Aquitard*—A saturated but poorly permeable stratum that impedes groundwater movement and does not yield water freely to wells, but that may transmit appreciable water to or from adjacent aquifers and, where sufficiently thick, may constitute an important groundwater storage zone; sandy clay is an example.

Table 6.1.1 lists the various aquifer types and their characteristics.

Table 6.1.1   Summary of the Characteristics of Principal Aquifer Types

Aquifer type	Lithology	Groundwater flow regime	Aquifer properties Porosity (percent)	Permeability (m/day)	Specific yield (percent)	Natural Flow rates (m/day)	Residence times
Shallow alluvium	Gravel	Intergranular	25–35	100–1000	12–25	2–10	Could be very short;
	Sand	Intergranular	30–42	1–50	10–25	0.05–1	a few months to years,
	Silt	Intergranular	40–45	0.0005–0.1	5–10	0.001–0.1	depending on volume
Deep sedimentary formations	Sand and silts	Intergranular	30–40	0.1–5	2–10	0.00.1–0.01	Many thousands of years
Sandstone	Cemented quartz grains	Intergranular and fissure	10–30	0.1–10	8–20	0.001–0.1	Tens to hundreds of years
Limestone	Cemented carbonate	Mainly fissure	5–30	0.1–5.0	5–15	0.001–1	Tens to hundreds of years
Karstic limestone	Cemented carbonate	Fissures and channels	5–25	100–10,000	5–15	10–2000	A few hours to days
Volcanic rock							
Basalt	Fine-grained crystalline	Fissure	2–15	0.1–1000	1–5	1–500	Very wide range; can be very short
Tuff	Cemented grains	Intergranular and fissure	15–30	0.1–5	10–20	0.001–1	Wide range
Igneous and metamorphic rocks (granites and gneisses):							
Fresh	Crystalline	Fissure	0.1–2	$10^{-7} - 5 \times 10^{-5}$		10^{-6}–10^{-2}	Thousands of years, but can be rapid where fractured
Weathered	Disaggregated crystalline	Intergranular and fissure	10–20	0.1–2	1–5	0.001–0.1	Tens to hundreds of years

Source: P. J. Chilton cited in M. Meybeck, D. V. Chapman, and R. Helmer, *Global Freshwater Quality: A First Assessment*, Global Environmental Monitoring System (GEMS), World Health Organization and United Nations Environment Programme, Basil Blackwell, Oxford, 1990, as presented by Gleick (1993).

Aquifers are classified as *unconfined* or *confined* depending upon the presence or absence of a water table (Figure 6.1.3). An *unconfined aquifer* is one in which a water table serves as the upper surface of the *zone of saturation*, also known as a *free, phreatic*, or *non-artesian aquifer*. Changes in the water table (rising or falling) correspond to changes in the volume of water in storage within an aquifer. A *confined aquifer* is one in which the groundwater is confined under pressure greater than atmospheric by overlying, relatively impermeable strata. Confined aquifers are also known as *artesian* or *pressure aquifers*. Water enters such aquifers in an area where the confining bed rises to the surface or ends underground, and such an area is known as a *recharge area*. Changes of the water levels in wells penetrating confined aquifers result primarily from changes in pressure rather than changes in storage volumes.

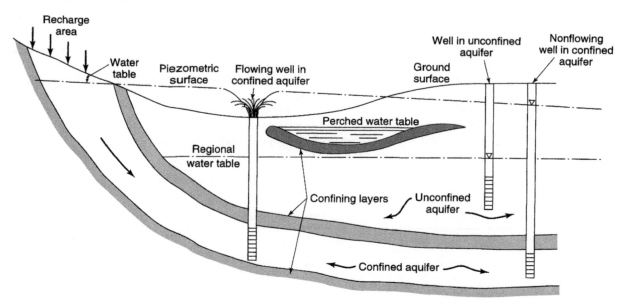

Figure 6.1.3 Types of aquifers (from U.S. Bureau of Reclamation (1981)).

Aquifer Properties

Aquifers perform two important functions—a *storage function* and a *conduit function*. In other words, aquifers store water and also function as a pipeline. When water is drained from a saturated material under the influence of gravity, only a portion of the total saturated volume in the pores is released. Part of the water is retained in interstices due to the losses of the molecular attraction, adhesion, and cohesion. The *specific yield* S_y, which is the storage term for unconfined aquifers, is the volume of water drained from a saturated sample of unit volume (1 ft^3 or 1 m^3) with a unit decrease in the water table. *Specific retention* S_r is the quantity of water that is retained in the unit volume after gravity drainage. The sum of the specific yield and the specific retention for saturated aquifers is the porosity, $\alpha = S_y + S_r$. *Porosity* is the pore volume divided by the total volume, expressed as a percent. Porosity represents the potential storage of an aquifer but does not indicate the amount of water a porous material will yield. Figure 6.1.4 illustrates the various types of porosities.

The *storativity* (or *storage coefficient*) of an aquifer is the volume of water the aquifer releases from or takes into storage per unit surface area of the aquifer per unit decline or rise of head. This is illustrated in Figure 6.1.5, considering a vertical column of unit area extending through a confined aquifer and an unconfined aquifer. In both cases, the storage coefficient S equals the volume of water (ft^3 or m^3) released from the aquifer when the piezometric surface or water table declines one unit of distance (1 ft or 1 m). The storativity then has dimensions of ft^3/ft^3 or m^3/m^3. In the case of unconfined aquifers, the storativity corresponds to the specific yield. Confined aquifers have storativities in the range of $10^{-5} \leq S \leq 10^{-3}$. These small values indicate that large pressure changes are required to produce substantial water yields. Storativity can be determined in the field by pump tests. The *specific storage* S_s of a saturated aquifer is the volume of water that a unit volume of aquifer releases from storage under a unit decline in hydraulic head, i.e., $S = S_s b$ where b is the thickness of a confined aquifer.

Hydraulic conductivity (also referred to as *coefficient of permeability*) is the property related to the conduit function of an aquifer. It is the measure of ease of moving groundwater through aquifers, with dimensions of (L/T). The hydraulic conductivity K is the rate of flow of water through a cross-section of unit area of the aquifer under a unit hydraulic gradient. The *hydraulic gradient* is the

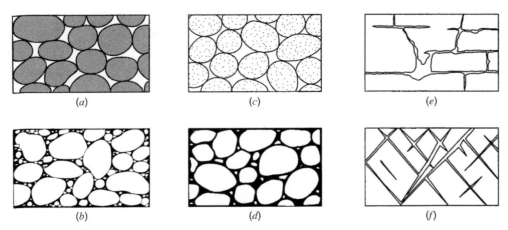

Figure 6.1.4 The various types of porosity: (*a*) Well-sorted sedimentary deposit possessing high porosity; (*b*) Poorly sorted sedimentary deposit of low porosity; (*c*) Similar to *a* but consisting of pebbles that are themselves porous, so the porosity of the deposit is very high; (*d*) Also similar to *a*, but the porosity has been diminished by the deposition of mineral matter in the interstices; (*e*) Rocks rendered porous by solution; and (*f*) Rocks rendered porous by fracturing (from Meinzer (1923)).

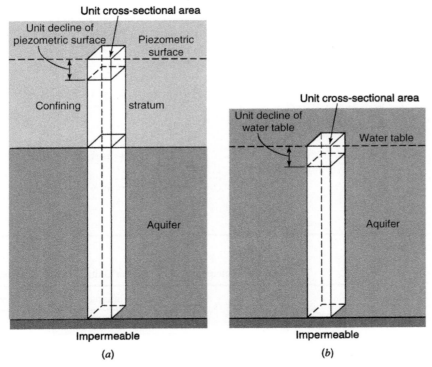

Figure 6.1.5 Illustrative sketches for defining storage coefficient of (*a*) confined and (*b*) unconfined aquifers (from Todd and Mays (2005)).

headloss divided by the distance between the two points. The hydraulic conductivity is commonly expressed in gallons per day/ft^2 or in ft/day in U.S. customary units or cm/d or m/d in SI units. Table 6.1.2 presents a range of hydraulic conductivity. Table 6.1.3 lists, in both the SI and the FPS systems, the dimensions and common units for some of the basic groundwater parameters. The

Table 6.1.2 Range of Values of Hydraulic Conductivity and Permeability

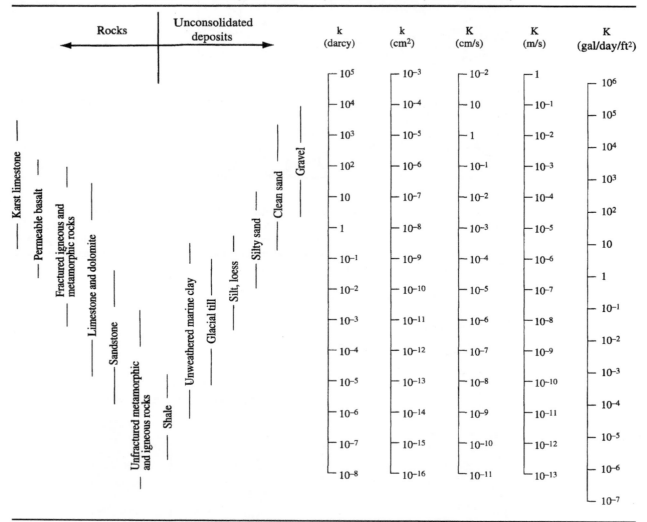

Source: Freeze and Cherry (1979).

Table 6.1.3 Dimensions and Common Units for Basic Groundwater Parameters

Parameter	Symbol	Systeme International*		Foot-Pound-Second System**	
		SI		FPS	
		Dimension	Units	Dimension	Units
Hydraulic head	h	$[L]$	m	$[L]$	ft
Elevation head	z	$[L]$	m	$[L]$	ft
Fluid pressure	p	$[M/LT^2]$	N/m^2 or Pa	$[F/LT^2]$	lb/ft^2
Mass density	ρ	$[M/L^3]$	kg/m^3	—	—
Weight density	γ	—	—	$[F/L^3]$	lb/ft^3
Specific discharge	v	$[L/T]$	m/s	$[L/T]$	ft/s
Hydraulic conductivity	K	$[L/T]$	m/s	$[L/T]$	ft/s

*Basic dimensions are length L, mass M, and time T.
**Basic dimensions are length L, force F, and time T.

hydraulic conductivity is a function of both the porous medium and the fluid properties. Table 6.1.1 summarizes the characteristics of principal aquifer types.

A *specific* or *intrinsic* permeability, which is a function of the medium alone and not the fluid properties, is $k = Cd^2$, where C is a constant of proportionality and d is the grain size diameter. This k has simply been referred to as permeability (Freeze and Cherry, 1979). Table 6.1.2 also presents ranges of k values, which have dimensions of $[L^2]$. The *darcy* is also a unit of permeability, where 1 darcy is the permeability that leads to a specific discharge of 1 cm/s for a fluid with a viscosity of 1 cP (1 centipoise, $cP = N = s/m^2 \times 10^{-3}$) under a hydraulic gradient of 1 cm/cm. One darcy is approximately equal to 10^{-8} cm^2 (Freeze and Cherry, 1979). Refer to Table 6.1.2 for the range of values of k.

Closely related to the hydraulic conductivity is the *transmissivity* (or *transmissibility*), which indicates the capacity of an aquifer to transmit water through its entire thickness. The transmissivity, T, is the flow rate (ft^2/s or m^2/s) through a vertical strip of the aquifer 1 unit wide (1 ft or 1 m) and extending through the saturated thickness under a unit hydraulic gradient. The transmissivity is equal to the hydraulic conductivity multiplied by the saturated thickness of the aquifer: $T = Kb$, in which b is the saturated thickness of the aquifer.

Heterogeneity and Anisotropy of Hydraulic Conductivity

In geologic formations the hydraulic conductivity usually varies through space, referred to as *heterogeneity*. A geologic formation is *homogeneous* if the hydraulic conductivity is independent of position in the formation, i.e., $K(x,y,z) =$ constant. A geologic formation is *heterogeneous* if the hydraulic conductivity is dependent on position in the formation, $K(x,y,z) \neq$ constant.

Hydraulic conductivity may also show variations with the direction of measurement at a given point in the formation. A geologic formation is *isotropic* at a point if the hydraulic conductivity is independent of the direction of measurement at the point, $K_x = K_y = K_z$. A geologic formation is *anisotropic* at a point if the hydraulic conductivity varies with the direction of measurement at that point, $K_x \neq K_y \neq K_z$.

Groundwater Basins

Aquifers exist in both *consolidated* (mainly bedrock) and *unconsolidated* (mainly surface deposits) *materials*. Figure 6.1.6 illustrates the distribution of major aquifers of different types. *Groundwater basins* are a group of interrelated bodies of groundwater linked together in a larger flow system (Marsh, 1987). These basins are typically complex, three-dimensional systems. The spatial configurations of groundwater basins are determined by regional geology.

Groundwater Movement

Groundwater in its natural state is invariably moving, and this movement is governed by hydraulic principles. The flow through an aquifer is expressed by *Darcy's law*, which is the foundation of groundwater hydraulics. This law states that the flow rate through porous media is proportional to the headloss and inversely proportional to the length of the flow path, expressed mathematically as

$$Q = -KA\frac{dh}{dL} \tag{6.1.1}$$

Figure 6.1.7 illustrates Darcy's law where $h = dh$ and $L = dL$ so that $Q = KAh/L$.

The Reynolds number for flow in porous media is

$$R_e = \frac{\rho VD}{\mu} \tag{6.1.2}$$

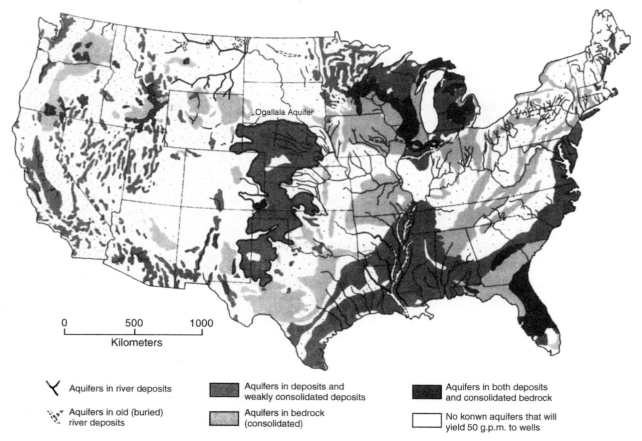

Figure 6.1.6 Distribution of major aquifers of different types (river deposits, other deposits, and consolidated bedrock) in the coterminous United States. A major aquifer is defined as one composed of material capable of yielding 50 gallons per minute or more to an individual well and having water quality generally not containing more than 2000 parts per million of dissolved solids (from Marsh (1987)).

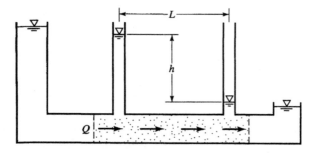

Figure 6.1.7 Illustration of Darcy's law (from U.S. Bureau of Reclamation (1981)).

where ρ is fluid density, V is the apparent velocity, D is the average grain diameter (an approximation to average pore diameter), and μ is viscosity of the fluid. The upper limit of validity for Darcy's law is for R_e, ranging between 1 and 10, and there is really no lower limit. In almost all natural groundwater motion $R_e < 1$; therefore, Darcy's law is applicable to natural groundwater problems. In summary, this law is valid when the flow is laminar or without turbulence.

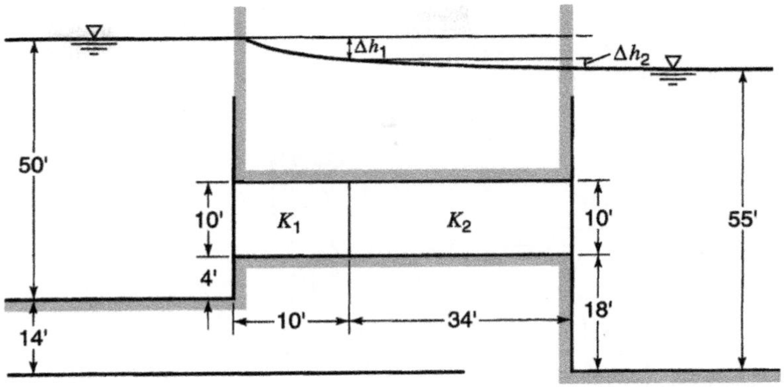

Figure 6.1.8 Two reservoirs connected by conduit of two porous medium for example 6.1.1.

EXAMPLE 6.1.1

Two very large reservoirs are connected by a conduit of two types of porous medium as shown in Figure 6.1.8. Compute the magnitude of flow per unit width between the two reservoirs if $K_1 = 1$ gpd/ft^2 and $K_2 = 2$ gpd/ft^2.

SOLUTION

Darcy's law for sections 1 and 2 where $Q_1 = Q_2$ is

$$K_1 A_1 \frac{dh_1}{dL_1} = K_2 A_2 \frac{dh_2}{dL_2}$$

Because $A_1 = A_2$,

$$\frac{\Delta h_1}{\Delta L_1} = 2 \frac{\Delta h_2}{\Delta L_2}$$

and

$$\Delta h_1 + \Delta h_2 = 9$$

The two equations above have two unknowns, Δh_1 and Δh_2, so they can be solved to obtain $\Delta h_2 = 5.67$ ft and $\Delta h_1 = 3.33$ ft. The flow rate is computed using Darcy's equation,

$$Q_1 = 1.10 \cdot \frac{3.33}{10} = 3.33 \text{ gpd}$$

6.2 SATURATED FLOW

6.2.1 Governing Equations

The control volume for a saturated flow is shown in Figure 6.2.1. The sides have lengths dx, dy, and dz in the coordinate directions. The total volume of the control volume is $dxdydz$, the volume of water in control volume is $\theta dxdydz$, where θ is the moisture content. With the control volume approach, the extensive property B is the mass of the groundwater, the intensive property is $\beta = dB/dm = 1$, and $d\beta/dt = 0$ because no phase change occurs. The general control volume equation for continuity, equation (3.3.1), is applicable:

$$0 = \frac{d}{dt} \int_{CV} \rho \, d\forall + \int_{CS} \rho \mathbf{V} \cdot d\mathbf{A} \qquad (3.3.1)$$

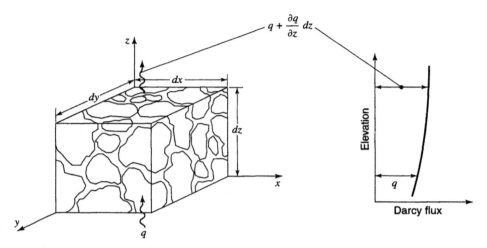

Figure 6.2.1 Control volume for development of the continuity equation in porous medium (from Chow et al. (1988)).

The time rate of change of mass stored in the control volume is defined as the time rate of change of fluid mass in storage, expressed as

$$\frac{d}{dt}\int_{CV} \rho\, d\forall = \rho S_s \frac{\partial h}{\partial t}(dxdydz) + \rho W(dxdydz) \tag{6.2.1}$$

where S_s is the specific storage and W is the flow out of the control volume, $W = Q/(dxdydz)$. The term $\rho S_s(\partial h/\partial t)(dxdydz)$ includes two parts: (1) the mass rate of water produced by an expansion of the water under change in density and (2) the mass rate of water produced by the compaction of the porous medium due to change in porosity (see Freeze and Cherry (1979) for more details).

The inflow of water through the control surface at the bottom of the control volume is $qdxdy$, and the outflow at the top is $[q + (\partial q/\partial z)dz]dxdy$, so that the net outflow in the vertical direction z is

$$\int_{CS(z)} \rho\mathbf{V}\cdot d\mathbf{A} = \rho\left(q + \frac{\partial q}{\partial z}dz\right)dxdy - \rho q dxdy = \rho dxdydz\frac{\partial q}{\partial z} \tag{6.2.2}$$

where q in $\partial q/\partial z$ is q_z, i.e., in the z direction.

Considering all three directions, we get

$$\int_{CS} \rho\mathbf{V}\cdot d\mathbf{A} = \rho dxdydz\frac{\partial q}{\partial x} + \rho dxdydz\frac{\partial q}{\partial y} + \rho dxdydz\frac{\partial q}{\partial z} \tag{6.2.3}$$

Substituting (6.2.1) and (6.2.3) into (3.3.1) results in

$$0 = \rho S_s \frac{\partial h}{\partial t}dxdydz + \rho W(dxdydz) + \rho dxdydz\frac{\partial q}{\partial x} + \rho dxdydz\frac{\partial q}{\partial y} + \rho dxdydz\frac{\partial}{\partial} \tag{6.2.4}$$

Dividing through by $\rho dxdydz$ gives

$$S_s \frac{\partial h}{\partial t} + \frac{\partial q}{\partial x} + \frac{\partial q}{\partial y} + \frac{\partial q}{\partial z} + W = 0 \tag{6.2.5}$$

Using Darcy's law, the Darcy flux in each direction is

$$q_x = -K_x \frac{\partial h}{\partial x} \tag{6.2.6a}$$

$$q_y = -K_y \frac{\partial h}{\partial y} \tag{6.2.6b}$$

$$q_z = -K_z \frac{\partial h}{\partial z} \tag{6.2.6c}$$

Substituting these definitions of q_x, q_y, and q_z into (6.2.5) and rearranging results in

$$\frac{\partial}{\partial x}\left(K_x \frac{\partial h}{\partial x}\right) + \frac{\partial}{\partial y}\left(K_y \frac{\partial h}{\partial y}\right) + \frac{\partial}{\partial z}\left(K_z \frac{\partial h}{\partial z}\right) = S_s \frac{\partial h}{\partial t} + W \tag{6.2.7}$$

This is the equation for *three-dimensional transient flow through a saturated anisotropic porous medium*. For a *homogeneous, isotropic medium* $(K_x = K_y = K_z)$, equation (6.2.7) becomes

$$\frac{\partial^2 h}{\partial x^2} + \frac{\partial^2 h}{\partial y^2} + \frac{\partial^2 h}{\partial z^2} = \frac{S_s}{K}\frac{\partial h}{\partial t} + \frac{W}{K} \tag{6.2.8}$$

For a *homogeneous, isotropic medium and steady-state flow*, equation (6.2.8) becomes

$$\frac{\partial^2 h}{\partial x^2} + \frac{\partial^2 h}{\partial y^2} + \frac{\partial^2 h}{\partial z^2} = \frac{W}{K} \tag{6.2.9}$$

For a horizontal confined aquifer of thickness b, $S = S_s b$ and transmissivity $T = Kb$, the two-dimensional form of (6.2.8) with $W = 0$ becomes

$$\frac{\partial^2 h}{\partial x^2} + \frac{\partial^2 h}{\partial y^2} = \frac{S}{T}\frac{\partial h}{\partial t} \tag{6.2.10}$$

The governing equation for radial flow can also be derived using the control volume approach. Alternatively, equation (6.2.10) can be converted into radial coordinates by the relation $r = \sqrt{x^2 + y^2}$. This is known as the *diffusion equation*, expressed as

$$\frac{1}{r}\frac{\partial}{\partial r}\left(r\frac{\partial h}{\partial r}\right) = \frac{\partial^2 h}{\partial r^2} + \frac{1}{r}\frac{\partial h}{\partial r} = \frac{S}{T}\frac{\partial h}{\partial t} \tag{6.2.11}$$

where r is the radial distance from a pumped well and t is the time since the beginning of pumping. For steady-state conditions, $\partial h / \partial t = 0$, so equation (6.2.11) reduces to

$$\frac{1}{r}\frac{\partial}{\partial r}\left(r\frac{\partial h}{\partial r}\right) = 0 \tag{6.2.12}$$

6.2.2 Flow Nets

Flow nets provide a graphical means to illustrate two-dimensional groundwater flow problems. For steady-state conditions, the two-dimensional flow equation (6.2.10) for a homogeneous, isotropic medium becomes

$$\frac{\partial^2 h}{\partial x^2} + \frac{\partial^2 h}{\partial y^2} = 0 \tag{6.2.13}$$

This is the classic partial differential equation form known as the *Laplace equation*. Equation (6.2.13) is linear in terms of the piezometric head h, and its solution depends entirely on the values of h on the boundaries of a flow field in the x–y plane. In other words, h at any point in a flow field can be determined uniquely in terms of h on the boundaries. Laplace's equation arises in other areas such as hydrodynamics and heat flow.

Velocity of flow is normal to lines of constant piezometric heads. This can be seen through examination of Darcy's equations $v_x = K_x(\partial h/\partial x)$ and $v_y = -K_y(\partial h/\partial y)$ and continuity equation $(\partial v_x/\partial x) + (\partial v_y/\partial y) = 0$. Figure 6.2.2 is a hypothetical flow net. The velocity vector **V** has components v_x and v_y with the resultant velocity vector in the direction of decreasing head h. The *streamlines* are a set of lines that are drawn tangent to the velocity vector and normal to the line of constant piezometric head. The family of streamlines is called the *stream function*, ψ. In steady flow, the streamlines define the paths of flowing particles of fluid. Because the streamlines are normal to the line of constant head, the velocities v_x and v_y can be expressed as

$$v_x = \frac{\partial \psi}{\partial x} \tag{6.2.14}$$

$$v_y = \frac{\partial \psi}{\partial y} \tag{6.2.15}$$

which can be substituted into the differential form of the two-dimensional equation of continuity:

$$\frac{\partial v_x}{\partial x} + \frac{\partial v_y}{\partial y} = 0 \tag{6.2.16}$$

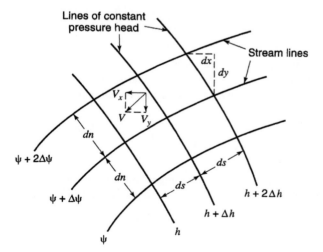

Figure 6.2.2 Hypothetical flow net.

resulting in

$$\frac{\partial^2 \psi}{\partial x^2} + \frac{\partial^2 \psi}{\partial y^2} = 0 \qquad (6.2.17)$$

which is also the *Laplace equation* (see equation (6.2.13)).

Because the streamlines are everywhere tangent to velocity vectors, no flow exists across them. The rate of flow is constant between any two streamlines. The flow lines in Figure 6.2.2 have been constructed so that $ds = dh$. The discharge through a cross-sectional area of unit depth perpendicular to the flow net is

$$dq = K\frac{dh}{ds}dh \qquad (6.2.18)$$

and since $ds = dh$ then

$$dq = Kdh \qquad (6.2.19)$$

and with m sections

$$q = mKdh \qquad (6.2.20)$$

If the total head drop across the region of flow is H and there are n divisions of head in the flow net ($H = ndh$), then

$$q = \frac{mKH}{n} \qquad (6.2.21)$$

This equation is applicable only to simple flow systems with one recharge boundary and one discharge boundary.

The rate of flow between streamlines can also be expressed as

$$dq = v_x dx - v_y dy \qquad (6.2.22)$$

Substituting (6.2.14) and (6.2.15) for v_x and v_y, we get

$$\frac{\partial \psi}{\partial x}dx + \frac{\partial \psi}{\partial y}dy = dq \qquad (6.2.23)$$

which implies that

$$dq = d\psi \qquad (6.2.31)$$

The value of the stream function is numerically equal to the unit discharge. Also, the increment in unit discharge between two streamlines is equal to the change in the value of the stream function between two streamlines.

Figure 6.2.3 illustrates the development of the flow distribution through the use of *equipotential lines* and *flow lines*. Flow lines represent the paths followed by molecules of water as they move through an aquifer in the direction of decreasing head. Equipotential lines intersect the flow lines at right angles, representing the piezometric-surface or water-table contours. The two families of lines or curves together are referred to as a *flow net*.

Figure 6.2.3 illustrates two example applications showing the flow lines and equipotential lines that form the orthogonal network of cells making up the flow net. Theoretically, the flow through any one cell equals the flow through any other cell. This figure shows contrasting flow nets for channel seepage through layered anisotropic media.

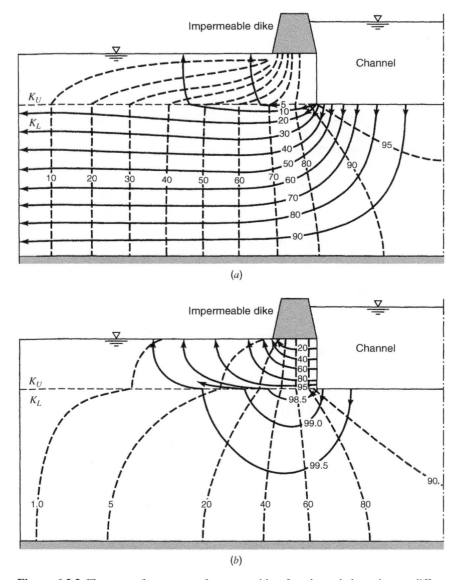

Figure 6.2.3 Flow nets for seepage from one side of a channel through two different anisotropic two-layer systems. (a) $K_u/K_L = 1/50$; (b) $K_u/K_L = 50$. The anisotropy ratio for all layers is $K_x/K_z = 10$ (after Todd and Bear (1961)).

6.3 STEADY-STATE ONE-DIMENSIONAL FLOW

Confined Aquifer

Consider steady-state groundwater flow in a confined aquifer of uniform thickness, as shown in Figure 6.3.1. For one-dimensional steady-state flow, $\partial^2 h/\partial y^2 = 0$ and $\partial h/\partial t = 0$, so equation

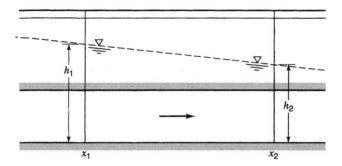

Figure 6.3.1 Flow in a one-dimensional confined aquifer.

(6.2.10) for $W = 0$ reduces to

$$\frac{\partial^2 h}{\partial x^2} = 0 \tag{6.3.1}$$

which has the solution

$$h = C_1 x + C_2 \tag{6.3.2}$$

where h is the head above a given datum and C_1 and C_2 are constants of integration. For $h = h_1$ at $x = 0$, $h = h_2$ at $x = L$, and $\partial h/\partial x = -(q/K)$ from Darcy's law (equation (6.2.6a)), then

$$h = q/K_x + \text{constant} \tag{6.3.3}$$

In other words, the head decreases linearly with flow in the x direction, as shown in Figure 6.3.1.

Unconfined Aquifer

For one-dimensional steady-state flow in an unconfined aquifer (Figure 6.3.2), a direct solution of equation (6.2.10) is not possible (see Todd and Mays, 2005). To obtain a solution, use Darcy's law, $q = -K(\partial h/\partial x)$, and by continuity define the discharge at any vertical section as

$$\sum_{CS} \rho \mathbf{V} \cdot d\mathbf{A} = Q = -Kh\frac{\partial h}{\partial x} \tag{6.3.4}$$

Equation (6.3.4) can be integrated over the distance from x_1 to x_2 where h_1 and h_2 are the respective heads; then

$$Q \int_{x_1}^{x_2} dx = K \int_{h_1}^{h_2} h \, dh \tag{6.3.5}$$

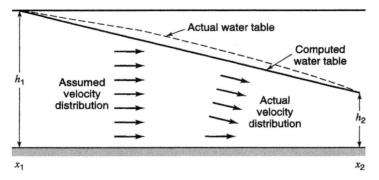

Figure 6.3.2 Flow in a one-dimensional unconfined aquifer.

which becomes

$$Q = K \frac{h_1^2 - h_2^2}{2(x_2 - x_1)} \qquad (6.3.6)$$

Equation (6.3.6) indicates a parabolic surface of the saturated portion of the aquifer between x_1 and x_2.

The assumptions in this derivation are (1) the velocity of flow is proportional to the tangent of the hydraulic gradient and (2) the flow is horizontal and uniform everywhere in a vertical section. The assumption of one-dimensional flow is valid, as velocities are small and dh/dx is small. Using the boundary conditions that $x_1 = 0$ and $h_1 = h_0$ and letting $x = X$ and $h = h_2$, then equation (6.3.6) reduces to the well-known *Dupuit equation*

$$Q = \frac{K}{2X} \left(h_0^2 - h^2 \right) \qquad (6.3.7)$$

Refer to Figure 6.3.3 for an example explanation.

EXAMPLE 6.3.1

A stratum of clean sand and gravel between two channels (see Figure 6.3.3) has a hydraulic conductivity $K = 10^{-1}$ cm/sec and is supplied with water from a ditch ($h_0 = 20$ ft deep) that penetrates to the bottom of the stratum. If the water surface in the second channel is 2 ft above the bottom of the stratum and its distance to the ditch is $x = 30$ ft, which is also the thickness of the stratum, what is the unit flow rate into the gallery?

SOLUTION

The flow is described using the Dupuit equation (6.3.7) for unit flow, where

$K = 10^{-1}$ cm/sec $(60 \text{ sec/min})(1 \text{ in}/2.54 \text{ cm})(1 \text{ ft}/12 \text{ in})(7.48 \text{ gal}/1 \text{ ft}^3) =$
$1.54 \text{ gpm/ft}^2 = 2.22 \times 10^3 \text{ gpd/ft}^2$

and

$$Q = \frac{1.5}{2(30)} \left(20^2 - 2^2 \right) = 10.16 \text{ gpm/ft} = 1.46 \times 10^4 \text{ gpd/ft}$$

or in SI units with $K = 10^{-1}$ cm/s $= 10^{-3}$ m/s and $h_0 = 20$ ft $= 6.10$ m, $h = 0.61$ m, and $x = 9.14$ m:

$$Q = \frac{1.0^{-3} \left(6.10^2 - 0.61^2 \right)}{2(9.14)} = 2.02 \times 10^{-3} \text{ m}^3\text{/s}$$

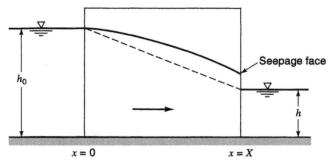

Figure 6.3.3 Steady flow in a one-dimensional unconfined aquifer between two bodies of water with vertical boundaries (application of Dupuit equation (6.3.7)).

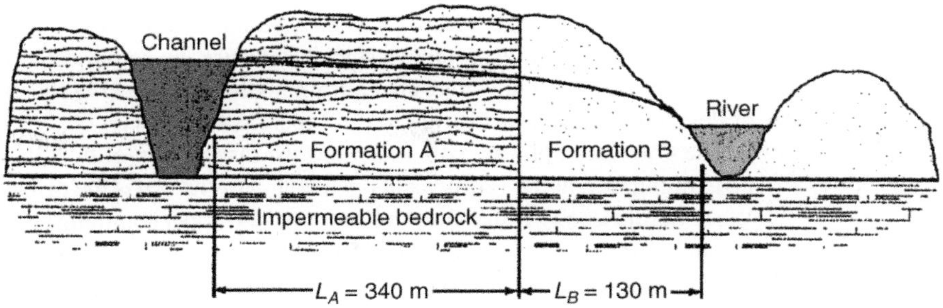

Figure 6.3.4 A channel and river for example 6.3.2.

EXAMPLE 6.3.2

For the channel and river shown in Figure 6.3.4, determine the volume of water that flows from the channel into the river. The water surface elevations in the channel and river with respect to the underlying bedrock are 13 and 10.5 m, respectively. The hydraulic conductivity of formation A is 5.6 m/day and that of formation B is 12.3 m/day.

SOLUTION

To account for multiple formations, first derive an expression by modifying the Dupuit equation (6.3.7). In developing the expression to account for multiple formations, subscript s denotes the separation, or divide, between the two formations so that h_s denotes the water table elevation at the separation. Under steady-state conditions, apply the Dupuit equation between the channel and the separation point, and between the separation point and the river in order to yield the same flow rate:

$$Q = \frac{K_A}{2L_A}\left(h_C^2 - h_S^2\right) = \frac{K_B}{2L_B}\left(h_S^2 - h_R^2\right)$$

Solving for h_S^2 gives

$$h_S^2 = \frac{\dfrac{K_A}{2L_A}h_C^2 + \dfrac{K_B}{2L_B}h_R^2}{\left(\dfrac{K_A}{2L_A} + \dfrac{K_B}{2L_B}\right)}$$

and substituting this expression for h_S^2 into the equation for the flow rate gives

$$Q = \frac{h_C^2 - h_R^2}{2\left[\dfrac{L_A}{K_A} + \dfrac{L_B}{K_B}\right]}$$

Now substitute
$h_C = 13$ m, $h_R = 10.5$ m, $K_A = 5.6$ m/day, $K_B = 12.3$ m/day, $L_A = 340$ m, and $L_B = 130$ m:

$$Q = \frac{(13\text{ m})^2 - (10.5\text{ m})^2}{2\left[\dfrac{340\text{ m}}{5.6\text{ m/day}} + \dfrac{130\text{ m}}{12.3\text{ m/day}}\right]} = 0.4121 \text{ m}^2/\text{day}$$

6.4   STEADY-STATE WELL HYDRAULICS

6.4.1   Flow to Wells

At initiation of discharge (pumpage) from a well, theoretically the water level or head in the well is lowered relative to the undisturbed condition of the piezometric surface or water table outside the well. In the aquifer surrounding the well, the water flows radially to the lower level in the well. For artesian conditions, the actual flow distribution of the flow conforms relatively close to the theoretical shortly after pumping starts. However, in non-artesian (free aquifer) conditions the actual distribution of flow may not conform to the theoretical as illustrated by the successive stages of development of flow distribution in Figure 6.4.1.

The flow net in Figure 6.4.2 illustrates the distribution of flow in an artesian aquifer for a fully penetrating well and a 100-percent open hole. *Drawdown* is the distance the water level is lowered. When the drawdown falls below the bottom of the upper confining bed, a mixed condition of artesian and non-artesian flow occurs. The flow net in Figure 6.4.3 illustrates the distribution of flow in an artesian aquifer for a well that penetrates through the upper confining bed but not into the artesian aquifer. A strong vertical component of flow is established out to a distance approximately equal to the thickness of the aquifer. *Drawdown curves (cones)* show the variation of drawdown with distance from the well.

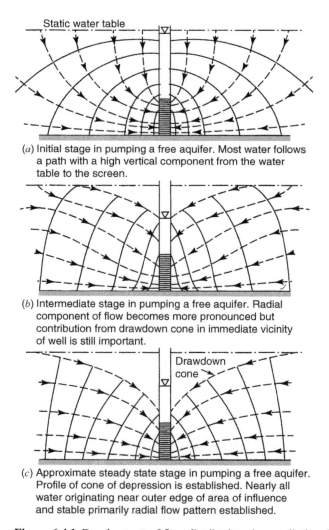

(*a*) Initial stage in pumping a free aquifer. Most water follows a path with a high vertical component from the water table to the screen.

(*b*) Intermediate stage in pumping a free aquifer. Radial component of flow becomes more pronounced but contribution from drawdown cone in immediate vicinity of well is still important.

(*c*) Approximate steady state stage in pumping a free aquifer. Profile of cone of depression is established. Nearly all water originating near outer edge of area of influence and stable primarily radial flow pattern established.

Figure 6.4.1 Development of flow distribution about a discharging well in a free aquifer—a fully penetrating and 33-percent open hole (from U.S. Bureau of Reclamation (1981)).

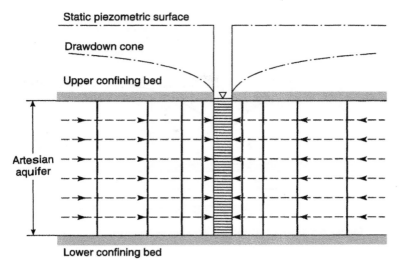

Figure 6.4.2 Distribution of flow to a discharging well in an artesian aquifer—a fully penetrating and 100-percent open hole (from U.S. Bureau of Reclamation (1981)).

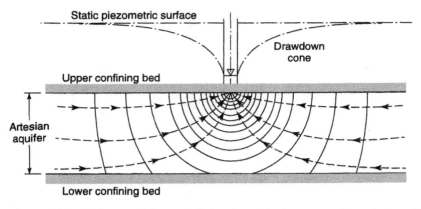

Figure 6.4.3 Distribution of flow to a discharging well—just penetrating to the top of an artesian aquifer (from U.S. Bureau of Reclamation (1981)).

6.4.2 Confined Aquifers

When pumping from a well, water is removed from the aquifer surrounding the well and the piezometric surface is lowered. *Drawdown* is the distance the piezometric surface is lowered (Figure 6.4.4). A radial flow equation can be derived to relate well discharge to drawdown. Consider the confined aquifer with steady-state radial flow to the fully penetrating well that is being pumped, as shown in Figure 6.4.5. For a homogenous, isotropic aquifer, the well discharge at any radial distance r from the pumped well is

$$\sum_{CS} \rho \mathbf{V} \cdot d\mathbf{A} = Q = 2\pi K r b \frac{dh}{dr} \tag{6.4.1}$$

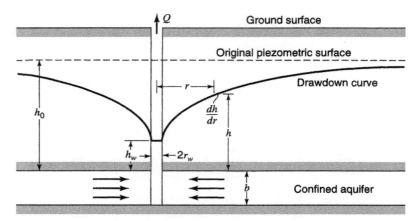

Figure 6.4.4 Well hydraulics for a confined aquifer.

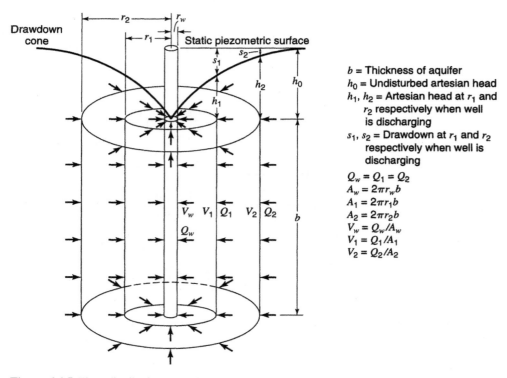

Figure 6.4.5 Flow distribution to a discharging well in an artesian aquifer—a fully penetrating and 100-percent open hole (from U.S. Bureau of Reclamation (1981)).

For boundary conditions of $h = h_1$ at $r = r_1$ and $h = h_2$ at $r = r_2$, then equation (6.4.1) can be integrated:

$$\int_{h_1}^{h_2} dh = \frac{Q}{2\pi Kb} \int_{r_1}^{r_2} \frac{dr}{r}$$

$$h_2 - h_1 = \frac{Q}{2\pi Kb} \ln \frac{r_2}{r_1} \tag{6.4.2}$$

Solving equation (6.4.2) for Q gives

$$Q = 2\pi Kb \left[\frac{h_2 - h_1}{\ln(r_2/r_1)} \right] \tag{6.4.3}$$

For the more general case of a well penetrating an extensive confined aquifer, there is no limit for r. Referring to Figure 6.4.4, equation (6.4.3) becomes

$$Q = 2\pi Kb \left[\frac{h - h_1}{\ln(r/r_1)} \right] \tag{6.4.4}$$

which shows that the head varies linearly with the logarithm of distance regardless of the rate of discharge.

Equation (6.4.4) is known as the *equilibrium* or *Thiem equation* and enables the aquifer permeability to be determined from a pumped well. Two points define the logarithmic drawdown curve, so drawdown can be measured at two observation wells at different distances from a well that is being pumped at a constant rate. The hydraulic conductivity can be computed using

$$K = \frac{Q \ln(r_2/r_1)}{2\pi b(h_2 - h_1)} \tag{6.4.5}$$

where r_1 and r_2 are the distances and h_1 and h_2 are the heads in the respective observation wells.

EXAMPLE 6.4.1

A well fully penetrates a 25-m thick confined aquifer. After a long period of pumping at a constant rate of 0.05 m^3/s, the drawdowns at distances of 50 m and 150 m from the well were observed to be 3 m and 1.2 m, respectively. Determine the hydraulic conductivity and the transmissivity. What type of unconsolidated deposit would you expect this to be?

SOLUTION

Use equation (6.4.5) to determine the hydraulic conductivity with $Q = 0.05$ m^3/s, $r_1 = 50$ m, $r_2 = 150$ m, $s_1 = h_0 - h_1$ and $s_2 = h_0 - h_2$, so $s_1 - s_2 = h_2 - h_1 = 3 - 1.2 = 1.8$ m:

$$K = \frac{Q \ln(r_2/r_1)}{2\pi b\,(s_1 - s_2)} = \frac{0.05 \ln(150/50)}{2\pi\,(25)(3 - 1.2)} = 1.94 \times 10^{-4} \text{ m/s}$$

Then the transmissivity is $T = Kb = (1.94 \times 10^{-4})(25) = 4.85 \times 10^{-3}$ m^2/s. From Table 6.1.1 for $K = 1.94 \times 10^{-4}$ m/s, this is probably a clean sand or silty sand.

EXAMPLE 6.4.2

A 2-ft diameter well penetrates vertically through a confined aquifer 50 ft thick. When the well is pumped at 500 gpm, the drawdown in a well 50 ft away is 10 ft and in another well 100 ft away is 3 ft. What is the approximate head in the pumped well for steady-state conditions, and what is the approximate drawdown in the well? Also compute the transmissivity. Take the initial piezometric level as 100 ft above the datum.

SOLUTION

First determine the hydraulic conductivity using equation (6.4.5):

$$K = \frac{Q \ln(r_2/r_1)}{2\pi b\,(h_2 - h_1)} = \frac{500 \ln(100/50)}{2\pi\,(50)(97 - 90)} = 0.158 \text{ gpm/ft}^2$$

Then compute the transmissivity:

$$T = Kb = 0.158 \times 50 = 7.90 \text{ gpm/ft}$$

Now compute the approximate head, h_w, in the pumped well:

$$h_2 - h_w = \frac{Q \ln(r_2/r_1)}{2\pi K b}$$

$$h_w = h_2 - \frac{Q \ln(r_2/r_w)}{2\pi K b}$$

$$h_w = 97 - \frac{500 \ln(100/1)}{2\pi(7.90)} = 97 - 46.4 = 50.6 \text{ ft}$$

Drawdown is then $s_w = 100 - 50.6 \text{ ft} = 49.6 \text{ ft}$.

6.4.3   Unconfined Aquifers

Now consider steady radial flow to a well completely penetrating an unconfined aquifer as shown in Figure 6.4.6. A concentric boundary of constant head surrounds the well. The well discharge is given by

$$\sum_{CS} \rho \mathbf{V} \cdot d\mathbf{A} = Q = 2\pi K h r \frac{dh}{dr} \tag{6.4.6}$$

Integrating equations (6.4.6) with the boundary conditions $h = h_1$ at $r = r_1$ and $h = h_2$ at $r = r_2$ yields

$$\int_{h_1}^{h_2} h \, dh = \frac{Q}{2\pi K} \int_{r_1}^{r_2} \frac{dr}{r} \tag{6.4.7}$$

$$Q = \pi K \left[\frac{h_2^2 - h_1^2}{\ln(r_2/r_1)} \right] \tag{6.4.8}$$

There are large vertical flow components near the well so that this equation fails to describe accurately the drawdown curve near the well, but can be defined for any two distances r_1 and r_2 away from the pumped well.

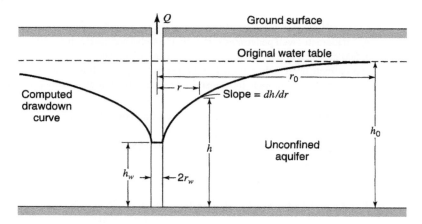

Figure 6.4.6 Well hydraulics for an unconfined aquifer.

EXAMPLE 6.4.3

A well penetrates an unconfined aquifer. Prior to pumping, the water level (head) is $h_0 = 25$ m. After a long period of pumping at a constant rate of 0.05 m³/s, the drawdowns at distances of 50 m and 150 m from the well were observed to be 3 m and 1.2 m, respectively. Determine the hydraulic conductivity. What type of deposit would the aquifer material probably be?

SOLUTION

Use equation (6.4.8) and solve for K:

$$K = \frac{Q \ln(r_2/r_1)}{\pi(h_2^2 - h_1^2)}$$

and solve for K with $Q = 0.05$ m³/s, $r_1 = 50$ m, $r_2 = 150$ m, $h_1 = 25 - 3 = 22$ m, and $h_2 = 25 - 1.2 = 23.8$ m:

$$K = \frac{0.05 \ln(150/50)}{\pi(23.8^2 - 22^2)} = \frac{1.7494 \times 10^{-2}}{(566.4 - 484)} = 2.12 \times 10^{-4} \text{ m/s}$$

The deposit is probably a silty sand or a clean sand.

EXAMPLE 6.4.4

A well 12-in in diameter penetrates 108 ft below the static water table. After a long period of pumping at a rate of 350 gpm, the drawdowns in wells 57 ft and 148 ft from the pumped well were found to be 12 ft and 7.4 ft respectively. What is the transmissivity of the aquifer? What is the approximate drawdown in the pumped well?

SOLUTION

Use equation (6.4.8) for steady-state radial flow to a well in an unconfined aquifer, where $h_1 = 108 - 12 = 96$ ft; $h_2 = 108 - 7.4 = 100.6$ ft; $r_2 = 148$ ft; and $r_1 = 57$ ft. Determine the hydraulic conductivity using equation (6.4.8):

$$K = \frac{Q \ln(r_2/r_1)}{\pi(h_2^2 - h_1^2)} = \frac{350 \cdot \ln(148/57)}{\pi[100.6^2 - 96^2]} = 0.118 \text{ gpm/ft}^2$$

$$T = Kb = 0.118 \times 108 = 12.74 \text{ gpm/ft}$$

Solve for the approximate head and approximate drawdown at the well:

$$h_w = \sqrt{100.6^2 - \frac{350 \cdot \ln(148/0.5)}{\pi \cdot (0.118)}} = 68.90 \text{ ft}$$

$$s_w = 108 - 68.90 = 39.10 \text{ ft.}$$

6.5 TRANSIENT WELL HYDRAULICS—CONFINED CONDITIONS

6.5.1 Nonequilibrium Well Pumping Equation

Consider the confined aquifer with transient flow conditions due to a constant pumping rate $Q(L^3/T)$ at the well shown in Figure 6.4.4. The initial condition is a constant head throughout the aquifer, $h(r, 0) = h_0$ for all r where h_0 is the constant initial head at $t = 0$. The boundary condition assumes (1) no drawdown in hydraulic head at the infinite boundary, $h(\infty, t) = h_0$ for all t with a constant pumping rate and (2) Darcy's law applies:

$$\lim_{r \to 0} \left(r \frac{\partial h}{\partial r} \right) = \frac{Q}{2\pi T} \text{ for } t > 0 \tag{6.5.1}$$

The governing differential equation is (6.2.11):

$$\frac{\partial^2 h}{\partial r^2} + \frac{1}{r}\frac{\partial h}{\partial r} = \frac{S}{T}\frac{\partial h}{\partial t} \tag{6.2.11}$$

The confined aquifer is nonleaky, homogeneous, isotropic, infinite in areal extent, and the same thickness throughout. The wells fully penetrate the aquifer and are pumped at a constant rate Q. During pumping of such a well, water is withdrawn from storage within the aquifer, causing the cone of depression to progress outward from the well. There is no stabilization of water levels, resulting in a continual decline of head, provided there is no recharge and the aquifer is effectively infinite in areal extent. The rate of decline of the head, however, continuously decreases as the cone of depression spreads. Water is released from storage by compaction of the aquifer and by the expansion of the water. Theis (1935) presented an analytical solution of equation (6.2.11) to solve for the *drawdown s* given as

$$s = h_0 - h = \frac{Q}{4\pi T}\int_u^\infty \left(\frac{e^{-u}}{u}\right)du \tag{6.5.2}$$

where h_0 is the piezometric surface before pumping started, Q is the constant pumping rate (L^3/T), T is the transmissivity (L^2/T), and

$$u = \frac{r^2 S}{4Tt} \tag{6.5.3}$$

in which r is in L and t is time in T.

The exponential integral can be expanded into a series expansion as

$$W(u) = \int \left(\frac{e^{-u}}{u}\right)du = -0.5772 - \ln u + u - \frac{u^2}{2\cdot 2!} + \frac{u^3}{3\cdot 3!} + \cdots \tag{6.5.4}$$

where $W(u)$ is called the dimensionless *well function* for nonleaky, isotropic, artesian aquifers fully penetrated by wells having constant discharge conditions. Values of this well function are listed in Table 6.5.1. The well function is also expressed in the form of a type curve, as shown in Figure 6.5.1. Both u and $W(u)$ are dimensionless.

Table 6.5.1   Values of $W(u)$ for Various Values of u

u	1.0	2.0	3.0	4.0	5.0	6.0	7.0	8.0	9.0
$\times 1$	0.219	0.049	0.013	0.0038	0.0011	0.00036	0.00012	0.000038	0.000012
$\times 10^{-1}$	1.82	1.22	0.91	0.70	0.56	0.45	0.37	0.31	0.26
$\times 10^{-2}$	4.04	3.35	2.96	2.68	2.47	2.30	2.15	2.03	1.92
$\times 10^{-3}$	6.33	5.64	5.23	4.95	4.73	4.54	4.39	4.26	4.14
$\times 10^{-4}$	8.63	7.94	7.53	7.25	7.02	6.84	6.69	6.55	6.44
$\times 10^{-5}$	10.94	10.24	9.84	9.55	9.33	9.14	8.99	8.86	8.74
$\times 10^{-6}$	13.24	12.55	12.14	11.85	11.63	11.45	11.29	11.16	11.04
$\times 10^{-7}$	15.54	14.85	14.44	14.15	13.93	13.75	13.60	13.46	13.34
$\times 10^{-8}$	17.84	17.15	16.74	16.46	16.23	16.05	15.90	15.76	15.65
$\times 10^{-9}$	20.15	19.45	19.05	18.76	18.54	18.35	18.20	18.07	17.95
$\times 10^{-10}$	22.45	21.76	21.35	21.06	20.84	20.66	20.50	20.37	20.25
$\times 10^{-11}$	24.75	24.06	23.65	23.36	23.14	22.96	22.81	22.67	22.55
$\times 10^{-12}$	27.05	26.36	25.96	25.67	25.44	25.26	25.11	24.97	24.86
$\times 10^{-13}$	29.36	28.66	28.26	27.97	27.75	27.56	27.41	27.28	27.16
$\times 10^{-14}$	31.66	30.97	30.56	30.27	30.05	29.87	29.71	29.58	29.46
$\times 10^{-15}$	33.96	33.27	32.86	32.58	32.35	32.17	32.02	31.88	31.76

Source: Wenzel (1942).

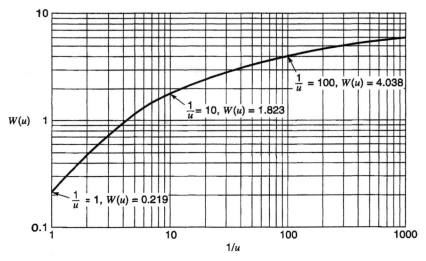

Figure 6.5.1 Type curve for use in solving the Theis's nonequilibrium equation graphically with values of $W(u)$ (well function of u) corresponding to values of $1/u$. The curve is plotted on a logarithmic scale.

The drawdown is then expressed as

$$s = \frac{Q}{4\pi T} W(u) \tag{6.5.5}$$

where Q is m³/day and T is m²/day for s in m; or Q is gal/day and T is gpd/ft for s in ft; or Q is ft³/day and T is ft²/day for s in ft. This equation is commonly referred to as the *nonequilibrium well pumping equation or Theis's equation.*

Equations (6.5.3) and (6.5.5) can be expressed in U.S. customary units (gallon-day-foot system) where s is in ft, Q is in gpm, T is in gpd/ft, r is in ft, and t is in days:

$$s = \frac{114.6Q}{T} W(u) \tag{6.5.6}$$

and

$$u = \frac{1.87r^2 S}{Tt} \tag{6.5.7}$$

or, for t in minutes,

$$u = \frac{2693r^2 S}{Tt} \tag{6.5.8}$$

EXAMPLE 6.5.1   Given that $T = 10{,}000$ gpd/ft, $S = 10^{-4}$, $t = 2693$ min, $Q = 1000$ gpm, and $r = 1000$ ft, compute the drawdown.

SOLUTION   Step 1.   Compute u: (for r in ft, T in gpd/ft, t in minutes)

$$u = \frac{2693r^2 S}{Tt} = \frac{2693 \times 10^6 \times 10^{-4}}{10^4 \times 2693} = 10^{-2}$$

Step 2.   Find $W(u)$ from Table 6.5.1: $W(u) = 4.04$

Step 3.   Compute drawdown: (for Q in gal/min and T in gpd/ft)

$$s = \frac{114.6\,Q}{T}W(u) = \frac{114.6 \times 10^3 \times 4.04}{10^4} = 46.30\,\text{ft}$$

6.5.2   Graphical Solution

Theis (1935) developed a graphical procedure for determining T and S, using time-drawdown data by expressing equation (6.5.6) as

$$\log s = \log\left(\frac{114.6Q}{T}\right) + \log W(u) \tag{6.5.9}$$

and equation (6.5.8) as

$$\log t = \log\left(\frac{2693r^2 S}{T}\right) + \log\frac{1}{u} \tag{6.5.10}$$

Because $(114.6Q/T)$ and $(2693r^2S/Tt)$ are constants for a given distance r from the pumped well, the relation between $\log(s)$ and $\log(t)$ must be similar to the relation between $\log W(u)$ and $\log 1/u$. Therefore, if s is plotted against t and $W(u)$ against $1/u$ on the same double-logarithmic paper, the resulting curves are of the same shape, but horizontally and vertically offset by the constants $(114.6Q/T)$ and $(2693r^2S/Tt)$. Plot each curve on a separate sheet, then match them by placing one graph on top of the other and moving it horizontally and vertically (keeping the coordinate axis parallel) until the curves are matched. This is further illustrated in Figure 6.5.2.

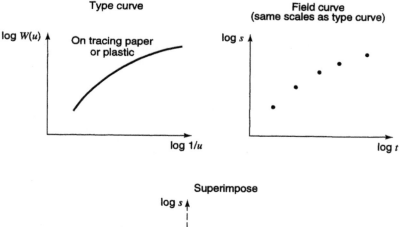

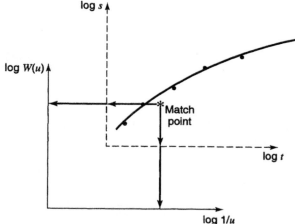

Figure 6.5.2 Graphical procedure to determine T and S from pump test data.

This graphical procedure is used to determine T and S from a field pumping test and requires measurements of a drawdown in at least one observation well. The observation of water levels should be made at proper intervals of time after the instant that pumping starts. Measurements from more than one observation well can also be used.

EXAMPLE 6.5.2

Pump test data (s vs. t) for an observation well are been plotted on log-log paper. The discharge was 150 gpm, and the observation well is 300 ft from the pumped well. The match points for the fitted data are at $W(u) = 1.0$ and $u = 1.0$, $t = 5.5$ min and $s = 0.112$ ft. Compute T and S.

SOLUTION

$$T = \frac{114.6Q}{s}W(u) = \frac{114.6(150)(1)}{0.112} = 153,482 \text{ gpd/ft}$$

and

$$S = \frac{Ttu}{2693r^2} = \frac{(153,482)(5.5)(1)}{2693(300)^2} = 0.00348$$

EXAMPLE 6.5.3

Drawdown was measured during a pumping test at frequent intervals in an observation well 200 ft from a well that was pumped at a constant rate of 500 gpm. The data for this pump test are listed below. These measurements show that the water level is still dropping after 4000 min of pumping; therefore, analysis of the test data requires use of the Theis's nonequilibrium procedure. Determine T and S for this aquifer.

Pump test data	
Time (min)	Drawdown (ft)
1	0.05
2	0.22
3	0.40
4	0.56
5	0.70
7	0.94
10	1.2
20	1.8
40	2.5
100	3.4
300	4.5
1000	5.6
4000	7.0

SOLUTION

Step 1.   Plot the time-drawdown data on log-log graph paper. The drawdown is plotted on the vertical axis and the time since pumping started on the horizontal axis.

Step 2.   Superimpose this plot on the type curve sheet so that the plotted points match the type curve. The axes of both graphs must be kept parallel.

Step 3.   Select a match point, which can be any point in the overlap area of the curve sheets. It is usually found most convenient to select a match point where the coordinates on the type curve are known in advance (e.g., $W(u) = 1$ and $1/u = 1$ or $W(u) = 1$ and $1/u = 10$, etc.). Then determine the value of s and t for this match point: $W(u) = 1$, $s = 1$, $1/u = 1$, and $t = 2$.

Step 4.   Determine T:

$$T = \frac{114.6\,Q}{s}W(u) = \frac{114.6 \times 500}{1} \times 1 = 57,300 \text{ gpd/ft}$$

Step 5.   Determine S:

$$S = \frac{Tt}{\dfrac{1}{u} \times 2693r^2} = \frac{57,300 \times 2}{1 \times 2693 \times 200^2} = 1.06 \times 10^{-3}$$

The above procedure can also be used to determine the drawdown in observation wells at rates of pumping different from those used in the pump test. The Theis nonequilibrium analysis can also be used to determine distance-drawdown information, once the T and S are known for a given time after pumping started. Equations (6.5.6) and (6.5.8) can be expressed as

$$\log s = \log\left(\frac{114.6Q}{T}\right) + \log W(u) \tag{6.5.9}$$

and

$$\log(r^2) = \log\left(\frac{Tt}{2693\,S}\right) - \log\frac{1}{u} \tag{6.5.11}$$

If the values of Q, T, and S are known (determined from a pump test), then the distance-drawdown curve can be determined for any time t. The terms $(114.6Q/T)$ and $(Tt/2693S)$ are known, and for an assumed match point $(W(u)$ and $1/u)$, s and r^2 can be determined.

EXAMPLE 6.5.4

For a pumping rate of 100 gpm in a confined aquifer with $T = 10^4$ gpd/ft and $S = 10^{-4}$, determine the distance-drawdown (r^2, s) curve after 269.3 min.

SOLUTION

Step 1.   Assume $W(u)$ and $1/u$ to compute match points (e.g., $W(u) = 1$ and $1/u = 1$).

Step 2.   Compute s:

$$s = \frac{114.6\,Q}{T}\,W(u) = \frac{114.6(100)}{10^4}(1) = 1.146 \text{ ft}$$

Compute r^2:

$$r^2 = \frac{Ttu}{2693\,S} = \frac{10^4(269.3)(1)}{2693\,S(10^{-4})} = 10^7 \text{ ft}^2$$

Step 3.   Use the computed match points to align the graph of s on the vertical scale and r^2 on the horizontal scale to the type curve. In other words, match points (r^2, s) with $(1/u, W(u))$ by superimposing the graph sheet onto the type curve. Keep all axes parallel. Once the points are matched, trace the type curve on the (r^2, s) graph sheet to form the distance-drawdown curve for t minutes after pumping started.

6.5.3   Cooper–Jacob Method of Solution

Time-Drawdown Analysis

Theis's method for nonequilibrium analysis was simplified by Cooper and Jacob (1946) for the conditions of small values of r and large values of t. From equation (6.5.3), it can be observed that, for small values of r or large values of t, u is small, so that the higher-order terms in equation (6.5.4) become negligible. Then the well function can be expressed as

$$W(u) = -0.5772 - \ln u \tag{6.5.12}$$

which has an error < 3% for $u < 0.1$. The drawdown for this approximation is expressed in U.S. customary units (gallon-day-foot) as

$$s = \frac{114.6\,Q}{T}[-0.5772 - \ln u] \tag{6.5.13}$$

and in SI units as

$$s = \frac{Q}{4\pi T}[-0.5772 - \ln u] \tag{6.5.14}$$

Substituting equation (6.5.13) in equation (6.5.8) yields

$$s = \frac{114.6\,Q}{T}\left[-0.5772 - \ln\frac{2693r^2 S}{Tt}\right] \tag{6.5.15}$$

This shows that the drawdown is a function of log t so that the equation plots as a straight line on semilog paper.

Consider the change in drawdown Δs a distance r from the pumped well over the time interval t_1 and t_2, which are one log cycle apart. The drawdowns at t_1 and t_2 are, respectively, s_1 and s_2. The change in drawdown Δs is then expressed as

$$\begin{aligned}
\Delta s &= s_2 - s_1 = \frac{264\,Q}{T}\left[\log\left(\frac{Tt_2}{2693r^2 S}\right) - \log\left(\frac{Tt_1}{2693r^2 S}\right)\right] \\
&= \frac{264\,Q}{T}\log\left(\frac{t_2}{t_1}\right)
\end{aligned} \tag{6.5.16}$$

If t_2 and t_1 are chosen one log cycle apart, $\log(t_2/t_1) = 1$, the above equation simplifies to

$$\Delta s = \frac{264\,Q}{T} \tag{6.5.17}$$

in which Q is in gpm and T is in gpd/ft.

Next consider that at time t_0, when pumping begins, the drawdown is 0 in the observation well r ft from the pumped well, so that equation (6.5.15) is expressed as

$$0 = \frac{114.6\,Q}{T}\left[-0.5772 + 2.3 \, \log\frac{Tt}{2693r^2 S}\right] \tag{6.5.18}$$

This equation then reduces to

$$S = \frac{Tt_0}{4790\,r^2} \tag{6.5.19}$$

The time t_0 is the time intercept on the zero-drawdown axis, keeping in mind that Jacob's method is a straight-line approximation on semilog paper. Refer to Figure 6.5.3.

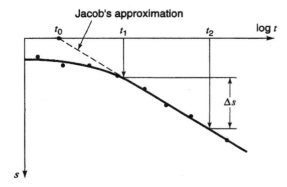

Figure 6.5.3 Illustration of Jacob's method showing the straight-line approximation on semilog paper.

EXAMPLE 6.5.5

For the time-drawdown data listed below, calculate T and S using Jacob's approximation. After computing T and S, check to see that the basic assumption of this approximation is satisfied. For the values of T and S that you computed, after how many minutes of pumping would Jacob's approximation be valid? The discharge is $Q = 1500$ gpm and the radius $r = 300$ ft.

Time after pumping started (min)	Drawdown (ft)
1	0.45
2	0.74
3	0.91
4	1.04
6	1.21
8	1.32
10	1.45
30	2.02
40	2.17
50	2.30
60	2.34
80	2.50
100	2.67
200	2.96
400	3.25
600	3.41
800	3.50
1000	3.60
1440	3.81

SOLUTION

Step 1. Plot the field data on semilog paper (s vs. $\log t$) as shown in Figure 6.5.4.

Step 2. Fit a straight line to the data (Figure 6.5.4).

Step 3. Find t_0 from the plot ($t_0 = 0.45$ min for $s = 0$).

Step 4. Find Δs for values of t_1 and t_2 one log cycle apart ($t_1 = 10$ min and $t_2 = 100$ min): $\Delta s = s_2 - s_1 = 2.67 - 1.45 = 1.08$ ft.

Step 5. Compute T (from equation (6.5.17)):

$$T = \frac{264\,Q}{\Delta s} = \frac{264(1500)}{1.08} = 366,700 \text{ gpd/ft}$$

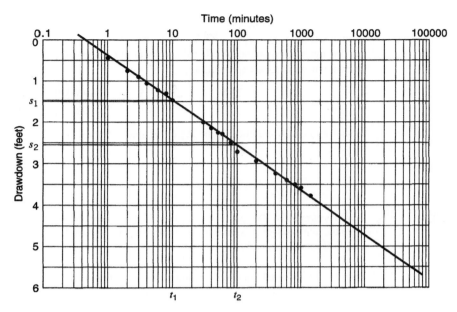

Figure 6.5.4 Time-drawdown for example 6.5.5.

Step 6. Compute S:

$$S = \frac{Tt_0}{4790\, r^2} = \frac{366,700(0.45)}{4790(300^2)} = 3.83 \times 10^{-4}$$

Step 7. Check to see if basic assumption is satisfied:

$$u = \frac{2693(300)^2(3.83 \times 10^{-4})}{366,700(1440)} = 0.00017 < 0.01 \quad \text{OK}$$

Jacob's approximation for this problem is valid for

$$t \geq \frac{2693r^2S}{Tu} = \frac{2693(300)^2(3.83 \times 10^{-4})}{366,700(0.01)} \quad \text{where } u = 0.01$$

$$t \geq 25.3 \text{ min}$$

Distance–Drawdown Analysis

A similar analysis using Jacob's approximation can be used to approximate T and S using distance-drawdown field data. The drawdown must be measured simultaneously in three or more observation wells, each at a different distance from the pumped well. The drawdowns in the observation wells plot as a straight line on semilog paper (s vs. $\log r$). By considering the distance r_0 from the pumped well, where the drawdown is 0, then equation (6.5.15) can be expressed as

$$0 = \frac{114.6\, Q}{T}\left[-0.5772 - \ln\frac{2693r_0^2S}{Tt}\right] \tag{6.5.20}$$

which can be simplified to

$$S = Tt/4790r_0^2 \tag{6.5.21}$$

where r_0 is the intercept at $s = 0$ of the extended straight line fitted to the field distance-drawdown data.

Following the same procedure as for the time-drawdown analysis, consider the change in drawdown $\Delta s = s_2 - s_1$ at a time t over the distance r_2 to r_1, where r_2 and r_1 are chosen to be one log cycle apart on the plot. The drawdowns at r_1 and r_2 are, respectively, s_1 and s_2. The change in drawdown Δs is then expressed as

$$\Delta s = s_2 - s_1 = \frac{264\, Q}{T}\left[\log\frac{2693 r_2^2 S}{Tt} - \log\frac{2693 r_1^2 S}{Tt}\right] = \frac{528\, Q}{T}\log\left(\frac{r_2}{r_1}\right) \tag{6.5.22}$$

Because r_2 and r_1 are chosen one log cycle apart, $(\log r_2/r_1 = 1)$, the above equation simplifies to

$$\Delta s = \frac{528\, Q}{T} \tag{6.5.23}$$

EXAMPLE 6.5.6

A well is pumped at 200 gpm for a period of 500 min. Three observation wells 1, 2, and 3 (50, 30, and 100 ft from the pumped well, respectively) have drawdowns of 10.6, 13.2, and 7.9 ft, respectively. Determine T and S. (From Gehm and Bregman, 1976.)

SOLUTION

Step 1.   Plot the field data on semilog paper (s vs. log r) as shown in Figure 6.5.5.

Step 2.   Fit a straight line to the data (Figure 6.5.5).

Step 3.   Find r_0 from the plot ($r_0 = 500$ ft).

Step 4.   Find Δs for values of r_1 and r_2 that are selected one log-cycle apart: $\Delta s = s_2 - s_1 = 10.6$ ft.

Step 5.   Compute T (from equation (6.5.23)):

$$T = \frac{528\, Q}{\Delta s} = \frac{528(200)}{10.6} = 9960 \text{ gpd/ft}$$

Step 6.   Compute S:

$$S = \frac{Tt}{4790 r_0^2} = \frac{9960(500)}{4790(500^2)} = 4.16 \times 10^{-3}$$

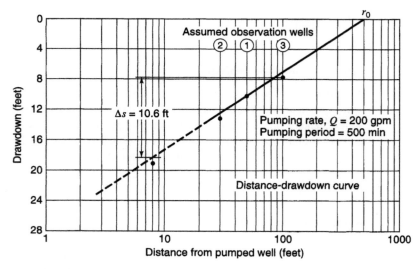

Figure 6.5.5 Trace of the cone of depression plotted on semilog coordinates becomes a straight line. Drawdown in each well was measured at 500 min after starting the pumping test.

6.6   TRANSIENT WELL HYDRAULICS—UNCONFINED CONDITIONS

Unconfined (water table) conditions differ significantly from those of the confined aquifer. For unconfined conditions, the pumped water is derived from storage by gravity drainage of the interstices above the cone of depression, by compaction of the aquifer and expansion of the water as pressure is reduced from pumping. For confined conditions, the nonequilibrium solution is based on the assumptions that the coefficient of storage is constant and water is released from storage instantaneously with a decline in head. The effects of gravity drainage are not considered in the nonequilibrium solution. Gravity drainage is not immediate and, for unsteady flow of water toward a well for unconfined conditions, is characterized by slow drainage of interstices.

There are three distinct segments of the time-drawdown curve for water table conditions, as shown in Figure 6.6.1. The first segment occurs for a short time after pumping begins: the drawdown reacts in the same manner as an artesian aquifer. In other words, the gravity drainage is not immediate, and the water is released instantaneously from storage. It is possible under some conditions to determine the coefficient of transmissivity by applying the nonequilibrium solution to the early time-drawdown data. The coefficient of storage computed using the early time-drawdown is in the artesian range and cannot be used to predict long-term drawdowns. The second segment represents an intermediate stage when the expansion of the cone of depression decreases because of the gravity drainage. The slope of the time-drawdown curve decreases, reflecting recharge. Pump test data deviate significantly from the nonequilibrium theory during the second segment. During the third segment (Figure 6.6.1), the time-drawdown curves conform closely to the nonequilibrium type curves, as shown in Figure 6.6.2. This segment may start from several minutes to several days after pumping starts, depending upon the aquifer condition. The coefficient of transmissibility of an aquifer can be determined by applying the nonequilibrium solution to the third segment of time-drawdown data. The coefficient of storage computed from these data will be in the unconfined range, which can be used to predict long-term effects.

Pricket (1965) and Neuman (1975) developed type curve solutions for water table conditions. The following equation for drawdown in an unconfined aquifer with fully penetrating wells and a constant discharge condition Q was presented by Neuman (1975):

$$s = \frac{Q}{4\pi T} W(u_a, u_y, \eta) \tag{6.6.1}$$

where

$$u_a = \frac{r^2 S}{Tt} \quad \text{(applicable for small values of } t\text{)} \tag{6.6.2}$$

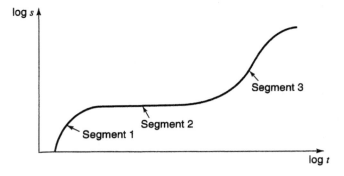

Figure 6.6.1 Three segments of time-drawdown curve for water table conditions.

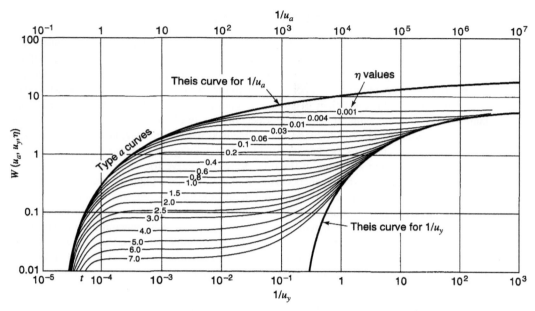

Figure 6.6.2 Theoretical curves of $W(u_a, u_y, \eta)$ versus $1/u_a$ and $1/u_y$ for an unconfined aquifer (after Neuman (1975)).

$$u_y = \frac{r^2 S_y}{Tt} \quad \text{(applicable for large values of } t\text{)} \tag{6.6.3}$$

where $\eta = r^2 K_z / b^2 K_r$.

$W(u_a, u_y, \eta)$ is referred to as the *unconfined well function*, and K_r and K_z are the horizontal and vertical hydraulic conductivities, respectively, for an anisotropic aquifer. Neuman (1975) tabulated the well function, as plotted in Figure 6.6.2. For an isotropic aquifer, $K_r = K_z$ and $\eta = r^2/b^2$. The u_a applies only to early-time response when the rate of drawdown is controlled by the elastic storage properties of the aquifer. The u_y applies to the late-time response when the rate of drawdown is controlled by the specific yield.

Distance-drawdown data for water table conditions can be used to compute T and S_y only after the effects of delayed gravity drainage have dissipated in observation wells. After the effects of delayed gravity drainage cease to influence the drawdown in observation wells, the time-drawdown field data conforms closely to the nonequilibrium solution, as illustrated above. This length of time was discussed in the previous section.

During the time when delayed gravity drainage is affecting the drawdown in observation wells, the cone of depression is distorted; therefore, the distance-drawdown data for this time cannot be analyzed. However, once the effects of delayed gravity drainage are negligible, then the distance-drawdown data can be analyzed using the nonequilibrium solution technique.

6.7 TRANSIENT WELL HYDRAULICS—LEAKY AQUIFER CONDITIONS

In leaky aquifer systems, water enters the aquifer from adjacent lower-permeability units. These conditions provide an additional source of water and reduce the drawdown predicted by the Theis solution. *Leaky-confined conditions* refer to wells penetrating a confined aquifer overlain by an aquitard, which in turn is overlain by a source bed having a water table. There is vertical leakage through the aquitard (confining layer). The aquifer is assumed to be homogeneous, isotropic, infinite

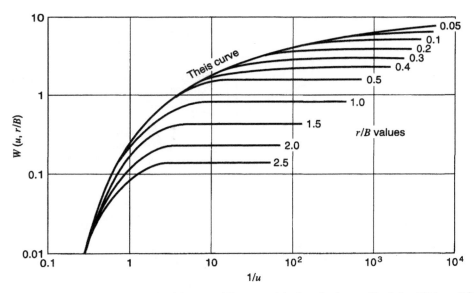

Figure 6.7.1 Theoretical curves of $W(u, r/B)$ versus $1/u$ for a leaky aquifer (after Walton (1960)).

in areal extent, and of constant thickness. Wells fully penetrate the aquifer to which there is radial flow. The confining layer is assumed to be incompressible, neglecting any storage in the layer. Discharge from a well is supplied from both storage within the aquifer and the vertical leakage. The rate of the vertical leakage is proportional to the difference in head between the water table for the source bed and the piezometric surface of the aquifer.

Hantush and Jacob (1955) developed a solution that describes the drawdown in a radially symmetric leaky confined aquifer separated from an overlying aquifer by a confining bed of lower permeability. Leakage into the pumped aquifer is a function of the vertical hydraulic conductivity of the aquitard, K', the aquitard thickness, b', and the difference in hydraulic head between the overlying aquifer and the aquifer being pumped. The drawdown is expressed as

$$s(r, \ t) = \frac{Q}{4\pi T} W[u, \ (r/B)] \tag{6.7.1}$$

where the *leaky well function*, $W[u, \ (r/B)]$, is defined as a function of two dimensionless parameters

$$u = \frac{r^2 S}{4Tt} \tag{6.7.2}$$

and

$$r/B = r\sqrt{\frac{K'}{Kbb'}} \tag{6.7.3}$$

The well function is plotted in Figure 6.7.1. The key simplification is that the hydraulic head in the upper aquifer remains constant and that water is not released from storage in the aquitard.

6.8 BOUNDARY EFFECTS: IMAGE WELL THEORY

The *nonequilibium Theis solution* for unsteady radial flow to a well is based on the assumption that the aquifer is infinite in areal extent; however, often aquifers are delimited by one or more boundaries, causing the time-drawdown data to deviate under the influence of these boundaries. The boundaries may be walls of impervious soil or rock, which are referred to as *barrier boundaries*,

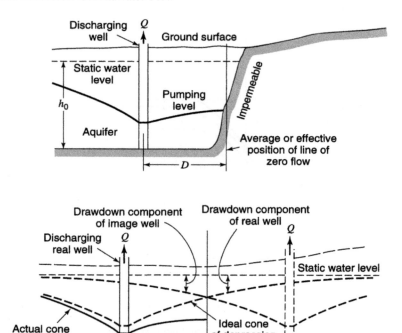

Figure 6.8.1 Relationship of an impermeable boundary and an image well (from U.S. Bureau of Reclamation (1981)).

or recharges of water from rivers, streams, or lakes, which are referred to as *recharge boundaries*. The effects of barrier and recharge boundaries on groundwater movement and storage can be described by image well theory.

Walton (1970) has stated the *image well theory* as "the effect of a barrier boundary on the drawdown in a well, as a result of pumping from another well, is the same as though the aquifer were infinite and a like discharging well were located across the real boundary on a perpendicular thereto and at the same distance from the boundary as the real pumping well. For a recharge boundary the principle is the same except that the image well is assumed to be discharging the aquifer instead of pumping from it." These concepts give rise to the use of imaginary wells, referred lo as *image wells*, that are introduced to simulate a groundwater system reflecting the effects of known physical boundaries on the system. Essentially, then, through the use of image wells, an aquifer of finite extent can be transformed to one of infinite extent. Once this is done, the unsteady radial flow equations can be applied to the transformed system. The concept of image wells is further illustrated in Figure 6.8.1 for a barrier boundary and Figure 6.8.2 for a recharge boundary.

6.8.1 Barrier Boundary

In the case of a *barrier boundary*, water cannot flow across the boundary, so that no water is being contributed to the pumped well from the impervious formation. The cone of depression that would exist for a pumped well in an aquifer of infinite areal extent is shown in Figure 6.8.1. Because of the barrier boundary, the cone of depression shown is no longer valid since there can be no flow across

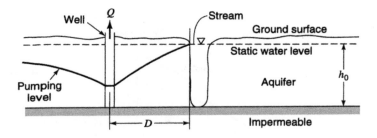

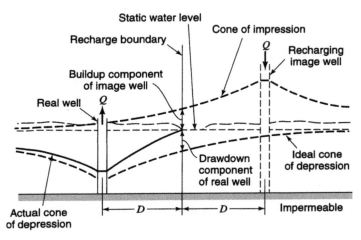

Figure 6.8.2 Relationship of a recharge boundary and an image well (from U.S. Bureau of Reclamation (1981)).

the boundary. Placing an image well, discharging in nature, across the barrier boundary creates the effect of no flow across the boundary. The image well must be placed perpendicular to the barrier boundary and at the same distance from the boundary as the real well. The resulting *real cone of depression* is the summation of the components of both the real and image well depression cones, as shown in Figure 6.8.1. Water levels in wells will decline at an initial rate due only to the influence of the pumped well. As pumping continues, the barrier boundary effects will begin as simulated by the image well affecting the real well. When the effects of the barrier boundary are realized, the time rate of drawdown will increase (Figure 6.8.3). When this occurs, the total rate of withdrawal from the aquifer is equal to that of the pumped well plus that of the discharging image well, causing the cone of depression of the real well to be deflected downward.

The total drawdown in the real well can be expressed as

$$s_B = s_p + s_i \qquad (6.8.1)$$

in which s_B is the total drawdown, s_p is the drawdown in an observation well due to pumping of the production well, and s_i is the drawdown due to the discharging image well (barrier boundary).

The total drawdown can be expressed as

$$s_B = \frac{Q}{4\pi T} W(u_p) + \frac{Q}{4\pi T} W(u_i) \qquad (6.8.2)$$

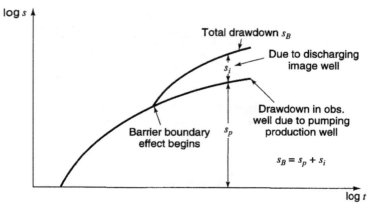

Figure 6.8.3 Barrier boundary effects on time-drawdown curve.

where Q is the constant pumping rate (L^3/T), T is the transmissivity (L^2/T), $W(u_p)$, and $W(u_i)$ are dimensionless, and u_p and u_i are

$$u_p = \frac{r_p^2 S}{4Tt_p} \tag{6.8.3}$$

$$u_i = \frac{r_i^2 S}{4Tt_i} \tag{6.8.4}$$

in which r_i and r_p are in L and t is time in T.

The drawdown equation (6.8.1) can also be expressed in U.S. customary units (gal-day-foot) system where S is in ft, Q is in gpm, T is in gpd/ft, r is in ft, and t is in days:

$$s_B = \frac{114.6Q}{T} W\left(u_p\right) + \frac{114.6Q}{T} W(u_i) = \frac{114.6Q}{T}\left[W\left(u_p\right) + W(u_i)\right] \tag{6.8.5}$$

where

$$u_p = \frac{1.87r_p^2 S}{Tt_p} \tag{6.8.6}$$

$$u_i = \frac{1.87r_i^2 S}{Tt_i} \tag{6.8.7}$$

Now suppose that we choose drawdowns at times t_p and t_i such that $s_p = s_i$, then $W\left(u_p\right) = W(u_i)$ and $u_p = u_i$, so that

$$\frac{r_i^2 S}{Tt_i} = \frac{r_p^2 S}{Tt_p} \tag{6.8.8}$$

which reduces to

$$\frac{r_i^2}{t_i} = \frac{r_p^2}{t_p} \tag{6.8.9}$$

Equation (6.8.9) defines the *law of times* (Ingersoll et al., 1948), which states that for a given aquifer, the times of occurrence of equal drawdown vary directly as the squares of distances from an observation well to a production well of equal discharge.

The law of times can be used to determine the distance from an image well to an observation well, using

$$r_i = r_p\sqrt{\frac{t_i}{t_p}} \tag{6.8.10}$$

in which r_i is the distance from the image well to the observation well in feet, r_p is the distance from the pumped well to the observation well in feet, t_p is the time after pumping started and before the barrier boundary is effective, and t_i is the time after pumping started and after the barrier boundary becomes effective, where $s_p = s_i$.

EXAMPLE 6.8.1

A well is pumping near a barrier boundary (see Figure 6.8.4) at a rate of 0.03 m³/s from a confined aquifer 20 m thick. The hydraulic conductivity of the aquifer is 27.65 m/day, and its storativity is 3×10^{-5}. Determine the drawdown in the observation well after 10 hours of continuous pumping. What is the fraction of the drawdown attributable to the barrier boundary?

SOLUTION

The following information is given in the above problem statement: $Q = 0.03$ m³/s, $b = 20$ m, $K = 27.65$ m/day $= 3.2 \times 10^{-4}$ m/s, $S = 3 \times 10^{-5}$, $t = 10$ hr $= 36,000$ s. An image well is placed across the boundary at the same distance from the boundary as the pumped well (as shown in Figure 6.8.4b). The drawdown in the observation well is due to the real well and the imaginary well (which accounts for the barrier boundary). Hence, using equation (6.8.5)

$$s = \frac{Q}{4\pi T} W(u_p) + \frac{Q}{4\pi T} W(u_i)$$

$$u_p = \frac{r_p^2 S}{4Tt} = \frac{(240)^2 (3 \times 10^{-5})}{4(20)(3.2 \times 10^{-4})(36,000)} = 1.88 \times 10^{-3}$$

Next, compute the distance from the observation well to the image well: $r_i^2 = 600^2 + 240^2 - 2(600)(300) \cos 30° = 168,185$ m² so $r_i = 410$ m. Using r_i, compute

$$u_i = \frac{168,185(3 \times 10^{-5})}{4(20)(3.2 \times 10^{-4})(36,000)} = 5.47 \times 10^{-3}$$

The well functions are now computed using equation (6.5.4) or obtained from Table 6.5.1 as $W(u_p) = 5.72$ for $u_p = 1.88 \times 10^{-1}$ and $W(u_i) = 4.64$ for $u_i = 5.47 \times 10^{-3}$.

The drawdown at the observation well is computed as

$$s = \frac{0.03}{4\pi(20)(3.2 \times 10^{-4})} = (5.72 + 4.64) = 3.86 \text{ m}.$$

The drawdown attributable to the barrier boundary is computed as

$$s_i = \frac{Q}{4\pi T} W(u_i) = \frac{0.03}{4\pi(20)(3.2 \times 10^{-4})} (4.64) = 1.73 \text{ m}$$

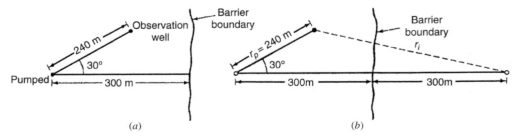

Figure 6.8.4 Example 6.8.1 system. (*a*) Well locations; (*b*) Image well location.

and the fraction of the drawdown attributable to the impermeable boundary is

$$\frac{s_i}{s} = \frac{1.73}{3.86} = 0.45 \ (45\%)$$

6.8.2 Recharge Boundary

The effect of recharge boundaries can also be simulated using the concepts of image well theory. For an aquifer bounded on one side by a recharge boundary, the cone of depression cannot extend beyond the stream, as shown in Figure 6.8.2. This results in no drawdown along the line of recharge. By placing an image well, recharging in nature, directly opposite and at the same distance from the stream as the real pumped well and also pumping at the same rate, the finite system can be simulated. The resulting cone of depression is the summation of the real well cone of depression without the recharge and the image well cone of depression, as shown in Figure 6.8.2.

The water level in the wells will draw down initially only under the influence of the pumped well. After a time the effects of the recharge boundary will cause the time rate of drawdown to decrease and eventually reach equilibrium conditions. This occurs when recharge equals the pumping rate, as illustrated in Figure 6.8.5.

The drawdown for equilibrium conditions can be expressed as

$$s_r = s_p - s_i \tag{6.8.11}$$

in which s_r is the drawdown in an observation well near a recharge boundary, s_p is the drawdown due to the pumped well, and s_i is the buildup due to the image well (recharge boundary). The drawdown equation can be written as

$$s_r = \frac{Q}{4\pi T} \left[W(u_p) - W(u_i) \right] \tag{6.8.12}$$

or in the gallon-day-foot system as

$$s_r = \frac{114.6Q}{T} \left[W(u_p) - W(u_i) \right] \tag{6.8.13}$$

For large values of time t, the well functions can be expressed as

$$W(u_p) = -0.5772 - \ln u_p \tag{6.8.14}$$

$$W(u_i) = -0.5772 - \ln u_i \tag{6.8.15}$$

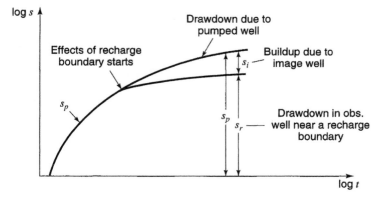

Figure 6.8.5 Recharge boundary effects on time-drawdown curve.

This allows equation (6.8.12) to be simplified to

$$s_r = \frac{Q}{4\pi T}\left[-\ln u_p + \ln u_i\right] \tag{6.8.16}$$

and equation (6.8.13) to

$$s_r = \frac{114.6Q}{T}\left[-\ln u_p + \ln u_i\right] \tag{6.8.17}$$

Now using the gallon-day-foot system with time in minutes, we get

$$u_p = \frac{2693r_p^2 S}{Tt} \tag{6.8.18}$$

and

$$u_i = \frac{2693r_i^2 S}{Tt} \tag{6.8.19}$$

The drawdown in the observation well from equation (6.8.17) is expressed as

$$s_r = \frac{114.6Q}{T}\left[-\ln\left(\frac{2693r_p^2 S}{Tt}\right) + \ln\left(\frac{2693r_i^2 S}{Tt}\right)\right] \tag{6.8.20}$$

which simplifies to

$$s_r = \frac{528}{T}Q\log\left(\frac{r_i}{r_p}\right) \tag{6.8.21}$$

Rorabaugh (1956) expressed this equation in terms of the distances between the pumped well and the line of recharges as

$$s_r = \frac{528Q\log\sqrt{\left(4a^2 + r_p^2 - 4ar_p\cos B_r\right)/r_p}}{T} \tag{6.8.22}$$

where a is the distance from the pumped well to the recharge boundary in feet, and B_r is the angle between a line connecting the pumped and image wells and a line connecting the pumped and observation wells. Refer to Figure 6.8.6 for an explanation of terms.

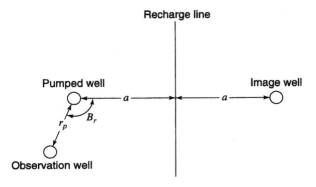

Figure 6.8.6 Definition of terms for equation (6.8.22).

6.8.3 Multiple Boundary Systems

Image well theory can also be applied to aquifer systems with multiple boundaries by considering successive reflections on the barrier and recharge boundaries. This is accomplished through a number of image wells. Placing a primary image well across each boundary balances the effect of the pumped well at each boundary. If a pair of converging boundaries is required, each primary image well then produces an unbalanced effect at the opposite boundary. This unbalanced effect is corrected by placing secondary image wells until the effects of the real and image wells are balanced at both boundaries. These concepts are illustrated in Figure 6.8.7.

A primary image well placed across a barrier boundary is discharging in character. A primary image well placed across a recharge boundary is recharging in character. A secondary image well placed across a barrier boundary has the same character as its parent image well. A secondary image

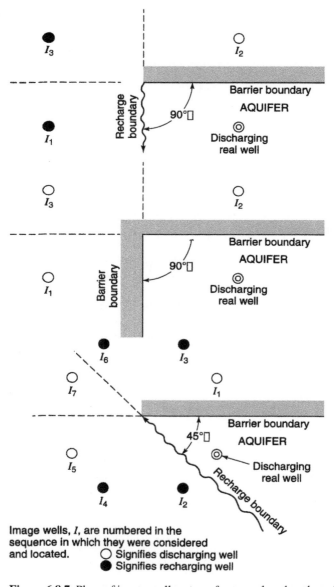

Image wells, I, are numbered in the sequence in which they were considered and located. ◯ Signifies discharging well
● Signifies recharging well

Figure 6.8.7 Plans of image-well systems for several wedge-shaped aquifers (from Ferris et al. (1946)).

well placed across a recharge boundary has the opposite character to its parent image well. Figure 6.8.7 shows image well systems for wedge-shaped aquifers. For parallel boundary systems, it is only necessary to add pairs of image wells until the next pair has negligible influence on the sum of all image well effects out to the point.

6.9 SIMULATION OF GROUNDWATER SYSTEMS

6.9.1 Governing Equations

Darcy's law relates the Darcy flux v with dimension L/T to the rate of headloss per unit length of porous medium $\partial h/\partial l$. The negative sign indicates that the total head is decreasing in the direction of flow because of friction. This law applies to a cross-section of porous medium that is large compared to the cross-section of individual pores and grains of the medium. At this scale, Darcy's law describes a steady uniform flow of constant velocity, in which the net force on any fluid element is zero. For unconfined saturated flow, the two forces are gravity and friction. Darcy's law can also be expressed in terms of the transmissivity for confined conditions as

$$v = -\frac{T}{b}\frac{\partial h}{\partial l} \tag{6.9.1}$$

or for unconfined conditions as

$$v = -\frac{T}{h}\frac{\partial h}{\partial l} \tag{6.9.2}$$

Considering two-dimensional (horizontal) flow, a general flow equation can be derived by considering flow through a rectangular element (control volume) shown in Figure 6.9.1. The flow components ($q = Av$) for the four sides of the element are expressed using Darcy's law where $A = \Delta x \cdot h$ for unconfined conditions and $A = \Delta x \cdot b$ for confined conditions, so that

$$q_1 = -T_{x_{i-1,j}}\Delta y_j \left(\frac{\partial h}{\partial x}\right)_1 \tag{6.9.3a}$$

$$q_2 = -T_{x_{i,j}}\Delta y_j \left(\frac{\partial h}{\partial x}\right)_2 \tag{6.9.3b}$$

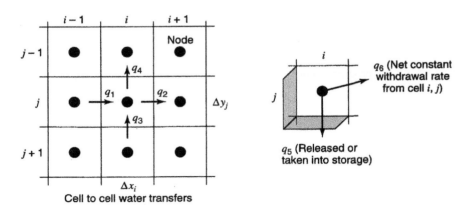

Cell to cell water transfers

Figure 6.9.1 Finite difference grid.

$$q_3 = -T_{y_{i,j+1}}\Delta x_i \left(\frac{\partial h}{\partial y}\right)_3 \tag{6.9.3c}$$

$$q_4 = -T_{y_{i,j}}\Delta x_i \left(\frac{\partial h}{\partial y}\right)_4 \tag{6.9.3d}$$

where $T_{x_{i,j}}$ is the transmissivity in the x flow direction from element (i, j) to element $(i+1, j)$. The terms $(\partial h/\partial x)_1$, $(\partial h/\partial x)_2$, ... define the hydraulic gradients at the element sides 1, 2, ...

The rate at which water is stored or released in the element over time is

$$q_5 = S_{i,j}\Delta x_i \Delta y_i \frac{\partial h}{\partial t} \tag{6.9.4}$$

in which $S_{i,j}$ is the storage coefficient for element (i, j). In addition, the flow rate q_6 for constant net withdrawal or recharge from the element over time interval Δt is considered:

$$q_6 = q_{i,j,t} \tag{6.9.5}$$

in which $q_{i,j,t}$ has a positive value for pumping whereas it has a negative value for recharge.

By continuity, the flow into and out of a grid or cell is

$$q_1 - q_2 + q_3 - q_4 = q_5 + q_6 \tag{6.9.6}$$

Substituting in equations (6.9.3) and (6.9.5) gives

$$-T_{x_{i-1,j}}\Delta y_j \left(\frac{\partial h}{\partial x}\right)_1 + T_{x_{i,j}}\Delta y_i \left(\frac{\partial h}{\partial x}\right)_2 - T_{y_{i,j+1}}\Delta x_i \left(\frac{\partial h}{\partial y}\right)_3 + T_{y_{i,j}}\Delta x_i \left(\frac{\partial h}{\partial y}\right)_4 = S_{i,j}\Delta x_i \Delta y_i \frac{\partial h}{\partial t} + q_{i,j,t} \tag{6.9.7}$$

Dividing equation (6.9.7) by $\Delta x \Delta y$ and simplifying for constant transmissivities in the x and y directions yields

$$-T_x \left[\frac{\left(\frac{\partial h}{\partial x}\right)_1 - \left(\frac{\partial h}{\partial x}\right)_2}{\Delta x_i}\right] - T_y \left[\frac{\left(\frac{\partial h}{\partial y}\right)_3 - \left(\frac{\partial h}{\partial y}\right)_4}{\Delta y_j}\right] = S_{i,j}\frac{\partial h}{\partial t} + \frac{q_{i,j,t}}{\Delta x_i \Delta y_i} \tag{6.9.8}$$

For Δx and Δy infinitesimally small, the terms in brackets become second derivatives of h: then equation (6.9.8) reduces to

$$T_x \frac{\partial^2 h}{\partial x^2} + T_y \frac{\partial^2 h}{\partial y^2} = S \frac{\partial h}{\partial t} + W \tag{6.9.9}$$

which is the general partial differential equation for unsteady flow in the horizontal direction in which $W = q_{i,j,t}/\Delta x_i \Delta y_j$ is a sink term with dimensions L/T.

In the more general unsteady, two-dimensional heterogeneous anisotropic case, equation (6.9.9) is expressed as

$$\frac{\partial}{\partial x}\left(T_x \frac{\partial h}{\partial x}\right) + \frac{\partial}{\partial y}\left(T_y \frac{\partial h}{\partial y}\right) = S \frac{\partial h}{\partial t} + W \tag{6.9.10a}$$

or more simply

$$\frac{\partial}{\partial x_i}\left(T_{i,j} \frac{\partial h}{\partial x_j}\right) = S \frac{\partial h}{\partial t} + W \quad i, j = 1, 2 \tag{6.9.10b}$$

6.9.2   Finite Difference Equations

The partial derivative expressions for Darcy's law, equations (6.9.3 a–d), can be expressed in finite difference form for time t in equation (6.9.7) using

$$\left(\frac{\partial h}{\partial x}\right)_1 = \left(\frac{h_{i-1,j,t} - h_{i,j,t}}{\Delta x_i}\right) \tag{6.9.11a}$$

$$\left(\frac{\partial h}{\partial x}\right)_2 = \left(\frac{h_{i,j,t} - h_{i+1,j,t}}{\Delta x_i}\right) \tag{6.9.11b}$$

$$\left(\frac{\partial h}{\partial y}\right)_3 = \left(\frac{h_{i,j+1,t} - h_{i,j,t}}{\Delta y_j}\right) \tag{6.9.11c}$$

$$\left(\frac{\partial h}{\partial y}\right)_4 = \left(\frac{h_{i,j,t} - h_{i,j-1,t}}{\Delta y_j}\right) \tag{6.9.11d}$$

and the time derivative in equation (6.9.7) is

$$\frac{\partial h}{\partial t} = \left(\frac{h_{i,j,t} - h_{i,j,t-1}}{\Delta t}\right) \tag{6.9.12}$$

Substituting equation (6.9.11) and (6.9.12) into equation (6.9.7) yields

$$-T_{x_{i-1,j}} \Delta y_j \left(\frac{h_{i-1,j,t} - h_{i,j,t}}{\Delta x_i}\right) + T_{x_{i,j}} \Delta y_j \left(\frac{h_{i,j,t} - h_{i+1,j,t}}{\Delta x_i}\right)$$

$$-T_{y_{i,j+1}} \Delta x_i \left(\frac{h_{i,j+1,t} - h_{i,t}}{\Delta y_i}\right) + T_{y_{i,j}} \Delta x_j \left(\frac{h_{i,j,t} - h_{i,j-1,t}}{\Delta y_j}\right)$$

$$-S_{i,j} \Delta x_i \Delta y_j \left(\frac{h_{i,j,t} - h_{i,j,t}}{\Delta t}\right) - q_{i,j,t} = 0 \tag{6.9.13}$$

which can be further simplified to

$$A_{i,j} h_{i,j,t} + B_{i,j} h_{i-1,j,t} + C_{i,j} h_{i+1,j,t} + D_{i,j} h_{i,j+1,t} + E_{i,j} h_{i,j-1,t} + F_{i,j,t} = 0 \tag{6.9.14}$$

where

$$A_{i,j} = \left[T_{x_{i-1,j}} \frac{\Delta y_j}{\Delta x_i} + T_{x_{i,j}} \frac{\Delta y_j}{\Delta x_i} + T_{y_{i,j+1}} \frac{\Delta x_i}{\Delta y_j} + T_{y_{i,j}} \frac{\Delta x_i}{\Delta y_j} - S_{i,j} \frac{\Delta x_i \Delta y_j}{\Delta t}\right] \tag{6.9.15a}$$

$$B_{i,j} = -T_{x_{i-1,j}} \frac{\Delta y_j}{\Delta x_i} \tag{6.9.15b}$$

$$C_{i,j} = -T_{y_{i,j}} \frac{\Delta y_j}{\Delta x_i} \tag{6.9.15c}$$

$$D_{i,j} = -T_{y_{i,j+1}} \frac{\Delta x_j}{\Delta y_j} \tag{6.9.15d}$$

$$E_{i,j} = -T_{y_{i,j}} \frac{\Delta x_i}{\Delta y_j} \tag{6.9.15e}$$

$$F_{i,j,t} = -S_{i,j} \frac{\Delta x_j \Delta y_j}{\Delta t} \left(\frac{h_{i,j,t} - h_{i,j,t-1}}{\Delta t}\right) - q_{i,j,t} \tag{6.9.15f}$$

The coefficients $A_{i,j}$, $B_{i,j}$, $C_{i,j}$, $D_{i,j}$, $E_{i,j}$, and $F_{i,j,t}$ are linear functions of the thickness of cell (i,j) and the thickness of one of the adjacent cells. For artesian conditions, this thickness is a known constant, so if cell (i,j) and its neighbors are artesian, equation (6.9.14) is linear for all t. For unconfined (water table) conditions, the thickness of cell (i,j) is $h_{i,j,t} - BOT_{i,j}$, where $BOT_{i,j}$ is the average elevation of the bottom of the aquifer for cell (i,j). Then for unconfined conditions, equation (6.9.14) involves products of heads and is nonlinear in terms of the heads.

An *iterative alternating direction implicit (IADI) procedure* can be used to solve the set of equations. The IADI procedure involves reducing a large set of equations to several smaller sets of equations. One such smaller set of equations is generated by writing equation (6.9.14) for each cell or element in a column but assuming that the heads for the nodes on the adjacent columns are known. The unknowns in this set of equations are the heads for the nodes along the column. The head for the nodes along adjoining columns are not considered unknowns. This set of equations is solved by Gauss elimination and the process is repeated until each column is treated. The next step is to develop a set of equations along each row, assuming the heads for the nodes along adjoining rows are known. The set of equations for each row is solved and the process is repeated for each row in the finite difference grid.

Once the sets of equations for the columns and the sets of equations for the row have been solved, one "iteration" has been completed. The iteration process is repeated until the procedure converges. Once convergence is accomplished, the terms $h_{i,j}$ represent the heads at the end of the time step. These heads are used as the beginning heads for the following time step. For a more detailed discussion of IADI procedure, see Peaceman and Rachford (1955), Prickett and Lonnquist (1971), Trescott et al., (1976), or Wang and Anderson (1982).

An example of the application of a two-dimensional finite-difference groundwater model is the Edwards (Balcones Fault Zone) aquifer. This aquifer has been modeled using the GWSIM groundwater simulation model developed by the Texas Water Development Board (1974). GWSIM is a finite-difference simulation model that uses the IADI method, similar to the model by Prickett and Lonnquist (1971). The finite-difference grid for the Edwards Aquifer is shown in Figure 6.9.2, which has 856 active cells to describe the aquifer. Wanakule (1989) has modeled only a small portion of the Edwards aquifer, called the Barton Springs-Edwards aquifer, Austin, Texas. This application used a finite difference grid system (Figure 6.9.3) containing 330 cells whose dimensions varied from 0.379×0.283 mi^2 to 0.95×1.51 mi^2. Figure 6.9.4 illustrates the 1981 water level contours.

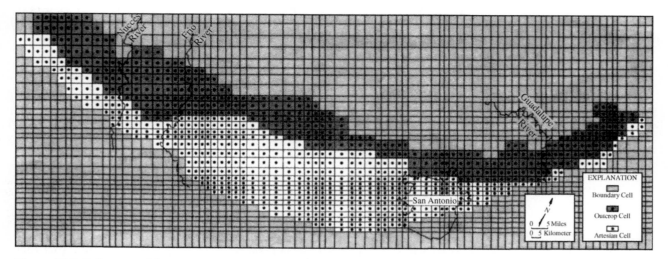

Figure 6.9.2 Cell map used for the digital computer model of the Edwards (Balcones Fault Zone) Aquifer (after Klemt et al. (1979)).

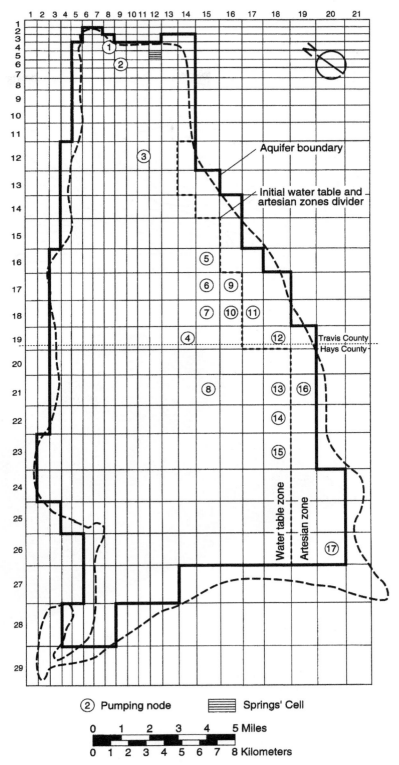

Figure 6.9.3 Pumping locations used in Barton Springs-Edwards Aquifer model (after Wanakule (1989)).

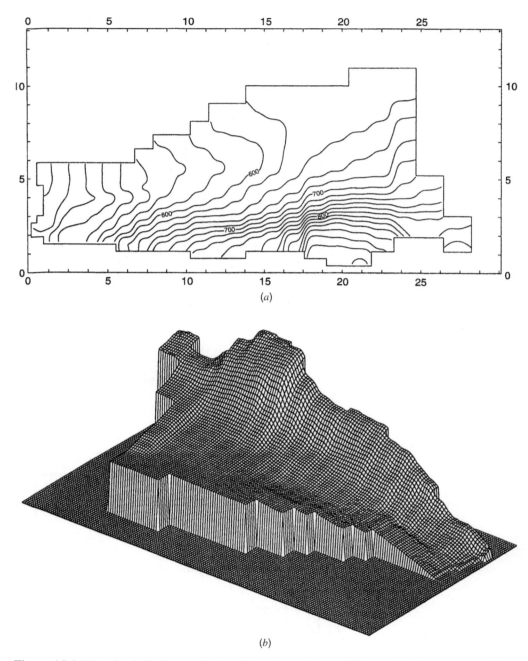

Figure 6.9.4 Water levels for Barton Springs-Edwards aquifer. (*a*) 1981 water-level contours at 20 ft intervals; (*b*) Perspective block diagram of 1981 water levels viewed from the east side of the aquifer (after Wanakule (1985)).

6.9.3   MODFLOW

One of the most widely used groundwater simulation models is the MODFLOW-2005 Model developed by the U.S. Geological Survey (Harbaugh, 2006). The World Wide Web (WWW) address from which this model can be downloaded is http://water.usgs.gov/gwsoftware/modflow2005/ modflow2005.html. MODFLOW-2005 is a finite difference model that can be used to solve groundwater flow problems in one, two, or three dimensions. The computer program is divided into a main program and a series of independent subroutines called *modules*. These modules are grouped into *packages*, each of which is a group of modules that deals with a single aspect of the simulation. A Basic Package handles tasks that are required for each simulation: the Well Package, which simulates the effects of injection and production wells; the River Package, which simulates the effect of rivers; and the Recharge Package, which simulates the effect of recharge. Others include the Evapotranspiration Package, the Drain Package, the General-Head Boundary Package, and the Solution Procedure Package. Individual packages may or may not be required, depending on the problem being simulated.

PROBLEMS

6.1.1 An experiment was conducted to determine the hydraulic conductivity of an artesian aquifer. The piezometric heads at two points 150 m apart were found to be 55 m and 48.5 m above a datum. A tracer injected into the first piezometer was observed after 32 hours in the second well. A test on porosity of a sample of the aquifer shows that $\alpha = 24\%$. What is the hydraulic conductivity of the aquifer? Suggest what the aquifer material may be and verify that your solution holds true. Take the subsurface temperature as 15°C.

6.1.2 Develop an inventory of wells in the county where you live using the USGS data sources for your state. Select a well that has a time history of water levels and print the hydrograph.

6.1.3 Perform a search of USGS publications for the topic "hydrologic budget and water budget." How many publications are listed?

6.1.4 Determine the water level rise in an unconfined aquifer produced by a seasonal precipitation of four inches. The aquifer's porosity is 20 percent and its specific retention is 9 percent.

6.1.5 The leakage from the artificially constructed Tempe Town Lake in Tempe, Arizona can be as low as 0.5 ft/day or as high as 3 ft/day. The lake covers 222 surface acres. If the specific yield of

the subsurface formation is 20 percent, estimate the average regional groundwater level rise assuming that the aerial extent of the effect of leakage is (a) 222 acres, (b) 1 square mile, (c) 5 square miles, and 25 square miles.

6.1.6 The coefficient of storage of a confined aquifer is found to be 6.8×10^{-4} as a result of a pumping test. The thickness of the aquifer is 50 m, and the porosity of the aquifer is 0.25. Determine the fractions of the expansibility of water and compressibility of the aquifer skeleton in making up the storage coefficient of the aquifer.

6.1.7 The specific storage of a 45-m thick confined aquifer is $3.0 \times 10^{-5} \, \text{m}^{-1}$. How much water would the aquifer produce if the piezometric surface is lowered by 10 m over an area of 1 km²?

6.2.1 Show that the steady state groundwater flow equation can be expressed in polar coordinate systems as $h = C_1 \ln r + C_2$, where C_1 and C_2 are constants.

6.2.2 Water flows through three confined aquifers in series as shown in Figure P6.2.2. For piezometric heads in the observation wells of 66.4 m and 60.6 m, determine the flow rate per unit width of the aquifer and the headlosses in each component of the aquifers between the observation wells. If the headlosses in each aquifer

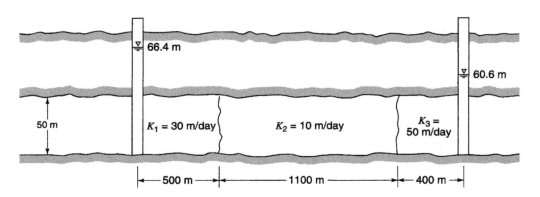

Figure P6.2.2

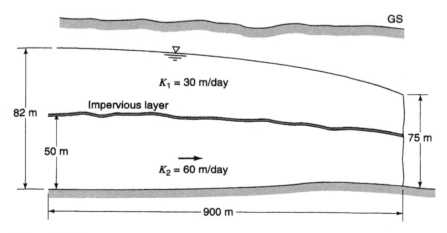

Figure P6.2.3

between the wells were to be equal, what would be the length of each aquifer?

6.2.3 Suppose an unconfined aquifer lies over a confined aquifer as shown in Figure P6.2.3. Determine the flow out of both aquifers.

6.3.1 Rework example 6.3.1 with $h_0 = 10$ ft.

6.3.2 Consider two strata of the same soil material that lie between two channels. The first stratum is confined, and the second one is unconfined, and the water surface elevations in the channels are 24 m and 16 m above the bottom of the unconfined aquifer. What should be the thickness of the confined aquifer for which

(1) the discharge through both strata are equal?
(2) the discharge through the confined aquifer is half of that through the unconfined aquifer?

6.3.3 Three monitoring wells are used (as shown in Figure P6.3.3) to determine the direction of groundwater flow in a confined aquifer. The piezometric heads in the wells are found to be 52 m in well 1, 49 m in well 2, and 56 m in well 3. Determine the direction of flow.

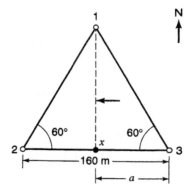

Figure P6.3.3

6.3.4 An earthen dam is 200 m across (i.e., the distance from the upstream face to the downstream face) and underlain by an impermeable bedrock. The average hydraulic conductivity of the material of which the dam is constructed is 0.065 m/day. If the water surface elevations in the reservoir and the tailwater are 25 m and 4.5 m, respectively, estimate the magnitude of leakage from the reservoir to the tailwater per 100-m width of the dam.

6.3.5 For the channel and river shown in Figure 6.3.4, determine the volume of water that flows from the channel into the river. The water surface elevations in the channel and river with respect to the underlying bedrock are 15 m and 11.5 m, respectively. The hydraulic conductivity of formation A is 5.6 m/day and that of formation B is 12.3 m/day.

6.3.6 For the channel and river shown in Figure 6.3.4, determine the volume of water that flows from the channel into the river. The water surface elevations in the channel and river with respect to the underlying bedrock are 15 m and 11.5 m, respectively. The hydraulic conductivity of formation A is 6.6 m/day and that of formation B is 10.3 m/day.

6.4.1 Rework example 6.4.1 with a constant pumping rate of 0.07 m^3/s and the same drawdowns.

6.4.2 Rework example 6.4.2 with a pumping rate of 400 gpm and the same drawdowns.

6.4.3 Rework example 6.4.3 with a constant pumping rate of 0.075 m^3/s and the same drawdowns.

6.4.4 Rework example 6.4.4 with the same pumping rate but drawdowns of 10 ft and 5 ft in the observation wells.

6.4.5 A 50-cm diameter well fully penetrates vertically through a confined aquifer 12 m thick. When the well is pumped at 0.035 m^3/s, the heads in the pumped well and the two other observation wells were found to be as shown in Figure P6.4.5. Does this test suggest that the aquifer material is fairly homogeneous in the directions of the observation wells?

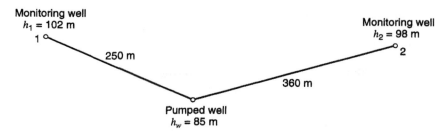

Figure P6.4.5

6.4.6 Three pumping wells along a straight line are spaced 200 m apart. What should be the steady state pumping rate from each well so that the drawdown in each well will not exceed 2 m: The transmissivity of the confined aquifer that all the wells penetrate fully is 2400 m²/day and all the wells are 40 cm in diameter. Take the thickness of the aquifer $b = 40$ m and the radius of influence of each well to be 800 m.

6.4.7 Reposition the wells in problem 6.4.6 such that they form an equilateral triangle (same spacings). For the same restrictions on the drawdown, will the discharge decrease or increase? If so, by what percent? If not, what difference do you perceive between the two problems?

6.4.8 It is required to dewater a construction site 80 m by 80 m. The bottom of the construction will be 1.5 m below the initial water surface elevation of 90 m. Four pumps are to be used in 0.5-m diameter wells at the four corners of the site. Determine the required steady-state pumping rate. The aquifer has $T = 1600$ m²/day and each well has a radius of influence of 600 m.

6.4.9 A 0.75-m radius well fully penetrates a 24-m thick unconfined aquifer and is pumped at a rate of 10 l/s (or 864 m³/day). The drawdown in an observation well 30 m from the pumped well is 1.6 m after a long period of pumping. The drawdown in another observation well 60 m from the pumped well is 1.1 m after the long period of pumping. Determine the (a) hydraulic conductivity of the aquifer, (b) the expected drawdown in the pumped well, and (c) the radius of influence of the pumped well.

6.5.1 Work example 6.5.1 with $T = 8500$ gpd/ft².

6.5.2 A 0.4-m diameter well is pumped continuously at a rate of 5.6 l/s from an aquifer of transmissivity 108 m²/day and storativity 2×10^{-5}. How long will it take before the drawdown in the well reaches 2 m?

6.5.3 In problem 6.5.2, how long will it take before a drawdown at a distance of 400 m from the well becomes 2 m?

6.5.4 In problem 6.5.2, determine the drawdown at a distance 200 m from the well after (a) 1 hr, (b) 1 day, and (c) 1 year of pumping.

6.5.5 A pumping test was performed to determine the transmissivity and the storativity of a confined aquifer. The drawdown versus time data in an observation well 100 m away from the pumping well are given in the following table. The discharge from the pumped well was at a constant rate of 3.6 m³/min throughout the test. Determine the aquifer properties.

Time (min)	Drawdown (m)
0.5	0.12
1	0.18
2	0.25
3	0.28
4	0.31
5	0.33
8	0.38
10	0.41
20	0.47
30	0.51
50	0.56
100	0.62
200	0.69
500	0.79
1000	0.89

6.5.6 The following data were obtained in an observation well 80 m away from the pumped well. The discharge from the pumped well was at 2.5 m³/min. Using Jacob's approximation, determine the aquifer properties. Assume a confined aquifer. Also verify your solution and determine the time after which Jacob's approximation will be valid.

Time (min)	Drawdown (m)
0.5	0.16
1	0.18
2	0.24
3	0.27
4	0.47
5	0.50
7	0.57
10	0.68
20	0.84
30	0.96
50	1.06
100	1.29
200	1.46
500	1.68
1000	1.86

6.5.7 A fully penetrating production well located in a confined aquifer is pumped at a constant rate of 64 l/s. Aquifer characteristics include a transmissibility of 1240 m²/day and coefficient of storage of 4×10^{-4}. Determine the drawdown at a distance of 200 m from the pumped well after 30 days of pumping. Compare results using the Theis and the Cooper–Jacobs procedures.

6.6.1 A fully penetrating well in an unconfined aquifer with a saturated thickness of 25 ft is pumped at a rate of 144.4 ft³/min. Time-drawdown data were collected from an observation well 73 feet from the pumped well and matched to type curves (such as Figure 6.6.2). The matching of early drawdown versus time best fit the type a curve for $\eta = 0.06$, with the selected match point: ($t = 0.17$ min, $s = 0.57$ ft) and ($1/u_a = 1.0$ and $W(u_a, \eta) = 1.0$). Determine the transmissivity and specific yield.

6.6.2 For the situation in problem 6.6.1, the time-drawdown data are now matched for the later in time drawdown data (segment 3) for the $\eta = 0.06$. The late-time match resulted in the match point ($t = 13$ min, $s = 0.57$ ft) and ($1/u_y = 0.1$ and $W(u_y, \eta) = 1.0$). Determine the transmissivity and the storativity. How does the transmissivity differ from that computed in problem 6.6.1?

6.6.3 Determine the vertical hydraulic conductivity of the aquifer described in problems 6.6.1 and 6.6.2.

6.8.1 Draw the generalized flow net showing the flow lines and potential lines in the vicinity of a discharging well near a recharge boundary.

6.8.2 Draw the generalized flow net showing the flow lines and potential lines in the vicinity of a discharging well near a barrier boundary.

6.8.3 A well is pumping near a barrier boundary at a rate of 0.03 m³/s from a confined aquifer 20 m thick (see Figure 6.8.3). The hydraulic conductivity of the aquifer is 3.2×10^{-4} m/s, and its storativity is 3×10^{-5}. Determine the drawdown in the observation well after 30 hours of continuous pumping. What is the fraction of the drawdown attributable to the barrier boundary?

6.8.4 A 0.5-m diameter well is pumping near a river at an unknown rate from a confined aquifer (see Figure 6.8.4). The aquifer properties are $T = 5.0 \times 10^{-3}$ m²/s and $S = 4.0 \times 10^{-4}$. After 8 hours of pumping, the drawdown in the observation well was 0.8 m. Determine the rate of pumping and the drawdown in the pumped well. How great was the effect of the river on the drawdown in the observation well and in the pumped well?

6.8.5 A production well fully penetrating a nonleaky isotropic artesian aquifer delimited by two barrier boundaries perpendicular to each other was continuously pumped at a constant rate of 1485 gpd for a period of 4 hours. The drawdowns in the table below were observed at a distance of 300 ft in a fully penetrating observation well. Compute the coefficients of transmissivity and storage of the aquifer and the distances to each image well from the observation well.

Time (min)	Drawdown (ft)
2	0.80
3	0.92
4	1.06
5	1.17
6	1.23
7	1.32
8	1.37
9	1.43
10	1.48
20	1.88
30	2.11
40	2.34
50	2.52
60	2.70
70	2.83
80	3.00
90	3.17
100	3.30
200	4.21
300	4.43

6.9.1 Search the Internet to find the various ASTM standards that apply to groundwater modeling.

6.9.2 Develop a two- to three-page description of the MODFLOW model.

REFERENCES

Batu, V., *Aquifer Hydraulics*, Wiley Interscience, New York, 1998.

Charbeneau, R. J., *Groundwater Hydraulics and Pollutant Transport*, Prentice-Hall, Upper Saddle River, NJ, 2000.

Chow, V. T., D. R. Maidment, and L. W. Mays, *Applied Hydrology*, McGraw-Hill, New York, 1988.

Cooper, H. H., Jr., and C. E. Jacob, "A Generalized Graphical Method for Evaluating Formation Constants and Summarizing Well Field History," *Trans. Amer. Geophys. Union*, vol. 27, pp. 526–534, 1946.

Delleur, J. W. (editor-in chief), *The Handbook of Groundwater Engineering*, CRC Press, Boca Raton, FL, 1999.

Ferris, J. G., D. B. Knowles, R. H. Brown, and R. W. Stallman, "Theory of Aquifer Tests," U.S. Geological Survey Water-Supply Paper 1536-E, 1946.

Freeze, R. A., and J. A. Cherry, *Groundwater*, Prentice-Hall, Englewood Cliffs, NJ, 1979.

Gehm, H. W., and J. I. Bregman, editors, *Handbook of Water Resources and Pollution Control*, Van Nostrand Reinhold Company, New York, 1976.

Gleick, P. H., *Water in Crisis*, Oxford University Press, Oxford, 1993.

Hantush, M. S., "Hydraulics of Wells," in *Advances in Hydroscience*, Academic Press, New York, 1964.

Hantush, M. S., and C. E. Jacob, "Non-Steady Radial Flow in an Infinite Leaky Aquifer," *Trans. Am. Geophys. Union*, vol. 36, no. 1, 1955.

Harbaugh, A. W., "MODFLOW-2005. The U.S. Geological Survey Modular Ground-Water Model—the Ground-Water Flow Process," *U.S. Geological Survey Techniques and Methods, B*ook 6 Modeling Techniques, Section A Groundwater, U.S. Geological Survey, 2006.

Ingersoll, L. R., O. J. Zobel, and A. C. Ingersoll, *Heat Conduction with Engineering and Geological Applications*, McGraw-Hill, New York, 1948.

Klemt, W. B., T. R. Knowles, G. R. Elder, and T. Sich,"Groundwater Resources and Model Applications for the Edwards (Balcones Fault Zone) Aquifer in the San Antonio Region, Texas," Report 239, Texas Department of Water Resources, Austin, TX, Oct. 1979.

Lohman. S. W., et al. "Definition of Selected Groundwater Terms-Revision and Conceptual Refinements," U.S. Geological Survey Water Supply Paper No. 1988 1972.

Marsh, W. M., *Earthscape: A Physical Geography*, John Wiley & Sons, New York, 1987.

Meinzer, O. E.,"The Occurrence of Groundwater in the United States," U.S. Geological Survey Water Supply Paper 489, 1923.

Neuman, S. P., "Analysis of Pumping Test Data from Anisotropic Unconfined Aquifers Considering Delayed Gravity Response," *Water Resources Research*, vol. 11, pp. 329–342, 1975.

Peaceman, D. W., and H. H. Rachford, Jr., "The Numerical Solutions of Parabolic and Elliptic Differential Equations," *Journal Soc. Industrial and Applied Mathematics*, vol. 3, pp. 28–41, 1955.

Prickett, T. A., "Type-Curve Solution to Aquifer Table Tests under Water-Table Conditions," *Groundwater*, vol. 3, no. 3, 1965.

Prickett, T. A., and C. G. Lonnquist, "Selected Digital Computer Techniques for Groundwater Resources Evaluation," *Bulletin No. 55*, Illinois State Water Survey, Urbana, IL, 1971.

Rorabaugh, M. I., "Ground-Water Resources of the Northeastern Part of the Louisville Area, Kentucky," U.S. Geological Survey Water Supply Paper 1360-B, 1956.

Texas Water Development Board, "GWS1M—Groundwater Simulation Program, Program Document and User's Manual," UMS7405, Austin, TX, 1974.

Theis, C. V., "The Relation Between the Lowering of Piezometric Surface and the Rate and Duration of Discharge of a Well Using Groundwater Storage," *Transactions American Geophysical Union*, vol. 2, pp. 519–524, 1935.

Todd, D. K., and J. Bear, "Seepage Through Layered Anisotropic Porous Media," *Journal of the Hydraulics Division, ASCE*, vol. 87, no. HY 3, pp. 31–57. 1961.

Todd, D. K. and L. W. Mays, *Groundwater Hydrology*, 3rd edition, John Wiley & Sons, New York, 2005.

Trescott, P. C., G. F. Pinder, and S. P. Larson, "Finite-Difference Model for Aquifer Simulation in Two Dimensions with Results of Numerical Experiments," in *U.S. Geological Survey Techniques of Water Resources Investigations*, Book 7, Cl, U.S. Geological Survey, Reston, VA, 1976.

U.S. Bureau of Reclamation, *Groundwater Manual*, U.S. Government Printing Office, Denver, CO, 1981.

U.S. Soil Conservation Service, *Soil Survey Manual*, Handbook no. 18, U.S. Department of Agriculture, 1951.

Walton, W. C, *Groundwater Resource Evaluation*, McGraw-Hill, New York, 1970.

Walton, W. E.,"Leaky Artesian Conditions in Illinois," Illinois State Water Survey, Dept. of Invest. Urbana, IL 39, 1960.

Wanakule, N.,"Optimal Groundwater Management Models for the Barton Springs-Edwards Aquifer," Edwards Aquifer Research and Data Center, San Marcos, TX, 1989.

Wang, H. F., and M. P. Anderson, *Introduction to Groundwater Modeling: Finite Difference and Finite Element Models*, W. H. Freeman, San Francisco, CA, 1982.

Wenzel, L. K., "Methods for Determining Permeability of Water-Bearing Materials with Special Reference to Discharging Well Methods," U.S. Geological Survey Water-Supply Paper 887, pp. 192, 1942.

Chapter 7

Hydrologic Processes

7.1 INTRODUCTION TO HYDROLOGY

7.1.1 What is Hydrology?

The U.S. National Research Council (1991) presented the following definition of hydrology:

Hydrology is the science that treats the waters of the Earth, their occurrence, circulation, and distribution, their chemical and physical properties, and their reaction with the environment, including the relation to living things. The domain of hydrology embraces the full life history of water on Earth.

For purposes of this book we are interested in the engineering aspects of hydrology, or what we might call *engineering hydrology*. From this point of view we are mainly concerned with quantifying amounts of water at various locations (spatially) as a function of time (temporally) for surface water applications. In other words, we are concerned with solving engineering problems using hydrologic principles. This chapter is not concerned with the chemical properties of water and their relation to living things.

Books on hydrology include: Bedient and Huber (1992); Bras (1990); Chow (1964); Chow, Maidment, and Mays (1988); Gupta (1989); Maidment (1993); McCuen (1998); Ponce (1989); Singh (1992); Viessman and Lewis (1996); and Wanielista, Kersten, and Eaglin (1997).

7.1.2 The Hydrologic Cycle

The central focus of hydrology is the *hydrologic cycle*, consisting of the continuous processes shown in Figure 7.1.1. Water *evaporates* from the oceans and land surfaces to become water vapor that is carried over the earth by atmospheric circulation. The *water vapor* condenses and *precipitates* on the land and oceans. The precipitated water may be *intercepted* by vegetation, become overland flow over the ground surface, *infiltrate* into the ground, flow through the soil as *subsurface flow,* and discharge as *surface runoff.* Evaporation from the land surface comprises evaporation directly from soil and vegetation surfaces, and *transpiration* through plant leaves. Collectively these processes are called *evapotranspiration*. Infiltrated water may percolate deeper to recharge groundwater and later become *springflow* or *seepage* into streams also to become streamflow.

The hydrologic cycle can be viewed on a global scale as shown in Figure 7.1.2. As discussed by the U.S. National Research Council (1991): "As a global average, only about 57 percent of the precipitation that falls on the land (P_l) returns directly to the atmosphere (E_l) without reaching the ocean. The remainder is runoff (R), which find its way to the sea primarily by rivers but also through subsurface (groundwater) movement and by the calving of icebergs from glaciers and ice shelves

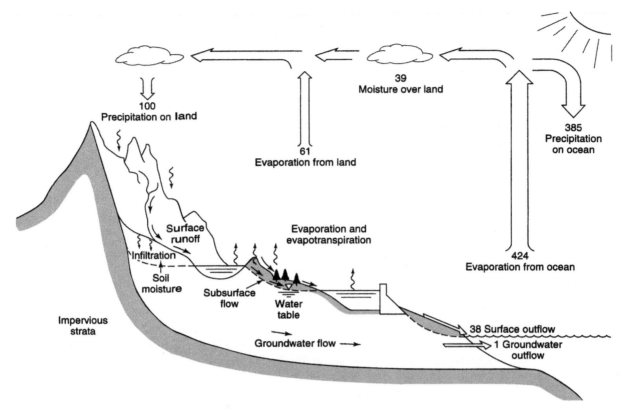

Figure 7.1.1 Hydrologic cycle with global annual average water balance given in units relative to a value of 100 for the rate of precipitation on land (from Chow et al. (1988)).

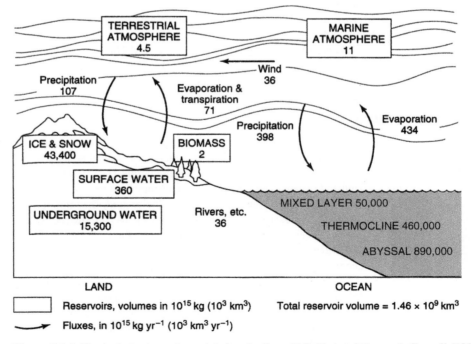

Figure 7.1.2 The hydrologic cycle at global scale (from U.S. National Research Council (1991)).

(*W*). In this gravitationally powered runoff process, the water may spend time in one or more natural storage reservoirs such as snow, glaciers, ice sheets, lakes, streams, soils and sediments, vegetation, and rock. Evaporation from these reservoirs short-circuits the global hydrologic cycle into subcycles with a broad spectrum of scale. The runoff is perhaps the best-known element of the global hydrologic cycle, but even this is subject to significant uncertainty."

7.1.3   Hydrologic Systems

Chow, Maidment, and Mays (1988) defined a *hydrologic system* as a structure or volume in space, surrounded by a boundary, that accepts water and other inputs, operates on them internally, and produces them as outputs. The structure (for surface or subsurface flow) or volume in space (for atmospheric moisture flow) is the totality of the flow paths through which the water may pass as throughput from the point it enters the system to the point it leaves. The boundary is a continuous surface defined in three dimensions enclosing the volume or structure. A *working medium* enters the system as input, interacts with the structure and other media, and leaves as output. Physical, chemical, and biological processes operate on the working media within the system; the most common working media involved in hydrologic analysis are water, air, and heat energy.

The *global hydrologic cycle* can be represented as a system containing three subsystems: the atmospheric water system, the surface water system, and the subsurface water system, as shown in Figure 7.1.3. Another example is the storm-rainfall-runoff process on a watershed, which can be represented as a hydrologic system. The input is rainfall distributed in time and space over the watershed, and the output is streamflow at the watershed outlet. The boundary is defined by the watershed divide and extends vertically upward and downward to horizontal planes.

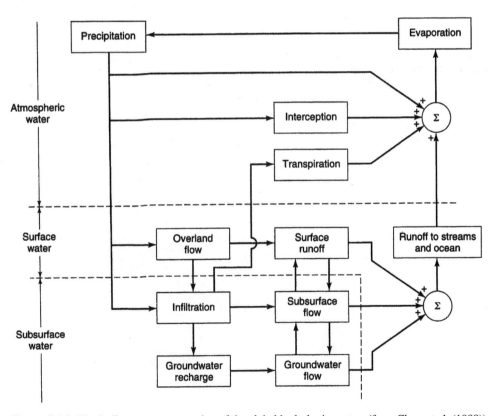

Figure 7.1.3 Block-diagram representation of the global hydrologic system (from Chow et al. (1988)).

Drainage basins, catchments, and *watersheds* are three synonymous terms that refer to the topographic area that collects and discharges surface streamflow through one outlet or mouth. Catchments are typically referred to as small drainage basins, but no specific area limits have been established. Figure 7.1.4 illustrates the drainage basin divide, watershed divide, or catchment divide, which is the line dividing land whose drainage flows toward the given stream from land whose drainage flows away from that stream. Think of drainage basin sizes ranging from the Mississippi

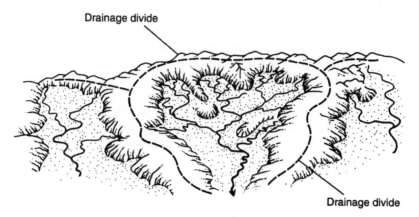

Figure 7.1.4 Schematic diagram of a drainage basin. The high terrain on the perimeter is the drainage divide (from Marsh (1987)).

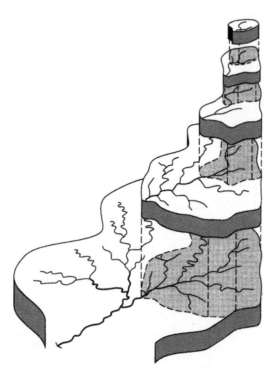

Figure 7.1.5 Illustration of the nested hierarchy of lower-order basins within a large drainage basin (from Marsh (1987)).

River drainage basin to small urban drainage basins in your local community or some small valley in the countryside near you.

As shown in Figure 7.1.5, drainage basins can be pictured in a pyramidal fashion as the runoffs from smaller basins (subsystems) combine to form larger basins (subsystem in system), and the runoffs from these basins in turn combine to form even larger basins, and so on. Marsh (1987) refers to this mode of organization as a *hierarchy* or *nested hierarchy*, as each set of smaller basins is set inside the next layer. A similar concept is that streams that drain small basins combine to form larger streams, and so on.

Figures 7.1.6–7.1.10 illustrate the hierarchy of the Friends Creek watershed located in the Lake Decatur watershed (drainage area upstream of Decatur, Illinois, on the Sangamon River with a drainage area of 925 mi^2 or 2396 km^2). Obviously, the Friends Creek watershed can be subdivided into much smaller watersheds. Figure 7.1.8 illustrates the Illinois River basin (29,000 mi^2) with the Sangamon River. Figure 7.1.9 illustrates the upper Mississippi River basin (excluding the Missouri River) with the Illinois River. Figure 7.1.10 illustrates the entire Mississippi River basin (1.15 million mi^2). This is the largest river basin in the United States, draining about 40 percent of the country. The main stem of the river is about 2400 miles long.

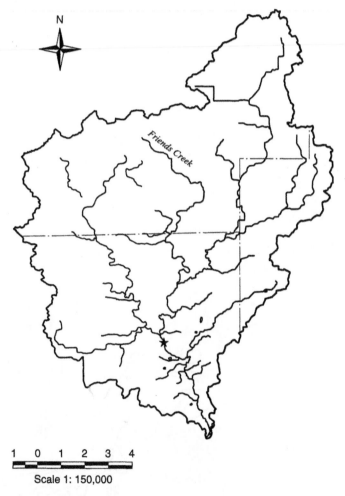

Figure 7.1.6 Friends Creek watershed, subwatershed of the Lake Decatur watershed. (Courtesy of the Illinois State Water Survey, compiled by Erin Hessler Bauer.)

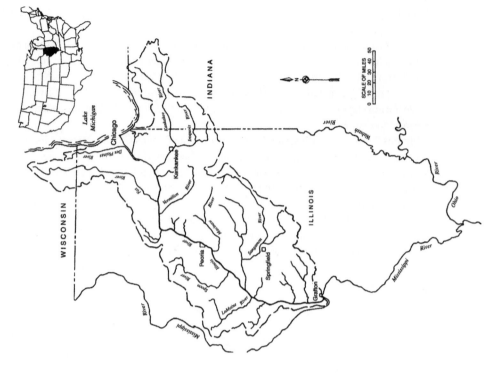

Figure 7.1.8 The Illinois River Basin showing the Sangamon River (from Bhowmik (1998)).

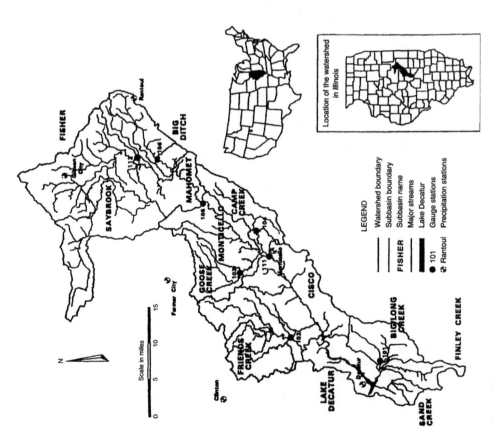

Figure 7.1.7 Lake Decatur watershed in the upper Sangamon River Basin (upstream of Decatur, Illinois). The location of Friends Creek watershed (Figure 7.1.6) is shown (from Demissee and Keefer (1998)).

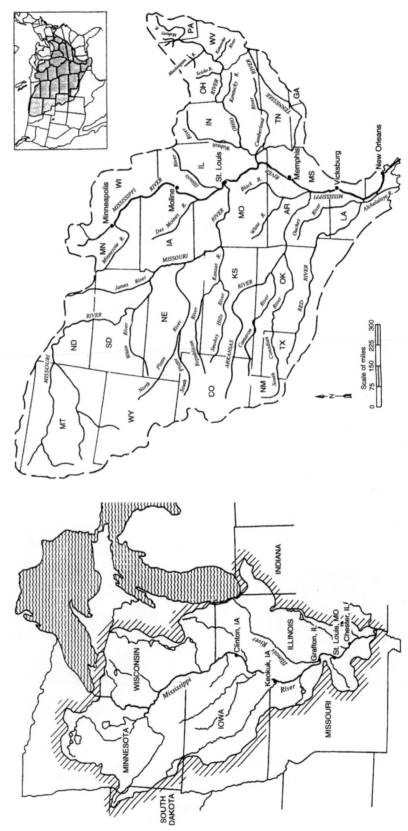

Figure 7.1.10 The Mississippi River Basin in the United States (from Bhowmik (1998)).

Figure 7.1.9 The Upper Mississippi River excluding the Missouri River. The location of the Illinois River (Figure 7.1.8) is shown (from Bhowmik (1998)).

7.1.4 Atmospheric and Ocean Circulation

Atmospheric circulation on Earth is a very complex process that is influenced by many factors. Major influences are differences in heating between low and high altitudes, the earth's rotation, and heat and pressure differences associated with land and water. The general circulation of the atmosphere is due to latitudinal differences in solar heating of the earth's surface and inclination of its axis, distribution of land and water, mechanics of the atmospheric fluid flow, and the Coriolis effect. In general, the atmospheric circulation is thermal in origin and is related to the earth's rotation and global pressure distribution. If the earth were a nonrotating sphere, atmospheric circulation would appear as in Figure 7.1.11. Air would rise near the equator and travel in the upper atmosphere toward the poles, then cool, descend into the lower atmosphere, and return toward the equator. This is called *Hadley circulation.*

The rotation of the earth from west to east changes the circulation pattern. As a ring of air about the earth's axis moves toward the poles, its radius decreases. In order to maintain angular momentum, the velocity of air increases with respect to the land surface, thus producing a westerly air flow. The converse is true for a ring of air moving toward the equator—it forms an easterly air flow. The effect producing these changes in wind direction and velocity is known as the *Coriolis force.*

The idealized pattern of atmospheric circulation has three cells in each hemisphere, as shown in Figure 7.1.12. In the *tropical cell,* heated air ascends at the equator, proceeds toward the poles at upper levels, loses heat, and descends toward the ground at latitude 30°. Near the ground it branches, one branch moving toward the equator and the other toward the pole. In the *polar cell,* air rises at 60° latitude and flows toward the poles at upper levels, then cools and flows back to 60° near the earth's surface. The *middle cell* is driven frictionally by the other two; its surface flows toward the pole, producing prevailing westerly air flow in the midlatitudes.

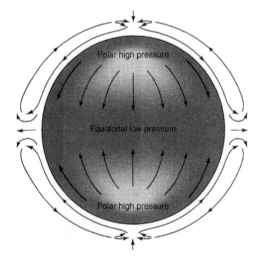

Figure 7.1.11 Atmospheric circulation pattern that would develop on a nonrotating planet. The equatorial belt would heat intensively and would produce low pressure, which would in turn set into motion a gigantic convection system. Each side of the system would span one hemisphere (from Marsh (1987)).

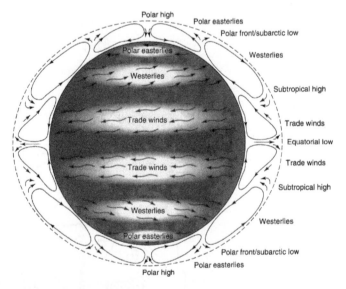

Figure 7.1.12 Idealized circulation of the atmosphere at the earth's surface, showing the principal areas of pressure and belts of winds (from Marsh (1987)).

The uneven distribution of ocean and land on the earth's surface, coupled with their different thermal properties, creates additional spatial variation in atmospheric circulation. The annual shifting of the thermal equator due to the earth's revolution around the sun causes a corresponding oscillation of the three-cell circulation pattern. With a larger oscillation, exchanges of air between adjacent cells can be more frequent and complete, possibly resulting in many flood years. Also, monsoons may advance deeper into such countries as India and Australia. With a smaller oscillation, intense high pressure may build up around 30° latitude, thus creating extended dry periods. Since the atmospheric circulation is very complicated, only the general pattern can be identified.

The atmosphere is divided vertically into various zones. The atmospheric circulation described above occurs in the *troposphere*, which ranges in height from about 8 km at the poles to 16 km at the equator. The temperature in the troposphere decreases with altitude at a rate varying with the moisture content of the atmosphere. For dry air, the rate of decrease is called the *dry adiabatic lapse rate* and is approximately 9.8°C/km. The *saturated adiabatic lapse rate* is less, about 6.5° C/km, because some of the vapor in the air condenses as it rises and cools, releasing heat into the surrounding air. These are average figures for lapse rates that can vary considerably with altitude. The *tropopause* separates the troposphere from the *stratosphere* above. Near the tropopause, sharp changes in temperature and pressure produce strong narrow air currents known as *jet streams*, with velocities ranging from 15 to 50 m/s (30 to 100 m/h). They flow for thousands of kilometers and have an important influence on air-mass movement.

The oceans exert an important control on global climate. Because water bodies have a high volumetric heat capacity, the oceans are able to retain great quantities of heat. Through wave and current circulation, the oceans redistribute heat to considerable depths and even large areas of the oceans. Redistribution is east-west or west-east and is also across the midaltitudes from the tropics to the subarctic, enhancing the overall poleward beat transfer in the atmosphere. Waves are predominately generated by wind. Ocean circulation is illustrated in Figure 7.1.13.

Oceans have a significant effect on the atmosphere; however, an exact understanding of the relationships and mechanisms involved is not known. The correlation between ocean temperatures and weather trends and midlatitude events has not been solved. One trend is the growth and decline of a warm body of water in the equatorial zone of the eastern Pacific Ocean, referred to as

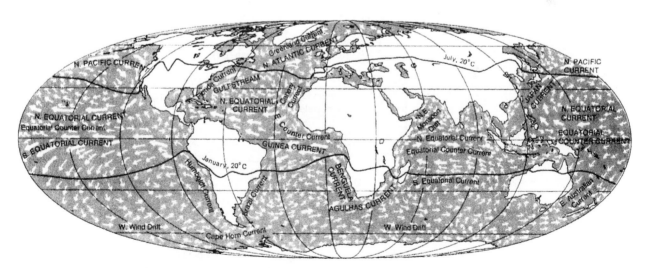

Figure 7.1.13 The actual circulation of the oceans. Major currents are shown with heavy arrows (from Marsh (1987)).

E1 Niño (meaning "The Infant" in Spanish, alluding to the Christ Child, because the effect typically begins around Christmas). The warm body of water develops and expands every five years or so off the coast of Peru, initiated by changes in atmospheric pressure resulting in a decline of the easterly trade winds. This reduction in wind reduces resistance, causing the eastwardly equatorial countercurrent to rise. As E1 Niño builds up, the warm body of water flows out into the Pacific and along the tropical west coast of the Americas, displacing the colder water of the California and Humboldt currents. One of the interesting effects of this weather variation is the South Oscillation, which changes precipitation patterns—resulting in drier conditions in areas of normally little precipitation.

7.1.5   Hydrologic Budget

A *hydrologic budget, water budget,* or *water balance* is a measurement of continuity of the flow of water, which holds true for any time interval and applies to any size area ranging from local-scale areas to regional-scale areas or from any drainage area to the earth as a whole. The hydrologists usually must consider an open system, for which the quantification of the hydrologic cycle for that system becomes a mass balance equation in which the change of storage of water (dS/dt) with respect to time within that system is equal to the inputs (I) to the system minus the outputs (O) from the system.

Considering the open system in Figure 7.1.14, the water balance equation can be expressed for the surface water system and the groundwater system in units of volume per unit time separately or, for a given time period and area, in depth.

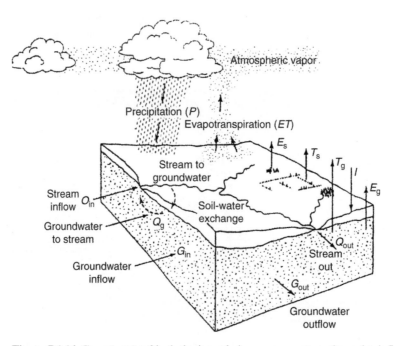

Figure 7.1.14 Components of hydrologic cycle in an open system: the major inflows and outflows of water from a parcel of land (from Marsh and Dozier (1986)) (Courtesy John Wiley & Sons, Inc.)

Surface Water System Hydrologic Budget

$$P + Q_{\text{in}} - Q_{\text{out}} + Q_g - E_s - T_s - I = \Delta S_s \tag{7.1.1}$$

where P is the precipitation, Q_{in} is the surface water flow into the system, Q_{out} is the surface water flow out of the system, Q_g is the groundwater flow into the stream, E_s is the surface evaporation, T_s is the transpiration, I is the infiltration, and ΔS_s is the change in water storage of the surface water system.

Groundwater System Hydrologic Budget

$$I + G_{\text{in}} - G_{\text{out}} - Q_g - E_g - T_g = \Delta S_g \tag{7.1.2}$$

where G_{in} is the groundwater flow into the system, G_{out} is the groundwater flow out of the system, and ΔS_g is the change in groundwater storage. The evaporation, E_g, and the transpiration, T_g, can be significant if the water table is near the ground surface.

System Hydrologic Budget

The system hydrologic budget is developed by adding the above two budgets together:

$$P - (Q_{\text{out}} - Q_{\text{in}}) - (E_s + E_g) - (T_s + T_g) - (G_{\text{out}} - G_{\text{in}}) = \Delta(S_s + S_g) \tag{7.1.3}$$

Using net mass exchanges, the above system hydrologic budget can be expressed as

$$P - Q - G - E - T = \Delta S \tag{7.1.4}$$

EXAMPLE 7.1.1

During January 1996, the water-budget terms for Lake Annie in Florida included precipitation (P) of 1.9 in, evaporation (E) of 1.5 in, surface water inflow (Q_{in}) of 0 in, surface outflow (Q_{out}) of 17.4 in, and change in lake volume (ΔS) of 0 in. Determine the net groundwater flow for January 1996 (the groundwater inflow minus the groundwater outflow).

SOLUTION

The water budget equation to define the net groundwater flow for the lake is

$$G = \Delta S - P + E - Q_{\text{in}} + Q_{\text{out}} = 0 - 1.9 + 1.5 - 0 + 17.4 = 17 \text{ in for January 1996}$$

7.2   PRECIPITATION (RAINFALL)

7.2.1   Precipitation Formation and Types

Even though precipitation includes rainfall, snowfall, hail, and sleet, our concern in this book will relate almost entirely to rainfall. The formation of water droplets in clouds is illustrated in Figure 7.2.1. *Condensation* takes place in the atmosphere on *condensation nuclei*, which are very small particles ($10^{-3} - 10\,\mu\text{m}$) in the atmosphere that are composed of dust or salt. These particles are called *aerosols*. During the initial occurrence of condensation, the droplets or ice particles are very small and are kept aloft by motion of the air molecules. Once droplets are formed they also act as condensation nuclei. These droplets tend to repel one another, but in the presence of an electric field in the atmosphere they attract one another and are heavy enough

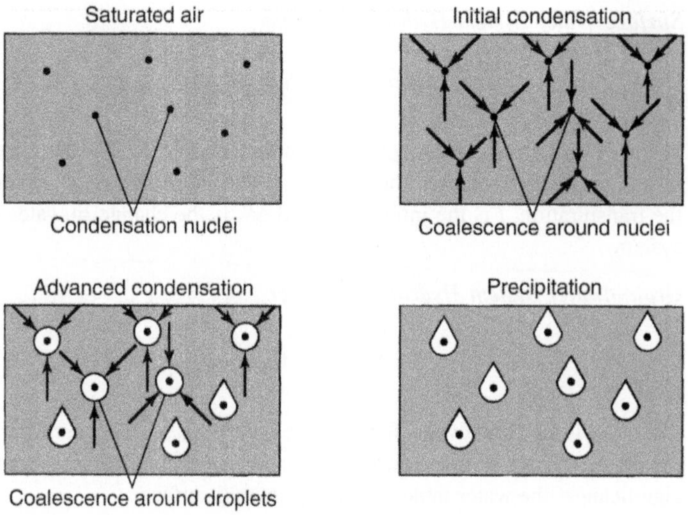

Saturated air

Condensation nuclei

Initial condensation

Coalescence around nuclei

Advanced condensation

Coalescence around droplets

Precipitation

Figure 7.2.1 Precipitation formation. Water droplets in clouds are formed by nucleation of vapor on aerosols, then go through many condensation-evaporation cycles as they circulate in the cloud, until they aggregate into large enough drops to fall through the cloud base (from Marsh (1987)).

($\sim$0.1 mm) to fall through the atmosphere. Some of the droplets evaporate in the atmosphere, some of the droplets decrease in size by evaporation, and some of the droplets increase in size by impact and aggregation.

Basically, the formation of precipitation requires lifting of an air mass in the atmosphere; it then cools and some of its moisture condenses. There are three main mechanisms of air mass lifting: *frontal lifting, orographic lifting*, and *convective lifting*. Frontal lifting (Figure 7.2.2) occurs when warm air is lifted over cooler air by frontal passage, orographic lifting (Figure 7.2.3) occurs when an air mass rises over a mountain range, and convective lifting (Figure 7.2.4) occurs when air is drawn upward by convective action such as a thunderstorm cell.

7.2.2 Rainfall Variability

In order to determine the runoff from a watershed and the resulting stream flow, precipitation is one of the primary inputs. Rainfall varies in space and time as a result of the general pattern of atmospheric circulation and local factors. Figure 7.2.5 shows the mean annual precipitation in the United States, and Figure 7.2.6 shows the normal monthly distribution of precipitation in the United States. Figure 7.2.7 shows the mean annual precipitation of the world.

Rainstorms can vary significantly in space and time. *Rainfall hyetographs* are plots of rainfall depth or intensity as a function of time. Figure 7.2.8*a* shows examples of two rainfall hyetographs. Cumulative rainfall hyetographs (rainfall mass curve) can be developed as shown in Figure 7.2.8*b*.

Isohyets (contours of constant rainfall) can be drawn to develop isohyetal maps of rainfall depth. *Isohyetal maps* are an interpolation of rainfall data recorded at gauged points. An example is shown in Figure 7.2.9 for the Upper Mississippi River Basin storm of January through July 1993. On a much smaller scale, shown in Figure 7.2.10, is the isohyetal map of the May 24–25, 1981 storm in Austin, Texas.

Figure 7.2.11 illustrates the three methods for determining areal average rainfall using rainfall gauge data. These are the *arithmetic-mean method*, the *Thiessen method*, and the *isohyetal method*.

Figure 7.2.6 Normal monthly distribution of precipitation in the United States in inches (1 in = 254 mm) (from U.S. Environmental Data Services (1968)).

illustrates the rate (as a function of time) at which water flows or is added to storage for each of the processes. At the beginning of a storm, a large proportion of rainfall contributes to *surface storage*, and as water infiltrates, the *soil moisture storage* begins. Both *retention storage* and *detention storage* prevail. Retention storage is held for a long period of time and is depleted by evaporation, whereas detention storage is over a short time and is depleted by flow from the storage location.

7.2.4   Design Storms

The determination of flow rates in streams is one of the central tasks of surface water hydrology. For most engineering applications, these flow rates are determined for specified events that are typically extreme events. A major assumption in these analyses is that a certain return period storm results in the same return period flow rates from a watershed. The return period of an event, whether the event is a storm or a flow rate, is the expected value or the average value measured over a very large number of occurrences. In other words, the return period refers to the time interval for which an event will occur once on the average over a very large number of occurrences.

Hershfield (1961), in a publication often referred to as TP-40, presented isohyetal maps of design rainfall depths for the United States for durations from 30 minutes to 24 hours and return periods from 1 to 100 years. The values of rainfall in these isohyetal maps are point precipitation values, which is precipitation occurring at a single point in space (as opposed to areal

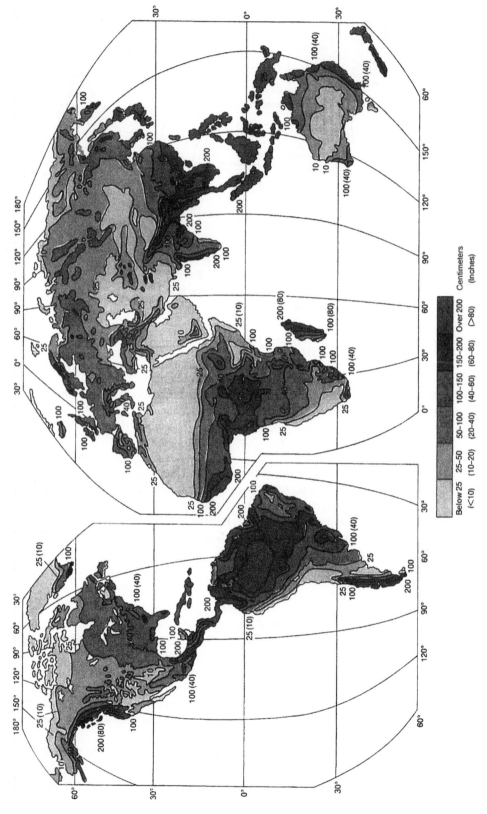

Figure 7.2.7 Average annual precipitation for the world's land areas, except Antarctica (from Marsh (1987)).

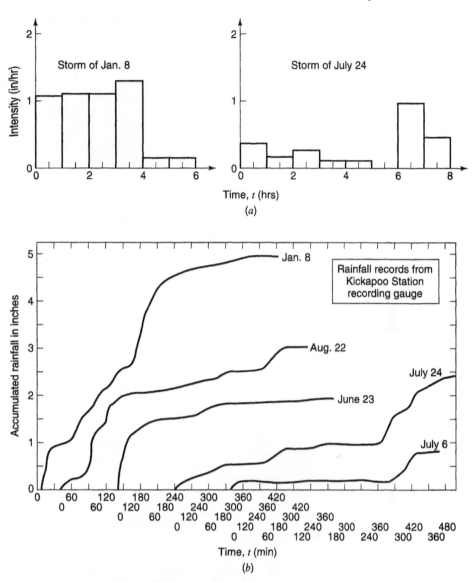

Figure 7.2.8 (*a*) Rainfall hyetographs for Kickapoo Station; (*b*) Mass rainfall curves (from Masch (1984)).

precipitation, which is over a larger area). Figure 7.2.13 is the isohyetal map for 100-year 24-hour rainfall. A later publication, U.S. Weather Bureau (1964), included maps for durations for 2 to 10 days, in what is referred to as TP-49. Miller et al. (1973) presented isohyetal maps for 6- and 24-hour durations for the 11 mountainous states in the western United States, which supersede the corresponding maps in TP-40.

Frederick et al. (1977), in a publication commonly referred to as HYDRO-35, presented isohyetal maps for events having durations from 5 to 60 minutes. The maps of precipitation depths for 5-, 15-, and 60-minutes durations and return periods of 2 and 100 years for the 37 eastern states are presented in Figures 7.2.14 *a–f*. Depths for a return period are obtained by interpolation from the 5-, 15-, and

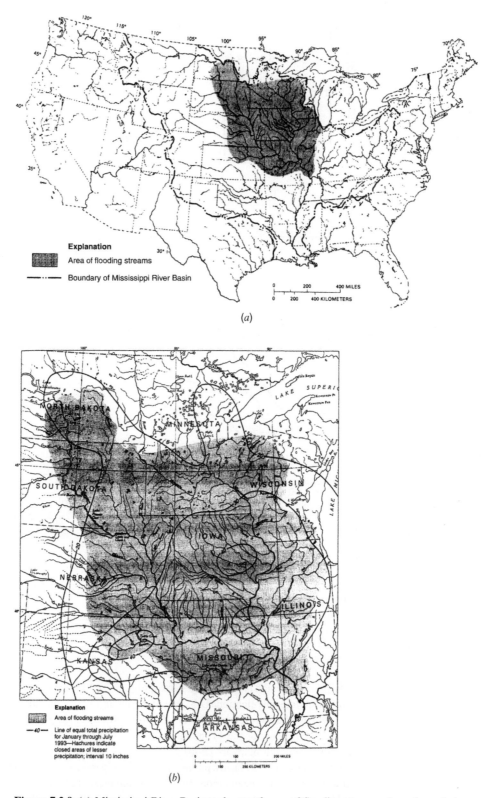

Figure 7.2.9 (*a*) Mississippi River Basin and general area of flooding streams, June through August 1993 (from Parrett et al. (1993)). (*b*) Areal distribution of total precipitation in the area of flooding in the upper Mississippi River Basin, January through May 1993 (from Parrett et al. (1993)).

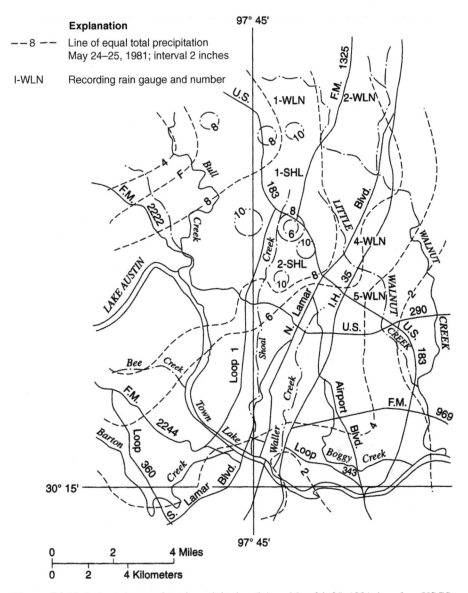

Figure 7.2.10 Isohyetal map of total precipitation (in) on May 24–25, 1981, based on USGS measurements, the City of Austin network, and unofficial precipitation reports (from Moore et al. (1982)).

60-minute data for the same return period:

$$P_{10min} = 0.41\, P_{5min} + 0.59\, P_{15min} \qquad (7.2.1a)$$

$$P_{30min} = 0.51\, P_{15min} + 0.49\, P_{60min} \qquad (7.2.1b)$$

To consider return periods other than 2 or 100 years, the following interpolation equation is used:

$$P_{T\,yr} = aP_{2yr} + bP_{100yr} \qquad (7.2.2)$$

where the coefficients a and b are found in Table 7.2.1.

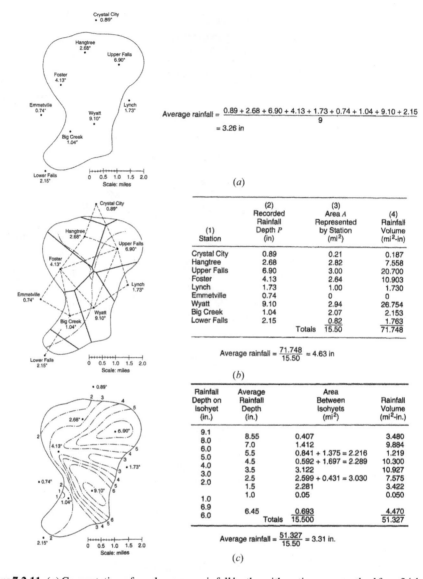

Average rainfall $= \dfrac{0.89 + 2.68 + 6.90 + 4.13 + 1.73 + 0.74 + 1.04 + 9.10 + 2.15}{9}$

$= 3.26$ in

(a)

(1) Station	(2) Recorded Rainfall Depth P (in)	(3) Area A Represented by Station (mi²)	(4) Rainfall Volume (mi²-in)
Crystal City	0.89	0.21	0.187
Hangtree	2.68	2.82	7.558
Upper Falls	6.90	3.00	20.700
Foster	4.13	2.64	10.903
Lynch	1.73	1.00	1.730
Emmetville	0.74	0	0
Wyatt	9.10	2.94	26.754
Big Creek	1.04	2.07	2.153
Lower Falls	2.15	0.82	1.763
Totals		15.50	71.748

Average rainfall $= \dfrac{71.748}{15.50} = 4.63$ in

(b)

Rainfall Depth on Isohyet (in.)	Average Rainfall Depth (in.)	Area Between Isohyets (mi²)	Rainfall Volume (mi²-in.)
9.1	8.55	0.407	3.480
8.0	7.0	1.412	9.884
6.0	5.5	0.841 + 1.375 = 2.216	1.219
5.0	4.5	0.592 + 1.697 = 2.289	10.300
4.0	3.5	3.122	10.927
3.0	2.5	2.599 + 0.431 = 3.030	7.575
2.0	1.5	2.281	3.422
1.0	1.0	0.05	0.050
6.9	6.45	0.693	4.470
6.0			
	Totals	15.500	51.327

Average rainfall $= \dfrac{51.327}{15.50} = 3.31$ in.

(c)

Figure 7.2.11 (*a*) Computation of areal average rainfall by the arithmetic-mean method for a 24-hr storm. This is the simplest method of determining areal average rainfall. It involves averaging the rainfall depths recorded at a number of gauges. This method is satisfactory if the gauges are uniformly distributed over the area and the individual gauge measurements do not vary greatly about the mean (after Roberson et al. (1998)); (*b*) Computation of areal average rainfall by the Thiessen method for 24-hr storm. This method assumes that at any point in the watershed the rainfall is the same as that at the nearest gauge, so the depth recorded at a given gauge is applied out to a distance halfway to the next station in any direction. The relative weights for each gauge are determined from the corresponding areas of application in a *Thiessen polygon* network, the boundaries of the polygons being formed by the perpendicular bisectors of the lines joining adjacent gauges for J gauges; the area within the watershed assigned to each is A_j and P_j is the rainfall recorded at the jth gauge. The areal average precipitation for the watershed is computed by dividing the total rainfall volume by the total watershed area, as shown in the table (after Roberson et al. (1998)). (*c*) Computation of areal average rainfall by the isohyetal method for 24-hr storm. This method connects isoyets, using observed depths at rain gauges and interpolation between adjacent gauges. Where there is a dense network of rain gauges, isohyetal maps can be constructed using computer programs for automated contouring. Once the isohyetal map is constructed, the area A_j between each pair of isohyets within the watershed, is measured and multiplied by the average P_j of the rainfall depths of the two boundary isohyets to compute the areal average precipitation (after Roberson et al. (1998)).

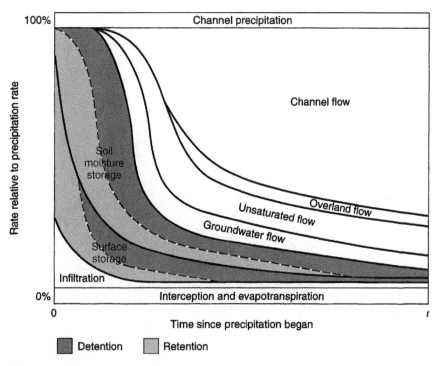

Figure 7.2.12 Schematic illustration of the disposal of precipitation during a storm on a watershed (from Chow et al. (1988)).

NOAA–Atlas 14

The new NOAA Atlas 14, Precipitation – Frequency Atlas of the United States, replaces the use of the old NOAA Atlas in the semi-arid region of the southwestern U.S., and replaces the use of the Hydro-35 and TP 40 in the Ohio River Valley including several surrounding states. The National Weather Service (NWS) Hydrometeorological Design Studies Center (HDSC) has issued the web page (http://hdsc.nws.noaa.gov/hdsc/pfds/index.html), which lists publications that should be used for each state in the U.S. and the time periods for which each should be used (5 min–60 min, 1 hr–24 hr, and 2-day–10-day periods). The two areas of the U.S. that are most impacted by the new NOAA Atlas 14 are the semi-arid southwest and the Ohio River Valley.

Semi-arid Southwest: NOAA-Atlas 2 is no longer valid for the semi-arid southwest, which includes: Arizona, Nevada, New Mexico, Utah, and Southeast California. For these states, NOAA-Atlas 2 (Volumes 4, 6, 7, 8, and part of 11) has been replaced by NOAA Atlas 14, Volume 1 which is available online: http://www.nws.noaa.gov/oh/hdsc/currentpf.htm

Ohio River Valley: Hydro-35 and Tech Paper No. 40 are no longer valid for most states in the Ohio River Valley and surrounding states, which includes: Delaware, Illinois, Indiana, Kentucky, Maryland, New Jersey, North Carolina, Ohio, Pennsylvania, South Carolina, Tennessee, Virginia, West Virginia, and Washington, DC. For these states, Hydro-35 and Tech Paper-40 have been replaced by NOAA Atlas 14, Volume 2, which is available online: http://www.nws.noaa.gov/oh/hdsc/currentpf.htm

Example point precipitation frequency estimates for two locations are presented in Figure 7.2.15. These are Chicago, Illinois and Phoenix, Arizona.

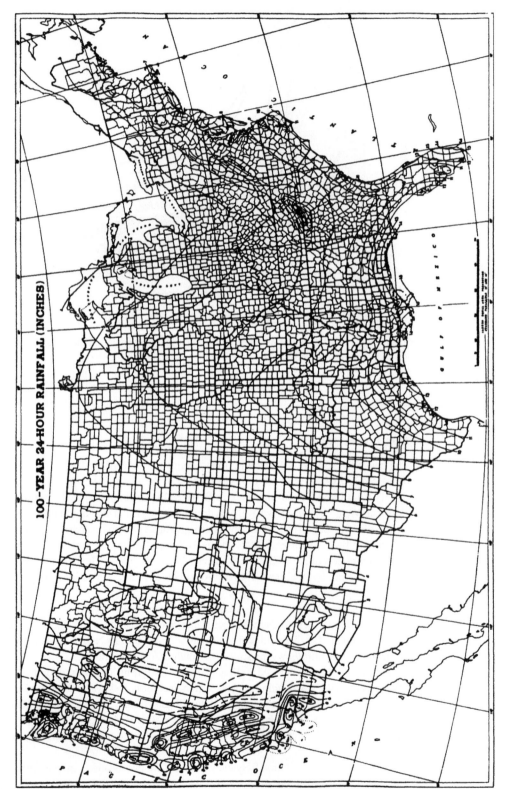

100-YEAR 24-HOUR RAINFALL (INCHES)

Figure 7.2.13 The 100-year 24-hr rainfall (in) the United States (from Hershfield (1961)).

248

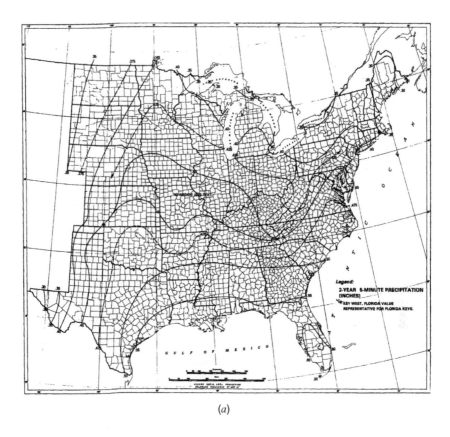

(a)

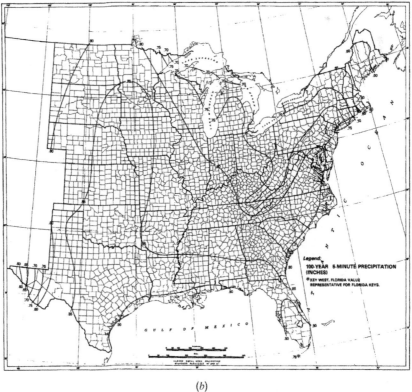

Figure 7.2.14 (*a*) 2-year 5-min precipitation (in) (from Frederick, Meyers, and Auciello (1977)). (*b*) 100-year 5-min precipitation (in) (from Frederick, Meyers, and Auciello (1977)). (*c*) 2-year 15-min precipitation (in) (from Frederick, Meyers, and Auciello (1977)). (*d*) 100-year 15-min precipitation (in) (from Frederick, Meyers, and Auciello (1977)). (*e*) 2-year 60-min precipitation (in) (from Frederick, Meyers, and Auciello (1977)). (*f*) 100-year 60-min precipitation (in) (from Frederick, Meyers, and Auciello (1977)).

(b)

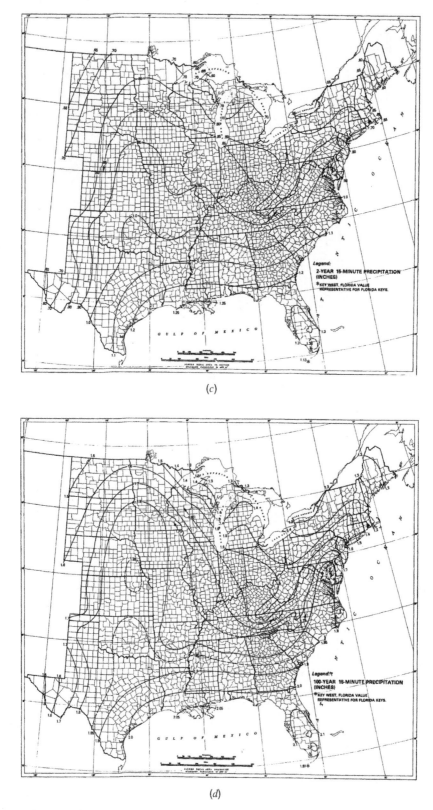

(c)

(d)

Figure 7.2.14 (*Continued*).

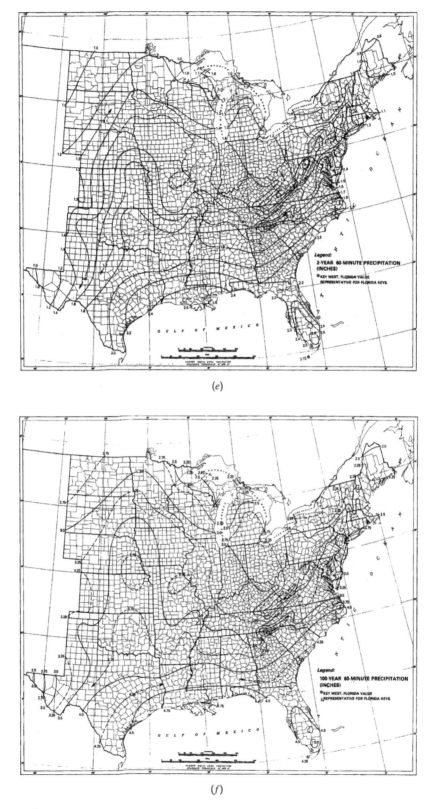

(e)

(f)

Figure 7.2.14 (*Continued*)

Partial duration based Point Precipitation Frequency Estimates - Version: 3
41.820 N 87.67 W 593 ft

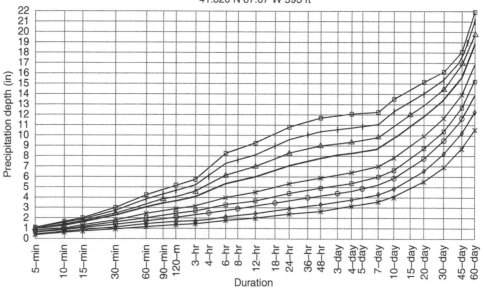

Mon Feb 02 21:45:19 2009

Average recurrence interval (years)		
1 —✳—		100 ——
2 —✦—		200 —△—
5 —⊖—		500 —+—
10 —⊟—		1000 —⊟—
25 —✕—		

					Precipitation Frequency Estimates (in)													
ARI (years)	5 min	10 min	15 min	30 min	60 min	120 min	3 hr	6 hr	12 hr	24 hr	48 hr	4 day	7 day	10 day	20 day	30 day	45 day	60 day
1	0.38	0.60	0.73	0.97	1.18	1.38	1.48	1.77	2.04	2.39	2.74	3.15	3.66	4.13	5.55	6.96	8.73	10.4
2	0.46	0.71	0.88	1.17	1.44	1.68	1.81	2.16	2.48	2.91	3.32	3.77	4.36	4.89	6.56	8.20	10.24	12.2
5	0.55	0.85	1.05	1.43	1.80	2.12	2.30	2.78	3.17	3.72	4.21	4.67	5.28	5.88	7.73	9.49	11.65	13.9
10	0.62	0.96	1.18	1.63	2.08	2.47	2.68	3.30	3.75	4.39	4.93	5.40	6.02	6.69	8.64	10.46	12.69	15.2
25	0.71	1.09	1.34	1.90	2.46	2.94	3.21	4.05	4.58	5.37	5.99	6.46	7.06	7.84	9.88	11.70	13.95	16.7
50	0.78	1.19	1.47	2.10	2.77	3.33	3.64	4.70	5.28	6.20	6.88	7.33	7.90	8.79	10.85	12.62	14.87	17.9
100	0.85	1.28	1.60	2.31	3.08	3.73	4.09	5.40	6.05	7.10	7.83	8.27	8.78	9.78	11.82	13.49	15.71	18.9
200	0.93	1.39	1.72	2.52	3.41	4.15	4.57	6.17	6.89	8.10	8.88	9.27	9.70	10.84	12.80	14.34	16.50	19.9
500	1.03	1.52	1.90	2.81	3.88	4.75	5.24	7.31	8.13	9.58	10.41	10.77	11.06	12.33	14.12	15.42	17.47	21.1
1000	1.11	1.62	2.02	3.04	4.26	5.22	5.78	8.29	9.18	10.83	11.71	12.05	12.26	13.56	15.15	16.20	18.15	21.9

(a) Chicago, Ilinois 41.820 N 87.67 W 593 feet

Figure 7.2.15 Point Precipitation Frequency Estimates from NOAA Atlas 14, (a) Chicago, Illinois (from Bonnin et al. (2004));
(b) Phoenix, Arizona (from Bonnin et al. (2004)).

Partial duration based Point Precipitation Frequency Estimates - Version: 4
33.482 N 112.05 W 1158 ft

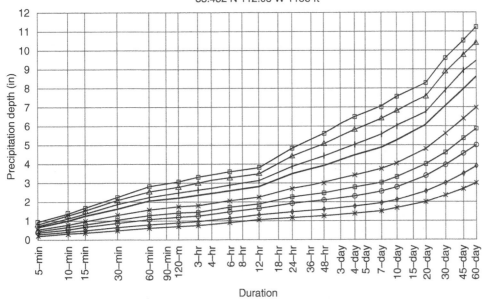

Mon Feb 02 22:31:58 2009

Average recurrence interval (years)		
1 —✳—		100 ——
2 —✦—		200 —△—
5 —⊙—		500 —+—
10 —▣—		1000 —☐—
25 —✕—		

Precipitation Frequency Estimates (in)

ARI (years)	5 min	10 min	15 min	30 min	60 min	120 min	3 hr	6 hr	12 hr	24 hr	48 hr	4 day	7 day	10 day	20 day	30 day	45 day	60 day
1	0.19	0.29	0.36	0.48	0.59	0.68	0.72	0.88	0.99	1.14	1.23	1.36	1.50	1.63	1.98	2.31	2.67	2.95
2	0.25	0.38	0.47	0.63	0.78	0.88	0.93	1.11	1.26	1.45	1.57	1.74	1.91	2.08	2.54	2.97	3.44	3.80
5	0.34	0.51	0.64	0.86	1.06	1.18	1.22	1.43	1.60	1.88	2.06	2.29	2.52	2.74	3.35	3.91	4.53	4.99
10	0.41	0.62	0.77	1.03	1.27	1.41	1.46	1.68	1.86	2.22	2.45	2.74	3.01	3.27	3.96	4.62	5.33	5.85
25	0.50	0.76	0.94	1.26	1.56	1.72	1.78	2.03	2.22	2.69	3.00	3.37	3.71	4.01	4.79	5.57	6.38	6.97
50	0.57	0.86	1.07	1.44	1.78	1.96	2.04	2.30	2.50	3.06	3.44	3.88	4.26	4.61	5.42	6.30	7.16	7.79
100	0.64	0.97	1.21	1.62	2.01	2.20	2.32	2.58	2.78	3.45	3.89	4.42	4.85	5.24	6.05	7.05	7.95	8.62
200	0.71	1.08	1.34	1.81	2.24	2.45	2.60	2.87	3.07	3.85	4.37	5.00	5.48	5.90	6.70	7.81	8.74	9.42
500	0.81	1.23	1.53	2.06	2.54	2.78	2.99	3.26	3.46	4.40	5.04	5.81	6.36	6.82	7.57	8.82	9.76	10.4
1000	0.88	1.34	1.67	2.24	2.78	3.04	3.31	3.57	3.75	4.84	5.57	6.47	7.07	7.57	8.23	9.60	10.53	11.2

(b) Arizona 33.482 N 112.05 W 1158 feet

Figure 7.2.15 (*Continued*)

Table 7.2.1 Coefficients for Interpolating
Design Precipitation Depths Using Equation (7.2.2)

Return period T years	a	b
5	0.674	0.278
10	0.496	0.449
25	0.293	0.669
50	0.146	0.835

Source: Frederick, Meyers, and Auciello (1997).

EXAMPLE 7.2.1

Determine the 2-, 10-, 25-, and 100-year rainfall depths for a 15-min duration storm at a location where $P_{2,15} = 0.9$ in and $P_{100,15} = 1.75$ in.

SOLUTION

$P_{10,15}$ and $P_{25,15}$ are determined using equation (7.2.2). From Table 7.2.1, $a = 0.496$ and $b = 0.449$ for 10 years and $a = 0.293$ and $b = 0.669$ for 25 years:

$$P_{10yr} = a\,P_{2yr} + b\,P_{100yr}$$

$$P_{10,15} = 0.496 \times 0.9 + 0.449 \times 1.75 = 0.446 + 0.786 = 1.23 \text{ in}$$

$$P_{25,15} = 0.293\,P_{2,15} + 0.669\,P_{100,15} = 0.293 \times 0.9 + 0.669 \times 1.75 = 1.43 \text{ in}$$

IDF Relationships

In hydrologic design projects, particularly urban drainage design, the use of *intensity-duration-frequency* relationships is recommended. *Intensity* refers to rainfall intensity (depth per unit time), and in some cases depths are used instead of intensity. *Duration* refers to rainfall duration, and *frequency* refers to *return periods*, which is the expected value of the *recurrence interval* (time between occurrences). See Chapter 10 for more details. The intensity-duration-frequency (IDF) relationships are also referred to as *IDF curves*. IDF relationships have also been expressed in equation form, such as

$$i = \frac{c}{T_d^e + f} \tag{7.2.3}$$

where i is the design rainfall intensity in inches per hour, T_d is the duration in minutes, and c, e, and f are coefficients that vary for location and return period. Other forms of these IDF equations include the return period, such as

$$i = \frac{cT^m}{T_d + f} \tag{7.2.4}$$

and

$$i = \frac{cT^m}{T_d^e + f} \tag{7.2.5}$$

where T is the return period. In Chapter 15, these IDF equations are used in urban drainage design. Chow et al. (1988) describe in detail how to derive the coefficients for these relationships using rainfall data.

Synthetic Storm Hyetograph

In many types of hydrologic analysis, such as *rainfall-runoff analysis*, to determine the runoff (discharge) from a watershed the time sequence of rainfall is needed. In such cases it is standard practice to use a *synthetic storm hyetograph*. The United States Department of Agriculture Soil Conservation Service (1973, 1986) developed synthetic storm hyetographs for 6- and 24-hr storms in

Table 7.2.2   SCS Rainfall Distributions

		24-hour storm				6-hour storm		
		P_t/P_{24}						
Hour t	$t/24$	Type I	Type IA	Type II	Type III	Hour t	$t/6$	P_t/P_6
0	0	0	0	0	0	0	0	0
2.0	0.083	0.035	0.050	0.022	0.020	0.60	0.10	0.04
4.0	0.167	0.076	0.116	0.048	0.043	1.20	0.20	0.10
6.0	0.250	0.125	0.206	0.080	0.072	1.50	0.25	0.14
7.0	0.292	0.156	0.268	0.098	0.089	1.80	0.30	0.19
8.0	0.333	0.194	0.425	0.120	0.115	2.10	0.35	0.31
8.5	0.354	0.219	0.480	0.133	0.130	2.28	0.38	0.44
9.0	0.375	0.254	0.520	0.147	0.148	2.40	0.40	0.53
9.5	0.396	0.303	0.550	0.163	0.167	2.52	0.42	0.60
9.75	0.406	0.362	0.564	0.172	0.178	2.64	0.44	0.63
10.0	0.417	0.515	0.577	0.181	0.189	2.76	0.46	0.66
10.5	0.438	0.583	0.601	0.204	0.216	3.00	0.50	0.70
11.0	0.459	0.624	0.624	0.235	0.250	3.30	0.55	0.75
11.5	0.479	0.654	0.645	0.283	0.298	3.60	0.60	0.79
11.75	0.489	0.669	0.655	0.357	0.339	3.90	0.65	0.83
12.0	0.500	0.682	0.664	0.663	0.500	4.20	0.70	0.86
12.5	0.521	0.706	0.683	0.735	0.702	4.50	0.75	0.89
13.0	0.542	0.727	0.701	0.772	0.751	4.80	0.80	0.91
13.5	0.563	0.748	0.719	0.799	0.785	5.40	0.90	0.96
14.0	0.583	0.767	0.736	0.820	0.811	6.00	1.0	1.00
16.0	0.667	0.830	0.800	0.880	0.886			
20.0	0.833	0.926	0.906	0.952	0.957			
24.0	1.000	1.000	1.000	1.000	1.000			

Source: U.S. Department of Agriculture Soil Conservation Service (1973, 1986).

the United States. These are presented in Table 7.2.2 and Figure 7.2.16 as cumulative hyetographs. Four 24-hr duration storms, Type I, IA, II, and III, were developed for different geographic locations in the U.S., as shown in Figure 7.2.17. Types I and IA are for the Pacific maritime climate, which has wet winters and dry summers. Type III is for the Gulf of Mexico and Atlantic coastal areas, which have tropical storms resulting in large 24-hour rainfall amounts. Type II is for the remainder of the United States.

In the midwestern part of the United States the Huff (1967) temporal distribution of storms is widely used for heavy storms on areas ranging up to 400 mi^2. Time distribution patterns were developed for four probability groups, from the most severe (first quartile) to the least severe (fourth quartile). Figure 7.2.18a shows the probability distribution of first-quartile storms. These curves are smooth, reflecting average rainfall distribution with time; they do not exhibit the burst characteristics of observed storms. Figure 7.2.18b shows selected histograms of first-quartile storms for 10-, 50-, and 90-percent cumulative probabilities of occurrence, each illustrating the percentage of total storm rainfall for 10 percent increments of the storm duration. The 50 percent histogram represents a cumulative rainfall pattern that should be exceeded in about half of the storms. The 90 percent histogram can be interpreted as a storm distribution that is equaled or exceeded in 10 percent or less of the storms.

EXAMPLE 7.2.2

Using equation (7.2.3), compute the design rainfall intensities for a 10-year return period, 10-, 20-, and 60-min duration storms for $c = 62.5$, $e = 0.89$, and $f = 9.10$.

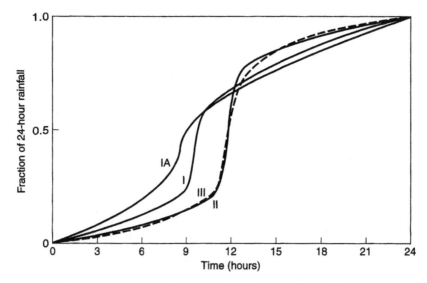

Figure 7.2.16 Soil Conservation Service 24-hour rainfall hyetographs (from U.S. Department of Agriculture Soil Conservation Service (1986)).

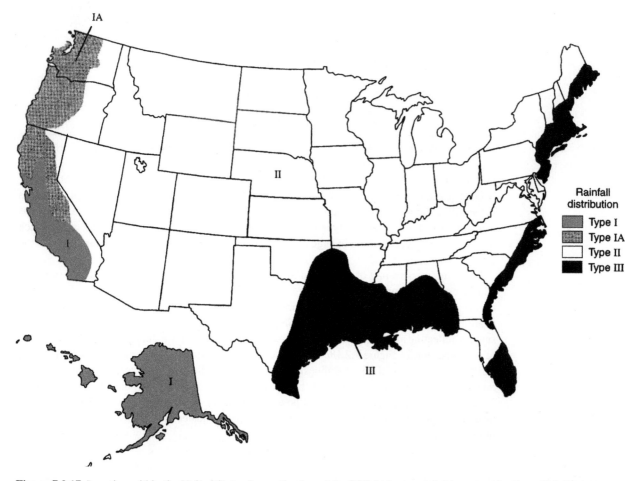

Figure 7.2.17 Location within the United States for application of the SCS 24-hour rainfall hyetographs (from U.S. Department of Agriculture Soil Conservation Service (1986)).

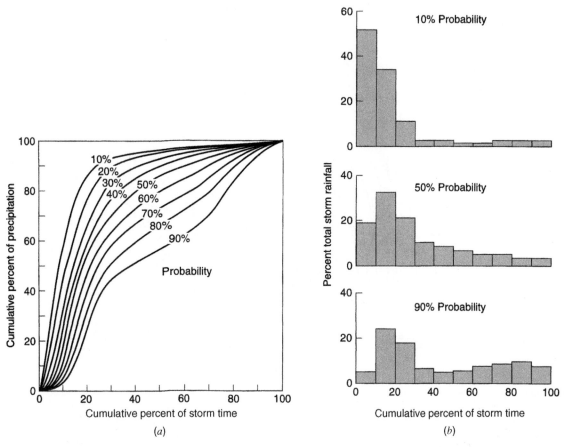

Figure 7.2.18 (*a*) Time distribution of first-quartile storms. The probability shown is the chance that the observed storm pattern will lie to the left of the curve; (*b*) Selected histograms for first-quartile storms (from Huff (1967)).

SOLUTION

The rainfall intensity duration frequency relationship is then

$$i = \frac{62.5}{T_d^{0.89} + 9.10}$$

For $T_d = 10$ min, $i = \dfrac{62.5}{10^{0.89} + 9.10} = 3.71$ in/hr. For $T_d = 20$ min, $i = 2.66$ in/hr, and for $T_d = 60$ min, $i = 1.32$ in/hr.

7.2.5   Estimated Limiting Storms

Estimated limiting values (ELVs) are used for the design of water control structures such as the spillways on large dams. Of particular interest are the *probable maximum precipitation* (PMP) and the *probable maximum storm* (PMS). These are used to derive a *probable maximum flood* (PMF). PMP is a depth of precipitation that is the estimated limiting value of precipitation, defined as the estimated greatest depth of precipitation for a given duration that is physically possible and reasonably characteristic over a particular geographical region at a certain time of year (Chow et al., 1988). Schreiner and Riedel (1978) presented generalized PMP charts for the United States east of the 105th meridian HMR 51. The all-seasons (any time of the year) estimates of PMP are presented in maps as a function of storm area (ranging from 10 to 20,000 mi^2) and storm durations ranging from 6 to 72 hours, as shown in Figure 7.2.19. For regions west of the 105th meridian, the diagram in Figure 7.2.20 shows the appropriate U.S. National Weather Service

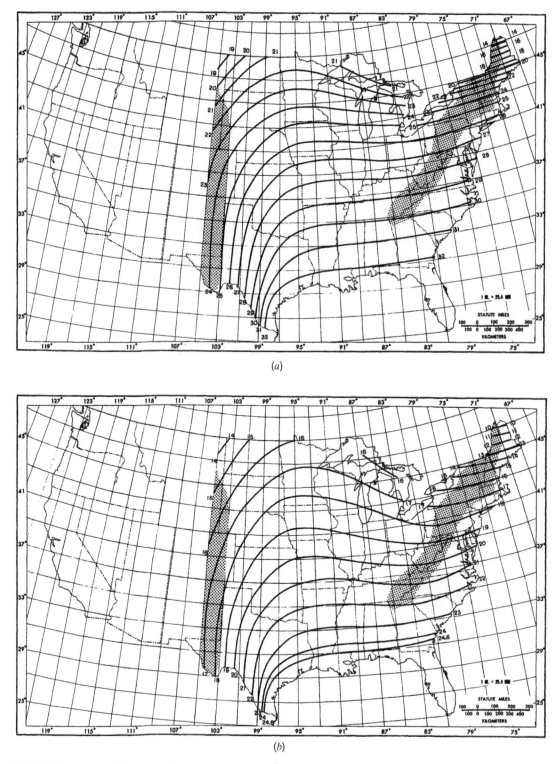

Figure 7.2.19 (*a*) Example of all-season PMP (in) for 6 hr, 10 mi^2 (from Schreiner and Riedel, 1978); (*b*) Example of all-season PMP (in) for 6 hr, 200 mi^2 (from Schreiner and Riedel, 1978).

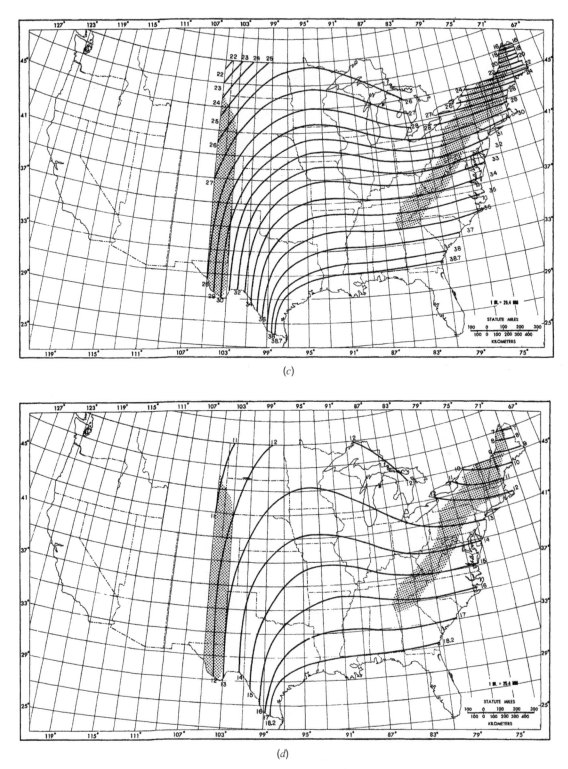

Figure 7.2.19 (*Continued*) (*c*) Example of all-season PMP (in) for 12 hr, 10 mi² (from Schreiner and Riedel, 1978); (*d*) Example of all-season PMP (in) for 6 hr, 1000 mi² (from Schreiner and Riedel, 1978).

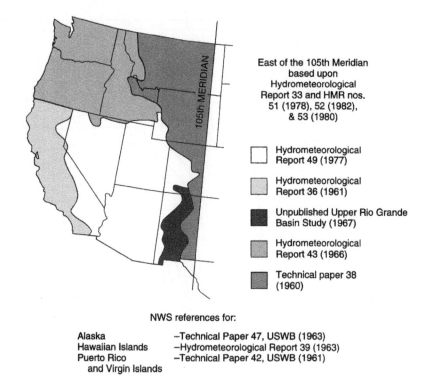

NWS references for:

Alaska	–Technical Paper 47, USWB (1963)
Hawaiian Islands	–Hydrometeorological Report 39 (1963)
Puerto Rico and Virgin Islands	–Technical Paper 42, USWB (1961)

Figure 7.2.20 Sources of information for probable maximum precipitation computation in the United States (from U.S. National Academy of Sciences (1983)).

publication. Hansen et al. (1982), in what is called HMR 52, present the procedure to determine the PMS for areas east of the 105th meridian.

7.3 EVAPORATION

Evaporation is the process of water changing from its liquid phase to the vapor phase. This process may occur from water bodies, from saturated soils, or from unsaturated surfaces. The evaporation process is illustrated in Figure 7.3.1. Above the water surface a number of things are happening. First

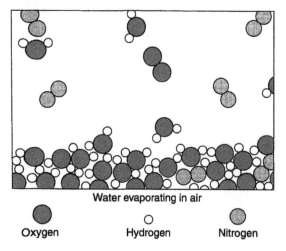

Figure 7.3.1 Evaporation (magnified one billion times) (from Feynman et al. (1963)).

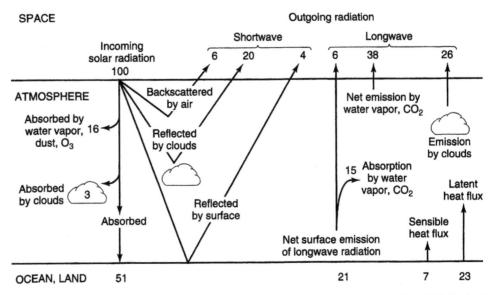

Figure 7.3.2 Radiation and heat balance in the atmosphere and at the earth's surface (from U.S. National Academy of Sciences (1975)).

of all, there are water molecules in the form of *water vapor*, which are always found above liquid water. In addition, there are some other molecules: (a) two oxygen atoms stuck together by themselves, forming an *oxygen molecule*, and (b) two nitrogen atoms stuck together, forming a nitrogen molecule. Above the water surface is the air, a gas, consisting almost entirely of nitrogen, oxygen, some water vapor, and lesser amounts of carbon dioxide, argon, and other substances. The molecules in the water are always moving around. From time to time, one on the surface gets knocked away. In other words, molecule by molecule, the water disappears or evaporates.

The computation of evaporation in hydrologic analysis and design is important in water supply design, particularly reservoir design and operation. The supply of energy to provide *latent heat of vaporization* and the *ability to transport water vapor away* from the evaporative surface are the two major factors that influence evaporation. *Latent heat* is the heat that is given up or absorbed when a phase (solid, liquid, or gaseous state) changes. *Latent heat of vaporization* (l_v) refers to the heat given up during vaporization of liquid water to water vapor, and is given as $l_v = 2.501 \times 10^6 - 2370T$, where T is the temperature in °C and l_v is in joules (J) per kilogram.

Three methods are used to determine evaporation: *the energy balance method, the aerodynamic method*, and *the combined aerodynamic and energy balance method*. The energy balance method considers the heat energy balance of a hydrologic system, and the aerodynamic method considers the ability to transport away from an open surface. Figure 7.3.2 illustrates the radiation and heat balance in the atmosphere and at the earth's surface along with relative values for the various components.

7.3.1 Energy Balance Method

Consider the evaporation pan shown in Figure 7.3.3 with the defined control volume. This control volume contains water in both the liquid phase and the vapor phase, with densities of ρ_w and ρ_a, respectively. The continuity equation must be written for both phases. For the liquid phase, the extensive property $B = m$ (mass of liquid water), the intensive property $\beta = 1$, and $dB/dt = \dot{m}_v$, which is the mass flow rate of evaporation. Continuity for the liquid phase is then

$$-\dot{m}_v = \frac{d}{dt} \int_{CV} \rho_w \, d\forall + \sum_{CS} \rho_w \mathbf{V} \cdot \mathbf{A} \qquad (7.3.1)$$

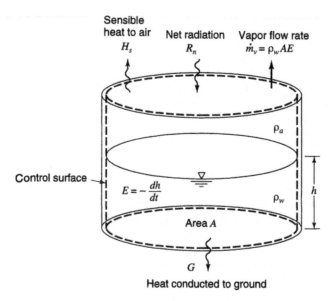

Sensible heat to air
H_s

Net radiation
R_n

Vapor flow rate
$\dot{m}_v = \rho_w A E$

ρ_a

Control surface

$E = -\dfrac{dh}{dt}$

ρ_w

h

Area A

G

Heat conducted to ground

Figure 7.3.3 Control volume defined for continuity and energy equation development for an evaporation pan (from Chow et al. (1988)).

Because the pan has impermeable sides, there is no flow of liquid water across the control surface, so

$$\sum_{CS} \rho_w \mathbf{V} \cdot \mathbf{A} = 0.$$

The rate of change of storage is

$$\frac{d}{dt} \int_{CV} \rho_w \, d\forall = \rho_w A \frac{dh}{dt} \tag{7.3.2}$$

where A is the cross-sectional area of the pan and h is the depth of water. Substituting equation (7.3.2) into (7.3.1) gives

$$-\dot{m}_v = \rho_w A \frac{dh}{dt} \tag{7.3.3a}$$

or

$$\dot{m}_v = \rho_w A E \tag{7.3.3b}$$

where $E = -dh/dt$ is the evaporation rate.

Considering the vapor phase, the extensive property is $B = m_v$ (mass of water vapor), the intensive property is $\beta = q_v$ (specific humidity), and $dB/dt = \dot{m}_v$. Continuity for the vapor phase is

$$\dot{m}_v = \frac{d}{dt} \int_{CV} q_v \rho_a \, d\forall + \sum_{CS} q_v \rho_a \mathbf{V} \cdot \mathbf{A} \tag{7.3.4}$$

Considering steady flow over the pan, the time derivative of water vapor in the control volume is zero, $\dfrac{d}{dt} \displaystyle\int_{CV} q_v \rho_a \, d\forall = 0$, and $\dot{m}_v$ from equation (7.3.3) is substituted into equation (7.3.4) to obtain

$$\rho_w A E \sum_{CS} q_v \rho_a \mathbf{V} \cdot \mathbf{A} \tag{7.3.5}$$

which is the continuity equation for the pan. Equation (7.3.5) can be rearranged to yield

$$E = \left(\frac{1}{\rho_w A}\right) \sum_{CS} q_v \rho_a \mathbf{V} \cdot \mathbf{A} \tag{7.3.6}$$

Next, consider the heat energy equation (3.4.19):

$$\frac{dH}{dt} - \frac{dW_s}{dt} = \frac{d}{dt}\int_{CV}\left(e_u + \frac{1}{2}V^2 + gz\right)\rho_w d\forall + \sum_{CS}\left(\frac{p}{\rho_w} + e_u + \frac{1}{2}V^2 + gz\right)\rho_w \mathbf{V} \cdot \mathbf{A} \tag{3.4.19}$$

using ρ_w to represent density of water; dH/dt is the rate of heat input from an external source, $dW_s/dt = 0$ because there is no work done by the system; e_u is the specific internal heat energy of the water; $V = 0$ for the water in the pan; and the rate of change of the water surface elevation is small $(dz/dt = 0)$. Then the above equation is reduced to

$$\frac{dH}{dt} = \frac{d}{dt}\int_{CV} e_u \rho_w d\forall \tag{7.3.7}$$

The rate of heat input from external sources can be expressed as

$$\frac{dH}{dt} = R_n - H_s - G \tag{7.3.8}$$

where R_n is the *net radiation flux* (watts per meter squared), H_s is the *sensible heat* to the air stream supplied by the water, and G is the *ground heat flux* to the ground surface. Net radiation flux is the net input of radiation at the surface at any instant. The net radiation flux at the earth's surface is the major energy input for evaporation of water, defined as the difference between the radiation absorbed, $R_i(1-\alpha)$ (where R_i is the *incident radiation* and α is the fraction of radiation reflected, called the *albedo*), and that emitted. R_e:

$$R_n = R_i(1-\alpha) - R_e \tag{7.3.9}$$

The amount of radiation emitted is defined by the *Stefan–Boltzmann law*

$$R_e = e\sigma T_p^4 \tag{7.3.10}$$

where e is the *emissivity* of the surface, σ is the *Stefan-Boltzmann constant* ($5.67 \times 10^{-8}\text{W/m}^2 \cdot \text{K}^4$), and T_p is the absolute temperature of the surface in degrees Kelvin.

Assuming that the temperature of the water in the control volume is constant in time, the only change in the heat stored within the control volume is the change in the internal energy of the water evaporated $l_v \dot{m}_v$, where l_v is the latent heat of vaporization, so that

$$\frac{d}{dt}\int_{CV} e_u \rho_w d\forall = l_v \dot{m}_v \tag{7.3.11}$$

Substituting equations (7.3.8) and (7.3.11) into (7.3.7) results in

$$R_n - H_s - G = l_v \dot{m}_v \tag{7.3.12}$$

From equation (7.3.3b), $\dot{m}_v = \rho_w AE$. Substituting in (7.3.12) with $A = 1\text{ m}^2$ and solving for E (to denote energy balance) gives the *energy balance equation for evaporation*,

$$E = \frac{1}{l_v \rho_w}(R_n - H_s - G) \tag{7.3.13}$$

Assuming the sensible heat flux H_s and the ground heat flux G are both zero, then an evaporation rate E_r, which is the rate at which all incoming net radiation is absorbed by evaporation, can be calculated as

$$E_r = \frac{R_n}{l_v \rho_w} \qquad (7.3.14)$$

EXAMPLE 7.3.1

For a particular location the average net radiation is 185 W/m^2, air temperature is 28.5°C, relative humidity is 55 percent, and wind speed is 2.7 m/s at a height of 2 m. Determine the open water evaporation rate in mm/d using the energy method.

SOLUTION

Latent heat of vaporization in joules (J) per kg varies with T (°C), or $l_v = 2.501 \times 10^6 - 2370T$, so $l_v = 2501 - 2.37 \times 28.5 = 2433$ kJ/kg, $\rho_w = 996.3$ kg/m^3. The evaporation rate by the energy balance method is determined using equation (7.3.14) with $R_n = 185$ W/m^2:

$E_r = R_n/(l_v \rho_w) = 185/(2433 \times 10^3 \times 996.3) = 7.63 \times 10^{-8}$ m/s $= 6.6$ mm/d.

7.3.2   Aerodynamic Method

As mentioned previously, the aerodynamic method considers the ability to transport water vapor away from the water surface; that is, generated by the humidity gradient in the air near the surface and the wind speed across the surface. These processes can be analyzed by coupling the equation for mass and momentum transport in the air. Considering the control volume in Figure 7.3.4, the vapor flux $\dot{m}_v$ passing upward by convection can be defined along with the momentum flux (as a function of the humidity gradient and the wind velocity gradient, respectively) (see Chow et al., 1988). The ratio of the vapor flux and the momentum flux can be used to define the *Thornthwaite–Holzman equation* for vapor transport (Thornthwaite and Holzman, 1939). Chow et al. (1988) present details of this derivation. The final form of the evaporation equation for the aerodynamic method expresses the evaporation rate E_a as a function of the difference of the vapor pressure at the surface e_{as}, which is the *saturation vapor pressure* at ambient air temperature (when the rate of evaporation and condensation are equal), and the vapor pressure at a height z_2 above the water surface, which is taken as the ambient vapor pressure in air e_a.

$$E_a = B(e_{as} - e_a) \qquad (7.3.15)$$

where E_a has units of mm/day and B is the vapor transfer coefficient with units of mm/day · Pa, given as

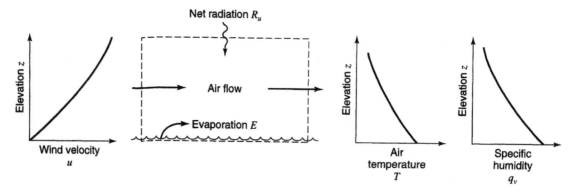

Figure 7.3.4 Evaporation from an open water surface (from Chow et al. (1988)).

$$B = \frac{0.102u_2}{[\ln(z_2/z_0)]^2} \qquad (7.3.16)$$

where u_2 is the wind velocity (m/s) measured at height z_2 (cm) and z_0 is the roughness height (0.01–0.06 cm) of the water surface. The vapor pressures have units of Pa (N/m^2). The saturation vapor pressure is approximated as

$$e_{as} = 611 \exp\left(\frac{17.27T}{237.3 + T}\right) \qquad (7.3.17)$$

and the vapor pressure is

$$e_a = R_h\, e_{as} \qquad (7.3.18)$$

where T is the air temperature in °C and R_h is the relative humidity ($R = e_a/e_{as}$) and ($0 \le R_h \le 1$).

EXAMPLE 7.3.2   Solve example 7.3.1 using the aerodynamic method, by using a roughness height $z_0 = 0.03$ cm.

SOLUTION   From equation (7.3.17), the saturated vapor pressure is

$$e_{as} = 611 \exp\left[17.27T/(237.3 + T)\right] = 611 \exp\left[17.27 \times 28.5/(237.3 + 28.5)\right]$$
$$= 3893 \text{ Pa}$$

The ambient vapor pressure e_a is determined from equation (7.3.18); for a relative humidity $R_h = 0.55$, $e_a = e_{as} R_h = 3893 \times 0.55 = 2141$ Pa. The vapor transfer coefficient B is given by equation (7.3.16) in which $u_2 = 2.7$ m/s, $z_2 = 2$ m, and $z_0 = 0.03$ cm for an open water surface, so that $B = 0.102u_2/[\ln(zz/z0)]^2 = 0.102 \times 2.7/[\ln(200/0.03)]^2 = 0.0036$ mm/d $\cdot$ Pa; then the evaporation rate by the aerodynamic method is given by equation (7.3.15):

$$E_a = B(e_{as} - e_a) = 0.0036(3893 - 2141) = 6.31 \text{ mm/d}.$$

7.3.3   Combined Method

When the energy supply is not limiting, the aerodynamic method can be used, and when the vapor transport is not limiting, the energy balance method can be used. However, both of these factors are not normally limiting, so a combination of these methods is usually required. The combined method equation is

$$E = \left(\frac{\Delta}{\Delta + \gamma}\right)E_r + \left(\frac{\gamma}{\Delta + \gamma}\right)E_a \qquad (7.3.19)$$

in which ( )E_r is the vapor transport term and ( )E_a is the aerodynamic term. γ is the *psychrometric constant* (approximately 66.8 Pa/°C) and Δ is the *gradient of the saturated vapor pressure curve* $\Delta = de_{as}/dT$ at air temperature T_α given as

$$\Delta = \frac{4098e_{as}}{(237.3 + T_\alpha)^2} \qquad (7.3.20)$$

in which e_{as} is the *saturated vapor pressure* (the maximum moisture content the air can hold for a given temperature).

The combination method is best for application to small areas with detailed climatological data including net radiation, air temperature, humidity, wind speed, and air pressure. For very large areas, energy (vapor transport) largely governs evaporation. Priestley and Taylor (1972) discovered that the aerodynamic term in equation (7.3.19) is approximately 30 percent of the energy term, so that equation (7.3.19) can be simplified to

$$E = 1.3\left(\frac{\Delta}{\Delta + \gamma}\right)E_r \qquad (7.3.21)$$

which is known as the *Priestley–Taylor evaporation equation.*

EXAMPLE 7.3.3   Solve example 7.3.1 using the combined method.

SOLUTION   The gradient of the saturated vapor pressure curve is, from equation (7.3.20),

$$\Delta = 4098e_{as}/(237.3 + T)^2 = 4098 \times 3893/(237.3 + 28.5)^2 = 225.8 \, \text{Pa/}^\circ\text{C}$$

The psychrometric constant γ is approximately 66.8 Pa/$^\circ$C; then E_r and E_a may be combined according to equation (7.3.19) to give

$$
\begin{aligned}
E &= \Delta/(\Delta + \gamma)E_r + \gamma/(\Delta + \gamma)E_a \\
&= (225.8/(225.8 + 66.8))6.6 + (66.8/(225.8 + 66.8))6.31 = 0.772(6.6) + 0.228(6.31) \\
&= 5.10 + 1.44 = 6.54 \, \text{mm/d}
\end{aligned}
$$

EXAMPLE 7.3.4   Solve example 7.3.1 using the Priestley–Taylor method.

SOLUTION   The evaporation is computed using equation (7.3.21):

$$E = 1.3\left(\frac{\Delta}{\Delta + \gamma}\right)E_r = 1.3\left(\frac{225.8}{225.8 + 66.8}\right)6.6$$

$$= 6.62 \, \text{mm/d}$$

7.4   INFILTRATION

The process of water penetrating into the soil is *infiltration.* The rate of infiltration is influenced by the condition of the soil surface, vegetative cover, and soil properties including porosity, hydraulic conductivity, and moisture content. In order to discuss infiltration, we must first consider the division of subsurface water (see Figure 6.1.2) and the various subsurface flow processes shown in Figure 7.4.1. These processes are infiltration of water to become *soil moisture, subsurface flow* (unsaturated flow) through the soil, and *groundwater flow* (saturated flow). *Unsaturated flow* refers to flow through a porous medium when some of the voids are occupied by air. *Saturated flow* occurs when the voids are filled with water. The *water table* is the interface between the saturated and unsaturated flow, where atmospheric pressure prevails. Saturated flow occurs below the water table and unsaturated flow occurs above the water table.

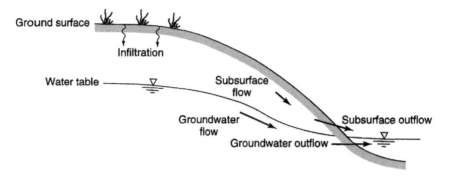

Figure 7.4.1 Subsurface water zones and processes (from Chow et al. (1988)).

7.4.1   Unsaturated Flow

The cross-section through an unsaturated porous medium (Figure 7.4.2*a*) is now used to define *porosity*, η:

$$\eta = \frac{\text{volume of voids}}{\text{total volume}} \tag{7.4.1}$$

in which η is $0.25 < \eta < 0.40$, and the *soil moisture content, θ*,

$$\theta = \frac{\text{volume of water}}{\text{total volume}} \tag{7.4.2}$$

in which θ is $0 \le \theta \le \eta$. For saturated conditions, $\theta = \eta$.

Consider the control volume in Figure 7.4.2*b* for an unsaturated soil with sides of lengths dx, dy, and dz, with a volume of $dxdydz$. The volume of water contained in the control volume is $\theta dxdydz$. Flow through the control volume is defined by the *Darcy flux*, $q = Q/A$, which is the volumetric flow rate per unit of soil area. For this derivation, the horizontal fluxes are ignored and only the vertical (z) direction is considered, with z positive upward.

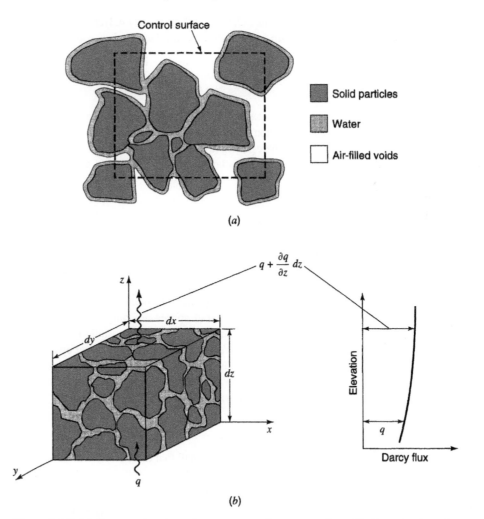

Figure 7.4.2 (*a*) Cross-section through an unsaturated porous medium; (*b*) Control volume for development of the continuity equation in an unsaturated porous medium (from Chow et al. (1988)).

With the control volume approach, the extensive property B is the mass of soil water, so the intensive property $\beta = dB/dm = 1$ and $dB/dt = 0$, because no phase changes are occurring in water. The general control volume equation for continuity, equation (3.3.1), is applicable:

$$0 = \frac{d}{dt} \int_{\text{CV}} \rho \, d\forall + \int_{\text{CS}} \rho \mathbf{V} \cdot d\mathbf{A} \tag{7.4.3}$$

The time rate of change of mass stored in the control volume is

$$\frac{d}{dt} \int_{\text{CV}} \rho \, d\forall = \frac{d}{dt} (\rho \theta dx dy dz) = \rho dx dy dz \frac{\partial \theta}{\partial t} \tag{7.4.4}$$

where the density is assumed constant. The net outflow of water is the difference between the volumetric inflow at the bottom $(q dx dy)$ and the volumetric outflow at the top $[q + (\partial q/\partial z)dz]dx dy$, so

$$\int_{\text{CV}} \rho \mathbf{V} \cdot d\mathbf{A} = \rho \left(q + \frac{\partial q}{\partial z} dz \right) dx dy - \rho q dx dy = \rho dx dy dz \frac{\partial q}{\partial z} \tag{7.4.5}$$

Substituting equations (7.4.4) and (7.4.5) into (7.4.3) and dividing by $\rho dx dy dz$ results in the following *continuity equation for one-dimensional unsteady unsaturated flow in a porous medium*:

$$\frac{\partial \theta}{\partial t} + \frac{\partial q}{\partial z} = 0 \tag{7.4.6}$$

Darcy's law relates the *Darcy flux q* to the rate of headloss per unit length of medium. For flow in the vertical direction the headloss per unit length is the change in total head ∂h over a distance, ∂z, i.e., $-\partial h/\partial z$, where the negative sign indicates that total head decreases (as a result of friction) in the direction of flow. Darcy's law can now be expressed as

$$q = -K \frac{\partial h}{\partial z} \tag{7.4.7}$$

where K is the *hydraulic conductivity*. This law applies to areas that are large compared with the cross-section of individual pores and grains of the medium. Darcy's law describes a steady uniform flow of constant velocity with a net force of zero in a fluid element. In unconfined saturated flow, the forces are gravity and friction. For unsaturated flow the forces are gravity, friction, and the *suction force* that binds water to soil particles through surface tension.

In unsaturated flow the void spaces are only partially filled with water, so that water is attracted to the particle surfaces through electrostatic forces between the water molecule polar bonds and the particle surfaces. This in turn draws water up around the particle surfaces, leaving air in the center of the voids. The energy due to the soil suction forces is referred to as the *suction head* ψ in unsaturated flow, which varies with moisture content. Total head is then the sum of the suction and gravity heads:

$$h = \psi + z \tag{7.4.8}$$

Note that the velocities are so small that there is no term for velocity head in this expression for total head.

Darcy's law can now be expressed as

$$q = -K \frac{\partial(\psi + z)}{\partial z} \tag{7.4.9}$$

Darcy's law was originally conceived for saturated flow and was extended by Richards (1931) to unsaturated flow with the provision that the hydraulic conductivity is a function of the suction head, i.e., $K = K(\psi)$. Also, the hydraulic conductivity can be related more easily to the degree of saturation, so that $K = K(\theta)$. Because the soil suction head varies with moisture content and moisture content varies with elevation, the *suction gradient* can be expanded by using the chain rule to obtain

$$\frac{\partial \psi}{\partial z} = \frac{d\psi}{d\theta}\frac{\partial \theta}{\partial z} \tag{7.4.10}$$

in which $\partial \theta / \partial z$ is the *wetness gradient*, and the reciprocal of $d\psi/d\theta$, i.e., $d\theta/d\psi$, is the *specific water capacity*. Now equation (7.4.9) can be modified to

$$q = -K\left(\frac{\partial \psi}{\partial z} + \frac{\partial z}{\partial z}\right) = -K\left(\frac{\partial \psi}{\partial \theta}\frac{\partial \theta}{\partial z} + 1\right) = -\left(K\frac{d\psi}{d\theta}\frac{\partial \theta}{\partial z} + K\right) \tag{7.4.11}$$

The *soil water diffusivity* $D(L^2/T)$ is defined as

$$D = K\frac{d\psi}{d\theta} \tag{7.4.12}$$

so substituting this expression for D into equation (7.4.11) results in

$$q = -\left(D\frac{\partial \theta}{\partial z} + K\right) \tag{7.4.13}$$

Using the continuity equation (7.4.6) for one-dimensional, unsteady, unsaturated flow in a porous medium yields

$$\frac{\partial \theta}{\partial t} = -\frac{\partial q}{\partial z} = \frac{\partial}{\partial z}\left(D\frac{\partial \theta}{\partial z} + K\right) \tag{7.4.14}$$

which is a one-dimensional form of *Richards' equation*. This equation is the governing equation for unsteady unsaturated flow in a porous medium (Richards, 1931). For a homogeneous soil, $\partial K/\partial z = 0$, so that $\partial \theta/\partial t = \partial(D\partial \theta/\partial z)/\partial z$.

EXAMPLE 7.4.1   Determine the flux for a soil in which the hydraulic conductivity is expressed as a function of the suction head as $K = 250(-\psi)^{-2.11}$ in cm/d at depth $z_1 = 80$ cm, $h_1 = -145$ cm, and $\psi_1 = -65$ cm at depth $z_2 = 100$ cm, $h_2 = -160$ cm, and $\psi_2 = -60$ cm.

SOLUTION   The flux is determined using equation (7.4.7). First the hydraulic conductivity is computed using an average value of $\psi = [-65 + (-60)]/2 = -62.5$ cm. Then $K = 250(-\psi)^{-2.11} = 250(62.5)^{-2.11} = 0.041$ cm/d. The flux is then

$$q = -K\left(\frac{h_1 - h_2}{z_1 - z_2}\right) = -0.041\left[\frac{-145 - (-160)}{-80 - (-100)}\right] = -0.03 \text{ cm/d}$$

The flux is negative because the moisture is flowing downward in the soil.

EXAMPLE 7.4.2   Determine the soil water diffusivity for a soil in which $\theta = 0.1$ and $K = 3 \times 10^{-11}$ mm/s from a relationship of $\psi(\theta)$ at $\theta = 0.1$, $\Delta \psi = 10^7$ mm, and $\Delta \theta = 0.35$.

SOLUTION   Using equation (7.4.12), the soil water diffusivity is $D = Kd\psi/d\theta = (3 \times 10^{-11}$ mm/s$)$ $(10^7 \text{ mm}/0.35) = 8.57 \times 10^{-4}$ mm/s.

7.4.2   Green-Ampt Method

Figure 7.4.3 illustrates the distribution of soil moisture within a soil profile during downward movement. These moisture zones are the *saturated zone*, the *transmission zone*, a *wetting zone*, and a *wetting front*. This profile changes as a function of time as shown in Figure 7.4.4.

The *infiltration rate f* is the rate at which water enters the soil surface, expressed in in/hr or cm/hr. The *potential infiltrate rate* is the rate when water is ponded on the soil surface, so if no ponding occurs the actual rate is less than the potential rate. Most infiltration equations describe a potential infiltration rate. *Cumulative infiltration F* is the accumulated depth of water infiltrated, defined mathematically as

$$F(t) = \int_0^t f(\tau)d\tau \qquad (7.4.15)$$

and the infiltration rate is the time derivative of the cumulative infiltration given as

$$f(t) = \frac{dF(t)}{dt} \qquad (7.4.16)$$

Figure 7.4.5 illustrates a rainfall hyetograph with the infiltration rate and cumulative infiltration curves. (See Section 8.2 for further details on the rainfall hyetograph.)

Green and Ampt (1911) proposed the simplified picture of infiltration shown in Figure 7.4.6. The *wetting front* is a sharp boundary dividing soil with moisture content θ_i below from saturated soil

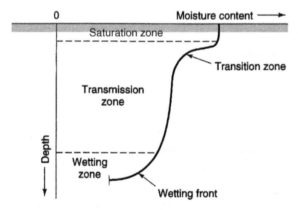

Figure 7.4.3 Moisture zones during infiltration (from Chow et al. (1988)).

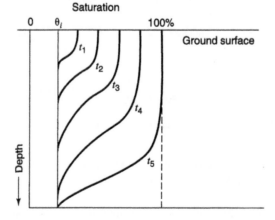

Figure 7.4.4 Moisture profile as a function of time for water added to the soil surface.

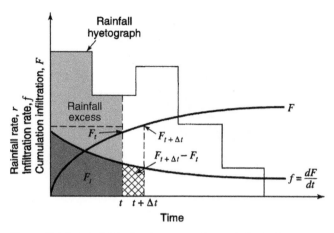

Figure 7.4.5 Rainfall infiltration rate and cumulative infiltration. The rainfall hyetograph illustrates the rainfall pattern as a function of time. The cumulative infiltration at time t is F_t or $F(t)$ and at time $t + \Delta t$ is $F_{t+\Delta t}$ or $F(t + \Delta t)$ and are computed using equation (7.4.15). The increase in cumulative infiltration from time t to $t + \Delta t$ is $F_{t+\Delta t} - F_t$ or $F(t + \Delta t) - F(t)$, as shown in the figure. Rainfall excess is defined in Chapter 8 as that rainfall that is neither retained on the land surface nor infiltrated into the soil.

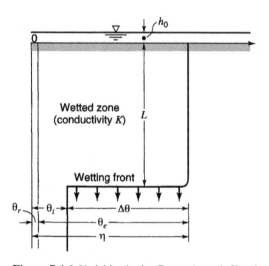

Figure 7.4.6 Variables in the Green–Ampt infiltration model. The vertical axis is the distance from the soil surface; the horizontal axis is the moisture content of the soil (from Chow et al. (1988)).

with moisture content η above. The wetting front has penetrated to a depth L in time t since infiltration began. Water is ponded to a small depth h_0 on the soil surface.

Consider a vertical column of soil of unit horizontal cross-sectional area (Figure 7.4.7) with the control volume defined around the wet soil between the surface and depth L. For a soil initially of moisture content θ_i throughout its entire depth, the *moisture content* will increase from θ_i to η (the porosity) as the wetting front passes. The moisture content θ is the ratio of the volume of water to the total volume within the control surface. $L(\eta - \theta_i)$ is then the increase in the water stored within the control volume as a result of infiltration, through a unit cross-section. By definition this quantity is equal to F, the cumulative depth of water infiltrated into the soil, so that

$$F(t) = L(\eta - \theta_i) = L\Delta\theta \tag{7.4.17}$$

where $\Delta\theta = (\eta - \theta_i)$.

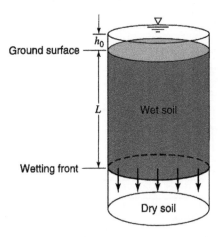

Figure 7.4.7 Infiltration into a column of soil of unit cross-sectional area for the Green–Ampt model (from Chow et al. (1988)).

Darcy's law may be expressed (using equation (7.4.7)) as

$$q = K\frac{\partial h}{\partial z} = -K\frac{\Delta h}{\Delta z} \tag{7.4.18}$$

The Darcy flux q is constant throughout the depth and is equal to $-f$, because q is positive upward while f is positive downward. If points 1 and 2 are located at the ground surface and just on the dry side of the wetting front, respectively, then equation (7.4.18) can be approximated by

$$f = K\left[\frac{h_1 - h_2}{z_1 - z_2}\right] \tag{7.4.19}$$

The head h_1 at the surface is equal to the ponded depth h_0. The head h_2, in the dry soil below the wetting front, equals $-\psi - L$. Darcy's law for this system is written as

$$f = K\left[\frac{h_0 - (-\psi - L)}{L}\right] \tag{7.4.20a}$$

and if the ponded depth h_0 is negligible compared to ψ and L,

$$f \approx K\left[\frac{\psi + L}{L}\right] \tag{7.4.20b}$$

This assumption $(h_0 = 0)$ is usually appropriate for surface water hydrology problems because it is assumed that ponded water becomes surface runoff.

From equation (7.4.17), the wetting front depth is $L = F/\Delta\theta$, and assuming $h_0 = 0$, substitution into equation (7.4.20) gives

$$f = K\left[\frac{\psi\Delta\theta + F}{F}\right] \tag{7.4.21}$$

Since $f = dF/dt$, equation (7.4.21) can be expressed as a differential equation in the one unknown F as

$$\frac{dF}{dt} = K\left[\frac{\psi\Delta\theta + F}{F}\right]$$

To solve for F, cross-multiply to obtain

$$\left[\frac{F}{F + \psi\Delta\theta}\right]dF = Kdt$$

Then divide the left-hand side into two parts

$$\left[\frac{F+\psi\Delta\theta}{F+\psi\Delta\theta} - \frac{\psi\Delta\theta}{F+\psi\Delta\theta}\right]dF = Kdt$$

and integrate

$$\int_0^{F(t)}\left[1 - \frac{\psi\Delta\theta}{F+\psi\Delta\theta}\right]dF = \int_0^t Kdt$$

to obtain

$$F(t) - \psi\Delta\theta\{\ln[F(t)+\psi\Delta\theta] - \ln(\psi\Delta\theta)\} = Kt \tag{7.4.22a}$$

or

$$F(t) - \psi\Delta\theta\ln\left(1 + \frac{F(t)}{\psi\Delta\theta}\right) = Kt \tag{7.4.22b}$$

Equation (7.4.22) is the *Green–Ampt equation* for cumulative infiltration. Once F is computed using equation (7.4.22), the infiltration rate f can be obtained from equation (7.4.21) or

$$F(t) = K\left[\frac{\psi\Delta\theta}{F(t)} + 1\right] \tag{7.4.23}$$

When the ponded depth h_0 is not negligible, the value of $\psi + h_0$ is substituted for ψ in equations (7.4.22) and (7.4.23).

Equation (7.4.22) is a nonlinear equation in F that can be solved by the method of successive substitution by rearranging (7.4.22)

$$F(t) = Kt + \psi\Delta\theta\ln\left(1 + \frac{F(t)}{\psi\Delta\theta}\right) \tag{7.4.24}$$

Given K, t, ψ, and $\Delta\theta$, a trial value F is substituted on the right-hand side (a good trial value is $F = Kt$), and a new value of F calculated on the left-hand side, which is substituted as a trial value on the right-hand side, and so on until the calculated values of F converge to a constant. The final value of cumulative infiltration F is substituted into (7.4.23) to determine the corresponding infiltration rate f.

Equation (7.4.22) can also be solved by Newton's method, which is more complicated than the method of successive substitution but converges in fewer iterations. Newton's iteration method is explained in Appendix A. Referring to equation (7.4.24), application of the Green–Ampt model requires estimates of the hydraulic conductivity K, the wetting front soil suction bead ψ (see Table 7.4.1), and $\Delta\theta$.

The *residual moisture content* of the soil, denoted by θ_r, is the moisture content after it has been thoroughly drained. The *effective saturation* is the ratio of the available moisture ($\theta - \theta_r$) to the maximum possible available moisture content ($\eta - \theta_r$), given as

$$s_e = \frac{\theta - \theta_r}{\eta - \theta_r} \tag{7.4.25}$$

where $\eta - \theta_r$ is called the *effective porosity* θ_e.

The effective saturation has the range $0 \le s_e \le 1.0$, provided $\theta_r \le \theta \le \eta$. For the initial condition, when $\theta = \theta_i$, cross-multiplying equation (7.4.25) gives $\theta_i - \theta_r = s_e\theta_e$, and the change in the moisture content when the wetting front passes is

$$\Delta\theta = \eta - \theta_i = \eta - (s_e\theta_e + \theta_r)$$
$$\Delta\theta = (1 - s_e)\theta_e \tag{7.4.26}$$

Table 7.4.1 Green–Ampt Infiltration Parameters for Various Soil Classes*

Soil class	Porosity η	Effective porosity θ_e	Wetting front soil suction head ψ (cm)	Hydraulic conductivity K (cm/h)
Sand	0.437	0.417	4.95	11.78
	(0.374–0.500)	(0.354–0.480)	(0.97–25.36)	
Loamy sand	0.437	0.401	6.13	2.99
	(0.363–0.506)	(0.329–0.473)	(1.35–27.94)	
Sandy loam	0.453	0.412	11.01	1.09
	(0.351–0.555)	(0.283–0.541)	(2.67–45.47)	
Loam	0.463	0.434	8.89	0.34
	(0.375–0.551)	(0.334–0.534)	(1.33–59.38)	
Silt loam	0.501	0.486	16.68	0.65
	(0.420–0.582)	(0.394–0.578)	(2.92–95.39)	
Sandy clay loam	0.398	0.330	21.85	0.15
	(0.332–0.464)	(0.235–0.425)	(4.42–108.0)	
Clay loam	0.464	0.309	20.88	0.10
	(0.409–0.519)	(0.279–0.501)	(4.79–91.10)	
Silty clay loam	0.471	0.432	27.30	0.10
	(0.418–0.524)	(0347–0.517)	(5.67–131.50)	
Sandy clay	0.430	0.321	23.90	0.06
	(0.370–0.490)	(0.207–0.435)	(4.08–140.2)	
Silty clay	0.479	0.423	29.22	0.05
	(0.425–0.533)	(0.334–0.512)	(6.13–139.4)	
Clay	0.475	0.385	31.63	0.03
	(0.427–0.523)	(0.269–0.501)	(6.39–156.5)	

* The numbers in parentheses below each parameter are one standard deviation around the parameter value given.

Source: Rawls, Brakensiek, and Miller (1983).

A logarithmic relationship between the effective saturation s_e and the soil suction head ψ can be expressed by the *Brooks–Corey equation* (Brooks and Corey, 1964):

$$s_e = \left[\frac{\psi_b}{\psi}\right]^{\lambda} \tag{7.4.27}$$

in which ψ_b and λ are constants obtained by draining a soil in stages, measuring the values of s_e and ψ at each stage, and fitting equation (7.4.27) to the resulting data.

Brakensiek et al. (1981) presented a method for determining the Green–Ampt parameters using the Brooks–Corey equation. Rawls et al. (1983) used this method to analyze approximately 5000 soil horizons across the United States and determined average values of the Green–Ampt parameters η, θ_e, ψ, and K for different soil classes, as listed in Table 7.4.1. As the soil becomes finer, moving from sand to clay, the wetting front soil suction head increases while the hydraulic conductivity decreases. Table 7.4.1 also lists typical ranges for η, θ_e, and ψ. The ranges are not large for η and θ_e, but ψ can vary over a wide range for a given soil. K varies along with ψ, so the values given in Table 7.4.1 for both ψ and K should be considered typical values that may show a considerable degree of variability in application (American Society of Agricultural Engineers, 1983; Devanrs and Gifford, 1986).

Table 7.4.2   Infiltration Computations Using the Green–Ampt Method

Time t (hr)	0.0	0.1	0.2	0.3	0.4	0.5	1.0	1.5	2.0	2.5	3.0	3.5	4.0	4.5	5.0	5.5	6.0
Infiltration rate f (cm/hr)	∞	1.78	1.20	0.97	0.84	0.75	0.54	0.44	0.39	0.35	0.32	0.30	0.28	0.27	0.26	0.25	0.24
Infiltration depth F (cm)	0.00	0.29	0.43	0.54	0.63	0.71	1.02	1.26	1.47	1.65	1.82	1.97	2.12	2.26	2.39	2.51	2.64

EXAMPLE 7.4.3

Use the Green–Ampt method to evaluate the infiltration rate and cumulative infiltration depth for a silty clay soil at 0.1-hr increments up to 6 hr from the beginning of infiltration. Assume an initial effective saturation of 20 percent and continuous ponding.

SOLUTION

From Table 7.4.1, for a silty clay soil, $\theta_e = 0.423$, $\psi = 29.22$ cm, and $K = 0.05$ cm/hr. The initial effective saturation is $s_e = 0.2$, so $\Delta\theta = (1 - s_e)$, $\theta_e = (1 - 0.20)0.423 = 0.338$, and $\psi\Delta\theta = 29.22 \times 0.338 = 9.89$ cm. Assuming continuous ponding, the cumulative infiltration F is found by successive substitution in equation (7.4.24):

$$F = Kt + \psi\Delta\theta \ln[1 + F/(\psi\Delta\theta)] = 0.05t + 9.89 \ln[1 + F/9.89]$$

For example, at time $t = 0.1$ hr, the cumulative infiltration converges to a final value $F = 0.29$ cm. The infiltration rate f is then computed using equation (7.4.23):

$$f = K(1 + \psi\Delta\theta/F) = 0.05(1 + 9.89/F)$$

As an example, at time $t = 0.1$ hr, $f = 0.05(1 + 9.89/0.29) = 1.78$ cm/hr. The infiltration rate and the cumulative infiltration are computed in the same manner between 0 and 6 hr at 0.1-hr intervals; the results are listed in Table 7.4.2.

EXAMPLE 7.4.4

Ponding time t_p is the elapsed time between the time rainfall begins and the time water begins to pond on the soil surface. Develop an equation for ponding time under a constant rainfall intensity i, using the Green–Ampt infiltration equation (see Figure 7.4.8).

SOLUTION

The infiltration rate f and the cumulative infiltration F are related in equation (7.4.22). The cumulative infiltration at ponding time t_p is $F_p = it_p$, in which i is the constant rainfall intensity (see Figure 7.4.8). Substituting $F_p = it_p$ and the infiltration rate $f = i$ into equation (7.4.23) yields

$$i = K\left(\frac{\psi\Delta\theta}{it_p} + 1\right)$$

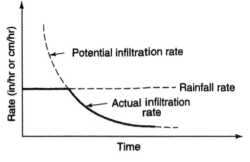
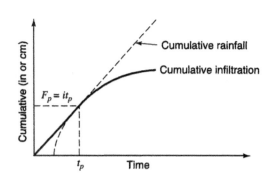

Figure 7.4.8 Ponding time. This figure illustrates the concept of ponding time for a constant intensity rainfall. Ponding time is the elapsed time between the time rainfall begins and the time water begins to pond on the soil surface.

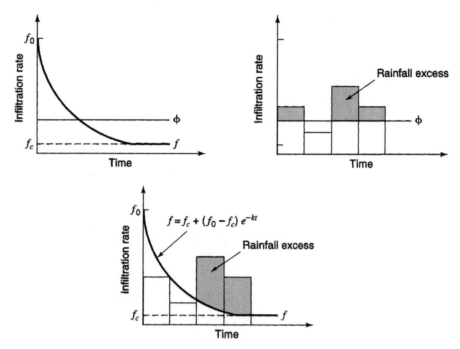

Figure 7.4.9 Φ-index and Horton's equation.

and solving, we get

$$t_p = \frac{K\psi\Delta\theta}{i(i-K)}$$

which is the ponding time for a constant rainfall intensity.

7.4.3 Other Infiltration Methods

The simplest accounting of abstraction is the Φ-*index* (refer to Figure 7.4.9 and Section 8.2), which is a constant rate of abstraction (in/h or cm/h). Other cumulative infiltration and infiltration rate equations include Horton's and the SCS method. Horton's equation (Horton, 1933) is an empirical relation that assumes infiltration begins at some rate f_o and exponentially decreases until it reaches a constant rate f_c (refer to Figure 7.4.9). The infiltration capacity is expressed as

$$f_t = f_c + (f_o - f_c)e^{-kt} \tag{7.4.28}$$

and the cumulative infiltration capacity is expressed as

$$F_t = f_c t + \frac{(f_0 - f_c)}{k}(1 - e^{-kt}) \tag{7.4.29}$$

where k is a decay constant. Many other empirical infiltration equations have been developed that can be found in the various hydrology texts.

Richard's equation can be solved (Philip, 1957, 1969) under less restrictive conditions by assuming that K and D (equation 7.4.14) can vary with moisture content. Philip used the Boltzmann transformation $B(\theta) = zt^{-1/2}$ to convert Richard's equation (7.4.14) into an ordinary differential equation in B and solved to obtain an intimate series for cumulative infiltration F_t, approximated as

$$F_t = St^{1/2} + Kt \qquad (7.4.30)$$

in which S is the sorptivity, a parameter that is a function of the soil suction potential, and K is the hydraulic conductivity. Differentiating $f(t) = df/dt$, the infiltration rate is defined as

$$f(t) = \frac{1}{2}St^{-1/2} + K \qquad (7.4.31)$$

As $t \rightarrow \infty$, $f(t) \rightarrow K$. The two terms S and K represent the effects of soil suction head and gravity head.

There are many other infiltration methods, including the SCS method described in Sections 8.6–8.8. In summary, the Green–Ampt and the SCS are both used in the U.S. Army Corps of Engineers HEC-1 and HEC-HMS models, and both are used widely in the United States.

PROBLEMS

7.2.1 Determine the 25-year return period rainfall depth for a 30-min duration in Chicago, Illinois.

7.2.2 Determine the 2-, 10-, 25-, and 100-year precipitation depths for a 15-min duration storm in Memphis, Tennessee.

7.2.3 Determine the 2- and 25-year intensity-duration-frequency curves for Memphis, Tennessee.

7.2.4 Determine the 10- and 50-year intensity-duration-frequency curves for Chicago, Illinois.

7.2.5 Determine the 2-, 5-, 10-, 25-, 50-, and 100-year depths for a 1-hr duration storm in Phoenix, Arizona. Repeat for a 6-hr duration storm.

7.2.6 Develop the Type I, IA, II, and III 24-hour storms for a 24-hour rainfall of 20 in. Plot and compare these storms.

7.2.7 Develop the SCS 6-hr storm for a 6-hr rainfall of 12 in.

7.2.8 Determine the design rainfall intensities (mm/hr) for a 25-year return period, 60-min duration storm using equation (7.2.5) with $c = 12.1$, $m = 0.25$, $e = 0.75$, and $f = 0.125$.

7.2.9 What is the all-season 6-hr PMP (in) for 200 mi^2 near Chicago, Illinois?

7.2.10 Tabulated are the data derived from a drainage basin of 21,100 hectares, given the areas covered with each of the rainfall isohyetal lines.

Interval of isohyets (cm)	0-2	2-4	4-6	6-8	8-10	10-12	12-14
Enclosed area (hectares $\times$ 1000)	5.3	4.4	3.2	2.6	2.3	1.9	1.4

(a) Determine the average depth of precipitation for the storm within the basin by the isohyetal method.

(b) Develop the depth-area relationship based on the data.

7.2.11 For a particular 24-hr storm event on a river basin, an isohyetal map was developed. The corresponding isohyets and area relationship is given in the table below. Based on the given information, (a) develop the depth-area relationship in tabular form for the basin and (b) assuming the maximum point rainfall

is 145 mm, estimate the parameters in the following dimensionless depth-area equation that best fit the existing data, $\frac{\bar{P}_A}{P_{max}} = \exp(-k \times A^n)$, where $\bar{P}_A$ = equivalent uniform rainfall depth for an area of A; P_{max} = maximum point rainfall depth; and k and n = parameters.

Isohyet (mm)	130	120	110	100	90
Incremental area between isohyets (km^2)	143	245	258	290	484

7.2.12 Based on the 100-year rainfall records at the Hong Kong Observatory, the rainfall intensities with an annual exceedance probability of 0.1 (i.e., 10-yr return period) of different durations are given in the table below.

Duration, t_d (min)	15	30	60	120
Intensity, i (mm/h)	161	132	103	74

(a) Determine the least-squares estimates of coefficients a and c in the following rainfall intensity-duration equation and the associated R^2 value, $i = \dfrac{a}{(t_d + 4.5)^c}$, in which i = rainfall intensity (in mm/h) and t_d = storm duration in (minutes).

(b) Estimate the total rainfall depth (in cm) for a 10-year, 4-hr storm event.

7.2.13 Based on the 100-year rainfall records at the Hong Kong Observatory, the rainfall intensities with an annual exceedance probability of 0.1 (i.e., 10-year return period) of different durations are given in the table below.

Return period, T (year)	5	10	50	100	200
Duration, t_d (hr)	6	8	12	18	24
Depth, d (mm/h)	192	255	392	500	622

Consider the following empirical model to be used to fit the above rainfall intensity-duration-frequency (IDF) data,

$$d(T, t_d) = \frac{aT^m}{(t_d + b)^c}, \text{ in which } d = \text{rainfall depth (in mm) corres-}$$

ponding to a storm event of return period T-year and duration t_d-hr.

(a) Describe a least-squares-based procedure to optimally estimate the coefficients a, b, c, and m in the above rainfall IDF model.

(b) Assuming $b = 4.0$, determine the least-squares estimates of coefficients a, c, and m in the above rainfall intensity-duration equation and the associated R^2 value.

(c) Estimate the average rainfall intensity (in mm/hr) for the 25-year, 4-hr storm event.

7.2.14 Consider a rainfall event having 5-min cumulative rainfall record given below:

Time (min)	0	5	10	15	20	25	30
Cumulative rainfall (mm)	0	7	14	23	34	45	58
Time (min)	35	40	45	50	55	60	65
Cumulative rainfall (mm)	70	81	91	100	110	119	125
Time (min)	70	75	80	85	90		
Cumulative rainfall (mm)	131	136	140	140	140		

(a) What is the duration of the entire rainfall event and the corresponding total rainfall amount?

(b) Find the rainfall depth hyetograph (in tabular form) with 10-min time interval for the storm event.

(c) Find the maximum 10-min and 20-min average rainfall intensities (in mm/hr) for the storm event.

7.2.15 Based on available rainfall records at a location, the rainfall intensity with an annual exceedance probability of 0.1 (i.e., 10-year return period) of different durations are given in the table below.

Duration, t_d (min)	15	30	60	120
Intensity, i (mm/h)	160	132	105	75

(a) Determine the least-squares estimates of coefficients a and c in the following rainfall intensity-duration equation and calculate the corresponding R^2 value, $i = \dfrac{a}{(t_d + 4.5)^c}$ in which i = rainfall intensity (in mm/h) and t_d = storm duration (in minutes).

(b) Estimate the total rainfall depth (in mm) for a 10-year, 90-min storm event.

7.2.16 Based on the available rainfall records at the Hong Kong Observatory, the rainfall intensities corresponding to different durations with an annual exceedance probability of 0.1 (i.e. 10-year return period) are given in the table below.

Duration, t_d (min)	15	30	60	120
Intensity, i (mm/h)	161	132	103	74

(a) Determine the least-squares estimates of coefficients a and c in the following rainfall intensity-duration equation, $i = \dfrac{a}{(t_d + 45)^c}$ in which i = rainfall intensity (in mm/h) and t_d = storm duration (in minutes).

(b) Estimate the total rainfall depth (in cm) for a 10-year, 4-hr storm event.

7.2.17 An experimental rectangular plot of 10 km × 12 km has five rain gauge stations as shown in the figure. The storm rainfall and coordinates of the stations are given in the table.

Station	(x, y)	Mean annual rainfall (cm)	Storm rainfall (cm)
A	(1, 3)	128	12.0
B	(8, 11)	114	11.4
C	(3, 10)	136	13.2
D	(5, 8)	144	14.6
E	(7, 5)	109	?

(a) Estimate the missing rainfall amount at station E.

(b) Based on the position of the five rain gauges, construct the Thiessen polygon for them.

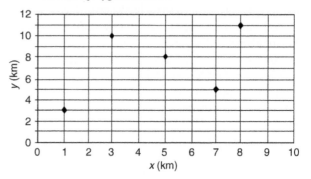

7.3.1 Solve example 7.3.1 for an average net radiation of 92.5 W/m^2. Compare the resulting evaporation rate with that in example 7.3.1.

7.3.2 Solve example 7.3.2 for a roughness height $z_0 = 0.04$ cm. Compare the resulting evaporation rate with that in example 7.3.2.

7.3.3 Solve example 7.3.3 for an average net radiation of 92.5 W/m^2. Compare the resulting evaporation rate with that in example 7.3.3.

7.3.4 Solve example 7.3.4 for an average net radiation of 92.5 W/m^2. Compare the resulting evaporation rate with that in example 7.3.4.

7.3.5 At a certain location during the winter, the average air temperature is 10°C and the net radiation is 40 W/m^2 and during the summer the net radiation is 200 W/m^2 and the temperature is 25°C Compute the evaporation rates using the Priestley–Taylor method.

7.3.6 The average weather conditions are net radiation = 40 W/m^2; air temperature = 28.5° C; relative humidity = 55 percent; and wind speed = 2.7 m/s at a height of 2 m. Calculate the open water evaporation rate in millimeters per day using the energy method, the aerodynamic method, the combination method, and the Priestley–Taylor method. Assume standard atmospheric pressure is 101 kPa and z_o is 0.03 cm.

7.3.7 A 600-hectare farm land receives annual rainfall of 2500 mm. There is a river flowing through the farm land with inflow rate of 5 m^3/s and outflow rate of 4 m^3/s. The annual water storage in the farm land increases by 2.5×10^6 m^3. Based on the hydrologic budget equation, determine the annual evaporation amount (in mm). [Note: 1 hectare = 10,000 m^2]

7.4.1 Determine the infiltration rate and cumulative infiltration curves (0 to 5 h) at 1-hr increments for a clay loam soil. Assume an initial effective saturation of 40 percent and continuous ponding.

7.4.2 Rework problem 7.4.1 using an initial effective saturation of 20 percent.

7.4.3 Rework example 7.4.3 for a sandy loam soil.

7.4.4 Compute the ponding time and cumulative infiltration at ponding for a sandy clay loam soil with a 30-percent initial effective saturation, subject to a rainfall intensity of 2 cm/h.

7.4.5 Rework problem 7.4.4 for a silty clay soil.

7.4.6 Determine the cumulative infiltration and the infiltration rate on a sandy clay loam after 1 hr of rainfall at 2 cm/hr if the initial effective saturation is 25 percent. Assume the ponding depth is negligible in the calculations.

7.4.7 Rework problem 7.4.6 assuming that any ponded water remains stationary over the soil so that the ponded depth must be accounted for in the calculations.

7.4.8 Derive the equation for cumulative infiltration using Horton's equation.

7.4.9 Use the Green–Ampt method to compute the infiltration rate and cumulative infiltration for a silty clay soil ($\eta = 0.479$, $\psi = 29.22$ cm, $K = 0.05$ cm/hr) at 0.25-hr increments up to 4 hr from the beginning of infiltration. Assume an initial effective saturation of 30 percent and continuous ponding.

7.4.10 The parameters for Horton's equation are $f_0 = 3.0$ in/h, $f_c = 0.5$ in/h, and $K = 4.0 \, h^{-1}$. Determine the infiltration rate and cumulative infiltration at 0.25-hr increments up to 4 hr from the beginning of infiltration. Assume continuous ponding.

7.4.11 Derive an equation for ponding time using Horton's equation.

7.4.12 Compute the ponding time and cumulative infiltration at ponding for a sandy clay loam soil of 25 percent initial effective saturation for a rainfall intensity of (a) 2 cm/h, (b) 3 cm/h, and (c) 5 cm/h.

7.4.13 Rework problem 7.4.12 considering a silt loam soil.

7.4.14 Consider a soil with porosity $\eta = 0.43$ and suction $\psi = 11.0$ cm. Before the rainfall event, the initial moisture content $\theta_i = 0.3$. It is known that after one hour of rainfall, the total infiltrated water is 2.0 cm.

(a) Determine the hydraulic conductivity in the Green–Ampt infiltration model.
(b) Estimate the potential total infiltration amount 1.5 hr after the beginning of the storm.

(c) Determine the instantaneous potential infiltration rate at $t = 1.5$ hr.

7.4.15 The following experimental data are obtained from an infiltration study. The objective of the study is to establish a plausible relationship between infiltration rate (f_t) and time (t).

Time, t (mm)	3	15	30	60	90
Infiltration rate, f_t (cm/hr)	8.5	7.8	7.0	6.1	5.6

From the scatter plot of infiltration rate and time, the following relationship between f_t and t is plausible, $f_t = 5.0 + (f_o - 5.0)\exp(-k \times t)$, where f_o is the initial infiltration rate (in cm/hr) and k is the decay constant (in min^{-1}).

(a) Determine the values of the two constants, i. e., f_o and k, by the least-squares method.
(b) Based on the result in part (a), estimate the infiltration rate at time $t = 150$ min.

7.4.16

(a) For a sandy loam soil, using Green–Ampt equation to calculate the infiltration rate (cm/h) and cumulative infiltration depth (cm) after 1, 60, and 150 min if the effective saturation is 40 percent. Assume a continuously ponded condition.
(b) Take the infiltration rates computed in Part (a) at 1 min and 150 min as the initial and ultimate infiltration rates in Horton's equation, respectively. Determine the decay constant, k.
(c) Use the decay constant, k, found in Part (b) to compute the cumulative infiltration at $t = 60$ min by Horton's equation.

7.4.17 Assume that ponded surface occurs at the beginning of a storm event.

(a) Calculate the potential infiltration rates (cm/hr) and potential cumulative infiltration depth (cm) by the Green–Ampt model for loamy sand at times 30 and 60 min. The initial effective saturation is 0.2.
(b) Based on the infiltration rates or cumulative infiltration depth computed in Part (a), determine the two parameters S and K in the following Philip two-term infiltration model.

$$F(t) = S \, t^{1/2} + K \, t; \quad f(t) = 0.5 \, S \, t^{0.5} + K$$

(c) Based on the S and K computed in Part (b), determine the effective rainfall hyetograph for the following total hyetograph of a storm event.

Time (min)	0-30	30-60	60-90	90-120
Intensity (cm/hr)	8.0	15.0	5.0	3.0

(d) Find the value of the Φ-index and the corresponding effective rainfall hyetograph having the total effective rainfall depth obtained in Part (c).

7.4.18 Consider the sandy loam soil with effective porosity $\theta_e = 0.43$, suction $\psi = 11.0$ cm, and hydraulic conductivity $K = 1.1$ cm/h. Before the rainfall event, the initial effective saturation is $S_e = 0.3$. It is known that, after 1 hr of rainfall, the total infiltrated water is 2.0 cm.

(a) Estimate the potential total infiltration amount 1.5 hr after the beginning of the storm event by the Green–Ampt equation.

(b) Also, determine the instantaneous potential infiltration rate at $t = 1.5$ hr.

7.4.19 Considering a plot of land with sandy loam soil, use the Green–Ampt equation to calculate the infiltration rate (cm/h) and cumulative infiltration depth (cm) at $t = 30$ min under the initial degree of saturation of 40 percent and continuously ponding condition. The relevant parameters are: suction head = 6.0 cm; porosity = 0.45; and hydraulic conductivity = 3.00 cm/h.

7.4.20 Suppose that the infiltration of water into a certain type of soil can be described by Horton's equation with the following parameters: initial infiltration rate = 50 mm/h; ultimate infiltration rate = 10 mm/h; and decay constant = 4 h^{-1}. A rain storm event has occurred and its pattern is given below.

Time (min)	0	15	30	45	60	75	90
Cumu. rain (mm)	0	15	25	30	32	33	33

(a) Determine the effective rainfall intensity hyetograph (in mm/h) from the storm event.

(b) What is the percentage of total rainfall infiltrated into the ground?

(c) What assumption(s) do you use in the Part (a) calculation? Justify them.

7.4.21 Consider a 2-hr storm event with a total rainfall amount of 80 mm. The measured direct runoff volume produced by the storm is 40 mm.

(a) Determine the decay constant k in Horton's infiltration model knowing that the initial and ultimate infiltration rates are 30 mm/hr and 2 mm/hr, respectively.

(b) Use the Horton equation developed in Part (a) to determine the effective rainfall intensity hyetograph for the following storm event.

Time (min)	0-10	10-20	20-30
Incremental rainfall depth (mm)	5	20	10

7.4.22 An infiltration study is to be conducted. A quick site investigation indicates that the soil has an effective porosity $\theta_e = 0.40$ and the initial effective saturation $S_e = 0.3$. Also, from the double-ring infiltrometer test, we learn that the cumulative infiltration amounts at $t = 1$ hr and $t = 2$ hr are 1 cm and 1.6 cm, respectively.

(a) It is decided that the Green–Ampt equation is to be used. Determine the parameters suction (ψ) and hydraulic conductivity (K) from the test data. (You can use an iterative procedure or apply simple approximation using $\ln(1 + x) = 2x/(2 + x)$ for a direct solution.)

(b) Also, determine the cumulative infiltration and the corresponding instantaneous potential infiltration rate at $t = 3$ hr. (You can use the Newton method to obtain the solution with accuracy within 0.1mm or the most accurate direct solution approach.)

7.4.23 You are working on a proposal to do some rainfall runoff modeling for a small city nearby. One of the things the city does not have is a hydrology manual. You must decide upon an infiltration methodology for the analysis. You are going to consider methods such as the Φ-index method, empirical methods (such as Horton's and Holtan's), the SCS method, and the Green–Ampt method. Would you choose the Green–Ampt method over the others? Why?

REFERENCES

American Society of Agricultural Engineers, "Advances in Infiltration," in *Proc. National Conf. on Advances in Infiltration*, Chicago, IL, ASAE Publication 11–83, St. Joseph, MI 1983.

Bedient, P. B., and W. C. Huber, *Hydrology and Floodplain Analysis*, second edition, Addison-Wesley, Reading, MA, 1992.

Bhowmik, N., "River Basin Management: An Integrated Approach" *Water International*, International Water Resources Association, vol. 23, no. 2, pp. 84–90, June 1998.

Bonnin, G. M., D. Martin, B. Lin, T. Parzybok, M. Yekta, and D. Riley, "Precipitation-Frequency Atlas of the United States," NOAA Atlas 14, vol. 2, version 3, NOAA, National Weather Service, Silver Spring, MD, 2004.

Bonnin, G. M., D. Martin, B. Lin, T. Parzybok, M. Yekta, and D. Riley, "Precipitation-Frequency Atlas of the United States," NOAA Atlas 14, vol. 1, version 4, NOAA, National Weather Service, Silver Spring, MD, 2006.

Brakensiek, D. L., R. L. Engleman, and W. J. Rawls, "Variation within Texture Classes of Soil Water Parameters," *Trans. Am. Soc. Agric. Eng.*, vol. 24, no. 2, pp. 335–339, 1981.

Bras, R. L., *Hydrology: An Introduction to Hydrologic Science*, Addison-Wesley, Reading, MA, 1990.

Brooks, R. H., and A. T. Corey, "Hydraulic Properties of Porous Media," *Hydrology Papers*, no. 3, Colorado State University, Fort Collins, CO, 1964.

Chow, V. T., *Handbook of Applied Hydrology*, McGraw Hill, New York 1964.

Chow, V. T., D. R. Maidment, and L. W. Mays, *Applied Hydrology*, McGraw-Hill, New York, 1988.

Demissee, M., and L. Keefer, "Watershed Approach for the Protection of Drinking Water Supplies in Central Illinois," *Water International*, IWRA, vol. 23, no. 4, pp. 272–277, 1998.

Devanrs, M., and G. F. Gifford, "Applicability of the Green and Ampt Infiltration Equation to Rangelands," *Water Resource Bulletin*, vol. 22, no. 1, pp. 19–27, 1986.

Feynman, R. P., R. B. Leighton, and M. Sands, *The Feynman Lecture Notes on Physics*, vol. I, Addison-Wesley, Reading, MA, 1963.

Frederick, R. H., V. A. Meyers, and E. P. Auciello, "Five to 60-Minute Precipitation Frequency for the Eastern and Central United States," NOAA Technical Memo NWS HYDRO-35, National Weather Service, Silver Spring, MD, June 1977.

Green, W. H., and G. A. Ampt, "Studies on Soil Physics. Part I: The Flow of Air and Water Through Soils," *J. Agric. Sci.*, vol. 4, no. 1, pp. 1–24, 1911.

Gupta, R. S., *Hydrology and Hydraulic Systems*, Prentice-Hall, Englewood Cliffs, NJ, 1989.

Hansen, E. M., L. C. Schreiner, and J. F. Miller, "Application of Probable Maximum Precipitation Estimates—United States East of 105th Meridian," NOAA Hydrometeorological Report 52, U.S. National Weather Service, Washington, DC, 1982.

Hershfield, D. M., "Rainfall Frequency Atlas of the United States for Durations from 30 Minutes to 24 Hours and Return Periods from 1 to 100 Years," Tech. Paper 40, U.S. Department of Commerce, Weather Bureau, Washington, DC, 1961.

Horton, R. E., "The Role of Infiltration in the Hydrologic Cycle," *Trans. Am. Geophysical Union*, vol. 14, pp. 446–460, 1933.

Huff, F. A., "Time Distribution of Rainfall in Heavy Storms," *Water Resources Research*, vol. 3, no. 4, pp. 1007–1019, 1967.

Maidment, D. R. (editor-in-chief), *Handbook of Hydrology*, McGraw-Hill, New York, 1993.

Marsh, W. M., *Earthscape: A Physical Geography*, John Wiley & Sons, New York, 1987.

Marsh, W. M., and J. Dozier, *Landscape: An Introduction to Physical Geography*, John Wiley & Sons, New York, 1986.

Masch, F. D., *Hydrology*, Hydraulic Engineering Circular No. 19, FHWA-10-84-15. Federal Highway Administration, U.S. Department of the Interior, McLean, VA, 1984.

McCuen, R. H., *Hydrologic Analysis and Design*, 2nd edition Prentice-Hall. Englewood Cliffs, NJ, 1998.

Miller J. F., R. H. Frederick, and R. J. Tracey, *Precipitation Frequency Atlas of the Coterminous Western United States (by States)*, NOAA Atlas 2, 11 vols., National Weather Service, Silver Spring, MD, 1973.

Moore, W. L., E. Cook, R. S. Gooch, and C. F. Nordin, Jr., "The Austin, Texas, Flood of May, 24-25, 1981," National Research Council, Committee on Natural Disasters, Commission on Engineering and Technical Systems National Academy Press, Washington, DC, 1982.

Parrett, C., N. B. Melcher, and R. W. James, Jr., "Flood Discharges in the Upper Mississippi River Basin," in *Floods in the Upper Mississippi River Basin*," U.S. Geological Survey Circular, 1120-A U.S. Government Printing Office, Washington, DC, 1993.

Philip, J. R., "Theory of Infiltration: 1. The Infiltration Equation and Its Solution," *Soil Science* 83, pp. 345–357, 1957.

Philip, J. R., "The Theory of Infiltration," in *Advances in Hydro Sciences*, edited by V. T. Chow, vol. 5, pp. 215–296, 1969.

Ponce, V. M., *Engineering Hydrology: Principles and Practices*, Prentice-Hall, Englewood Cliffs, NJ, 1989.

Priestley, C. H. B., and R. J. Taylor, "On the Assessment of Surface Heat Flux and Evaporation Using Large-Scale Parameter," *Monthly Weather Review*, vol. 100, pp. 81–92, 1972.

Rawls, W. J., D. L. Brakensiek, and N. Miller, "Green-Ampt Infiltration Parameters from Soils Data," *J. Hydraulic Div., ASCE*, vol. 109, no. 1, pp. 62–70, 1983.

Richards, L. A., "Capillary Conduction of Liquids Through Porous Mediums," *Physics*, vol. 1, pp. 318–333, 1931.

Roberson, J. A., J. J. Cassidy, and M. H. Chaudhry, *Hydraulic Engineering*, 2nd edition, John Wiley & Sons, New York, 1998.

Schreiner, L. C., and J. T. Riedel, Probable Maximum Precipitation Estimates, United States East of the 105th Meridian, NOAA Hydrometeorological Report no. 51, National Weather Service, Washington, DC, June 1978.

Singh, V. P., *Elementary Hydrology*, Prentice-Hall, Englewood Cliffs, NJ, 1992.

Thornthwaite, C. W., and B. Holzman, "The Determination of Evaporation from Land and Water Surface," *Monthly Weather Review*, vol. 67, pp. 4–11, 1939.

U.S. Army Corps of Engineers, Hydrologic Engineering Center, HEC-1, *Flood Hydrograph Package*, User's Manual, Davis, CA, 1990.

U.S. Department of Agriculture Soil Conservation Service, *A Method far Estimating Volume and Rate of Runoff in Small Watersheds*, Tech. Paper 149, Washington, DC, 1973.

U.S. Department of Agriculture Soil Conservation Service, *Urban Hydrology for Small Watersheds,* Tech. Release no. 55, Washington, DC, 1986.

U.S. Department of Commerce, *Probable Maximum Precipitation Estimates, Colorado River and Great Basin Drainages,* Hydrometeorological Report no 49, NOAA, National Weather Service, Silver Spring, MD, 1977.

U.S. Environmental Data Services, *Climate Atlas of the US.*, U.S. Environment Printing Office, Washington, DC, pp. 43–44, 1968.

U.S. National Academy of Sciences, *Understanding Climate Change*, National Academy Press, Washington, DC, 1975.

U.S. National Academy of Sciences, *Safety of Existing Dams; Evaluation and Improvement*, National Academy Press, Washington, DC, 1983.

U.S. National Research Council, *Global Change in the Geosphere-Biosphere*, National Academy Press, Washington, DC, 1986.

U.S. National Research Council, Committee on Opportunities in the Hydrologic Science, Water Science and Technology Board, *Opportunities in the Hydrologic Sciences*, National Academy Press, Washington, DC, 1991.

U.S. Weather Bureau, *Two- to Ten-Day Precipitation for Return Periods of 2 to 100 Years in the Contiguous United States*, Tech. Paper 49, Washington, DC, 1964.

Viessman, W. Jr., and G. L. Lewis, *Introduction to Hydrology*, 4th edition, Harper and Row, New York, 1996.

Wanielista, M., R. Kersten, and R. Eaglin, *Hydrology: Water Quantity and Quality Control*, John Wiley & Sons, New York, 1997.

Chapter **8**

Surface Runoff

8.1 DRAINAGE BASINS AND STORM HYDROGRAPHS

8.1.1 Drainage Basins and Runoff

As defined in Chapter 7, *drainage basins, catchments*, and *watersheds* are three synonymous terms that refer to the topographic area that collects and discharges surface streamflow through one outlet or mouth. The study of topographic maps from various physiographic regions reveals that there are several different types of drainage patterns (Figure 8.1.1). *Dendritic patterns* occur where rock and weathered mantle offer uniform resistance to erosion. Tributaries branch and erode headward in a random fashion, which results in slopes with no predominant direction or orientation. *Rectangular*

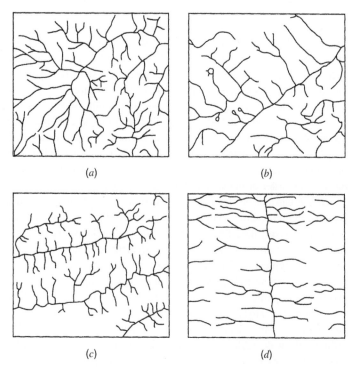

(a) (b)

(c) (d)

Figure 8.1.1 Common drainage patterns: (*a*) Dendritic; (*b*) Rectangular; (*c*) Trellis on folded terrain; (*d*) Trellis on mature, dissected coastal plain (from Hewlett and Nutter (1969)).

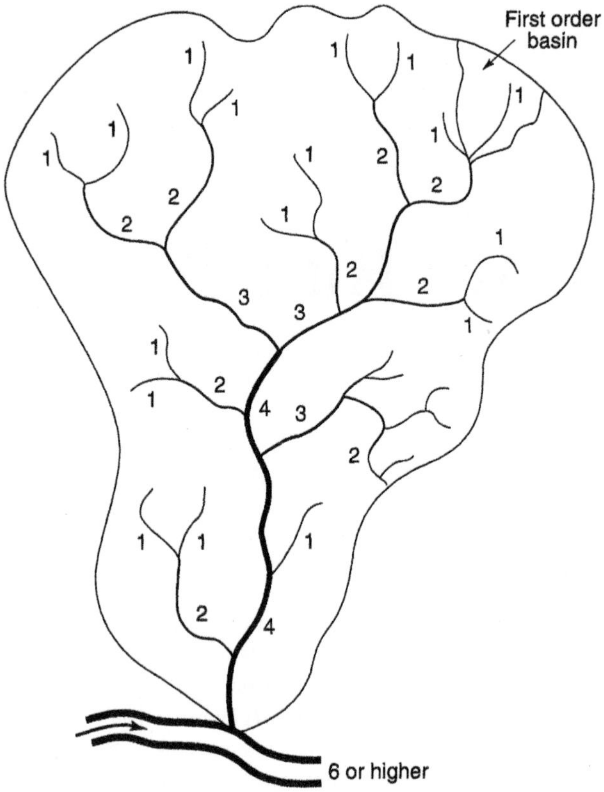

Figure 8.1.2 Stream orders (from Hewlett and Nutter (1969)).

patterns occur in faulted areas where streams follow a more easily eroded fractured rock in fault lines. *Trellis patterns* occur where rocks being dissected are of unequal resistance so that the extension and daunting of tributaries is most rapid on least resistant areas.

Streams can also be classified within a basin by systematically ordering the network of branches. Horton (1945) originated the quantitative study of stream networks by developing an ordering system and laws relating to the number and length of streams of different order (see Chow, et al. 1988). Strahler (1964) slightly modified Horton's stream ordering to that shown in Figure 8.1.2. Essentially, each non-branching channel segment is a *first-order stream*. Streams, which receive only first-order segments, are *second order*, and so on. When a channel of lower order joins a channel of higher order, the channel downstream retains the higher of the two orders. The order of a drainage basin is the order of the stream draining its outlet.

Streams can also be classified by the period of time during which flow occurs. *Perennial streams* have a continuous flow regime typical of a well-defined channel in a humid climate. *Intermittent streams generally* have flow occurring only during the wet season (50 percent of the time or less). *Ephemeral streams* generally have flow occurring during and for short periods after storms. These streams are typical of climates without very well-defined streams.

The *stream flow hydrograph* or *discharge hydrograph* is the relationship of flow rate (discharge) and time at a particular location on a stream (see Figure 8.1.3a). The *hydrograph* is "an integral expression of the physiographic and climatic characteristics that govern the relation between rainfall and runoff of a particular drainage basin" (Chow, 1964). Figures 8.1.3b and c illustrate the rising and falling of the water table in response to rainfall. Figures 8.1.3d and e illustrate that the flowing stream channel network expands and contracts in response to rainfall.

The spatial and temporal variations of rainfall and the concurrent variation of the abstraction processes define the runoff characteristics from a given storm. When the local abstractions have been

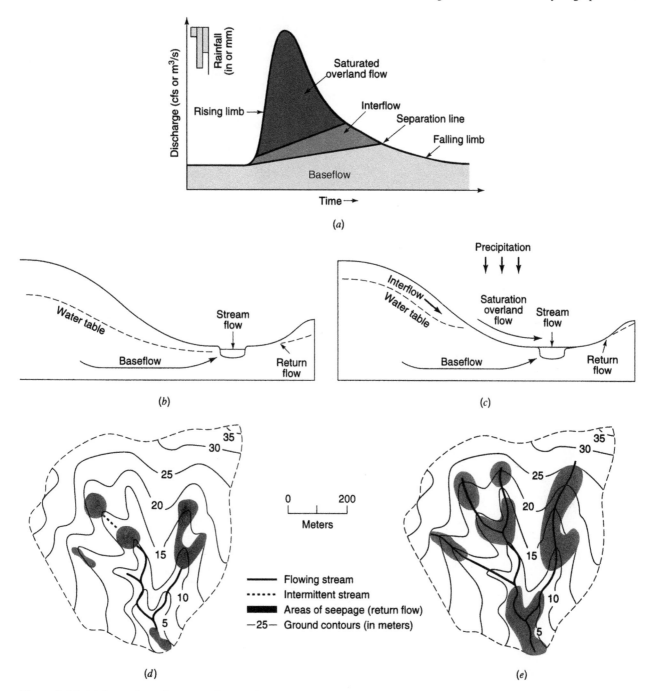

Figure 8.1.3 (*a*) Separatism of sources of streamflow on an idealized hydrograph; (*b*) Sources of streamflow on a hillslope profile during a dry period; (*c*) During a rainfall event; (*d*) Stream network during dry period; (*e*) Stream network extended during and after rainfall (from Mosley and McKerchar (1993)).

accomplished for a small area of a watershed, water begins to flow overland as *overland flow* and eventually into a drainage channel (in a gulley or stream valley). When this occurs, the hydraulics of the natural drainage channels have a large influence on the runoff characteristics from the watershed. Some of the factors that determine the hydraulic character of the natural drainage system include:

(a) drainage area, (b) slope, (c) hydraulic roughness, (d) natural and channel storage, (e) stream length, (f) channel density, (g) antecedent moisture condition, and (h) other factors such as vegetation, channel modifications, etc. The individual effects of each of these factors are difficult, and in many cases impossible, to quantify. Figure 8.1.4 illustrates the effects of some of the drainage

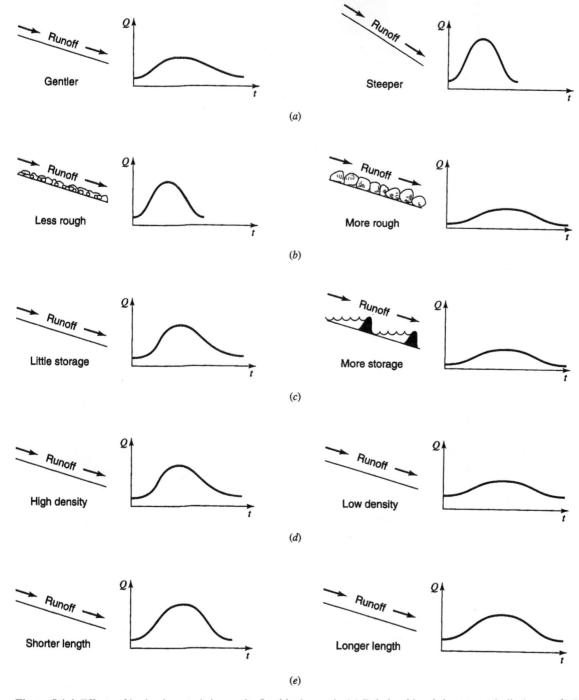

Figure 8.1.4 Effects of basin characteristics on the flood hydrograph. (*a*) Relationship of slope to peak discharge. (*b*) Relationship of hydraulic roughness to runoff. (*c*) Relationship of storage to runoff. (*d*) Relationship of drainage density to runoff. (*e*) Relationship of channel length to runoff (from Masch (1984)).

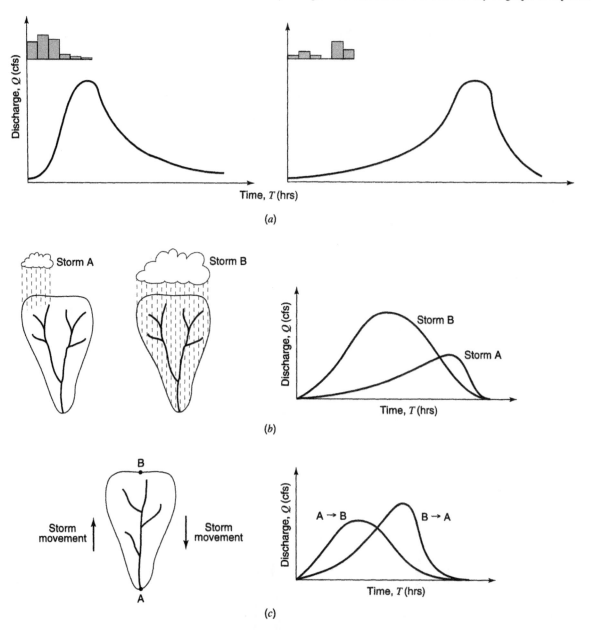

Figure 8.1.5 Effects of storm shape, size, and movement on surface runoff. (*a*) Effect of time variation of rainfall intensity on the surface runoff. (*b*) Effect of storm size on surface runoff. (*c*) Effect of storm movement on surface runoff (from Masch (1984)).

basin characteristics on the surface runoff (discharge hydrographs) and Figure 8.1.5 illustrates the effects of storm shape, size, and movement on surface runoff.

8.2   HYDROLOGIC LOSSES, RAINFALL EXCESS, AND HYDROGRAPH COMPONENTS

Rainfall excess, or *effective rainfall*, is that rainfall that is neither retained on the land surface nor infiltrated into the soil. After flowing across the watershed surface, rainfall excess becomes direct runoff at the watershed outlet. The graph of rainfall excess versus time is the rainfall excess hyetograph. As shown in Figure 8.2.1, the difference between the observed total rainfall hyetograph

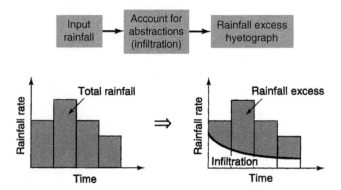

Figure 8.2.1 Concept of rainfall excess. The difference between the total rainfall hyetograph on the left and the total rainfall excess hyetograph on the right is the abstraction (infiltration).

and the rainfall excess hyetograph is the *abstractions*, or *losses*. Losses are primarily water absorbed by infiltration with some allowance for interception and surface storage. The relationships of rainfall, infiltration rate, and cumulative infiltration are shown in Figure 8.2.2. Figure 8.2.2 illustrates the relationships for rainfall and runoff data of an actual storm that can be obtained from data recorded by the U.S. Geological Survey. Using the rainfall data, rainfall hyetographs can be computed.

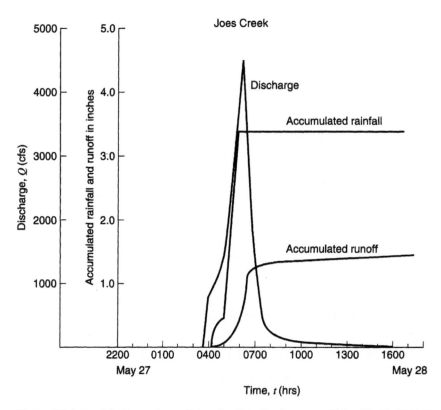

Figure 8.2.2 Precipitation and runoff data for Joes Creek, storm of May 27–28, 1978 (from Masch (1984)).

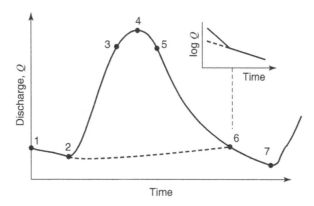

Figure 8.2.3 Components of a streamflow hydrograph: (1-2) baseflow recession; (2-3) rising limb; (3-5) crest segment; (4) peak; (5-6) falling limb; and (6-7) baseflow recession.

8.2.1 Hydrograph Components

There are several sources that make up a hydrograph (total runoff hydrograph) including direct surface runoff, interflow, baseflow (groundwater), and channel precipitation. Figure 8.1.3a illustrates the direct runoff (saturated overland flow), interflow, and baseflow. For hydrologic purposes, the total runoff consists of only two parts, direct runoff and baseflow. Baseflow is the result of water entering the stream from the groundwater that discharges from the aquifer. Figure 8.2.3 defines the components of the hydrograph, showing the baseflow recession (1-2) and (6-7), the rising limb (2-3), the crest segment (3-5), and the falling limb (5-6).

The process of defining the baseflow is referred to as baseflow separation. A number of baseflow separation methods have been suggested. Baseflow recession curves (Figure 8.2.3) can be described in the form of an exponential decay

$$Q(t_2) = Q(t_2)e^{-k(t_2 - t_1)}, \quad t_2 > t_1 \tag{8.2.1}$$

where k is the exponential decay constant having dimensions of $(time)^{-1}$. With a known streamflow runoff hydrograph, the decay constant can be determined by plotting the curve of log Q versus time as shown in Figure 8.2.3 or by using a least-squares procedure. Baseflow recession curves for particular streams can be superimposed to develop a normal depletion curve or master baseflow recession curve.

8.2.2 Φ-Index Method

The Φ-index is a constant rate of abstractions (in/hr or cm/hr) that can be used to approximate infiltration. Using an observed rainfall pattern and the resulting known volume of direct runoff, the Φ-index can be determined. Using the known rainfall pattern, Φ is determined by choosing a time interval Δt, identifying the number of rainfall intervals N of rainfall that contribute to the direct runoff volume, and then subtracting $\Phi \cdot \Delta t$ from the observed rainfall in each time interval. The values of Φ and N will need to be adjusted so that the volume of direct runoff (r_d) and excess rainfall are equal

$$r_d = \sum_{n=1}^{N} (R_n - \Phi \cdot \Delta t) \tag{8.2.2}$$

where R_n is the observed rainfall (in or cm) in time interval n.

EXAMPLE 8.2.1

Consider the following storm event given below for a small catchment of 120 hectares. For a baseflow of 0.05 m³/s, (a) compute the volume of direct runoff (in mm), (b) assuming the initial losses (abstractions) are 5 mm, determine the value of Φ-index (in mm/hr), and (c) the corresponding effective rainfall intensity hyetograph (in mm/hr).

Time (min)	0	5	10	15	20	25	30
Cumulative rainfall Depth (mm)	0	5	20	35	45		
Discharge (m³/s)	0.05	0.05	0.25	0.65	0.35	0.15	0.05

SOLUTION

The following table shows the analysis to obtain the incremental rainfalls and rainfall intensities. There are three remaining pulses of rainfall after eliminating the first 5 mm as initial abstractions. For each of the first two rainfall increments after the initial losses are accounted for, the incremental rainfall volume is $\Phi \cdot \Delta t = \Phi(5 \text{ min})(1 \text{ hr}/60 \text{ min}) = 15$ mm, so solving $\Phi = 180$ mm/hr. For the third interval, $\Phi \cdot \Delta t = \Phi(5 \text{ min})(1 \text{ hr}/60 \text{ min}) = 10$, so the rainfall intensity is 120 mm/hr.

Time (min)	Discharge (m³/sec)	Direct runoff (m³/sec)	Cumulative rainfall (mm)	Incremental rainfall (mm)	Rainfall intensity (mm/hr)
0	0.05	0	0		
5	0.05	0	5 (initial loss)	0	
10	0.25	0.20	20	15	180
15	0.65	0.60	35	15	180
20	0.35	0.30	45	10	120
25	0.15	0.10			
30	0.05	0			

(a) The direct runoff volume $= (0.2 + 0.6 + 0.3 + 0.1)$ m²/sec (5 min.) (60 sec/min) $= 360$ m³, which converts to $r_d = 360$ m³/(120 hectares $\times$ 10,000 m²/hectare) $= 0.3$ mm.

(b) There are three pulses of rainfall after eliminating the first 5 mm as initial abstractions. Considering the two largest rainfall pulses, the rainfall volume above the 120 mm/hr level is 10 mm, and

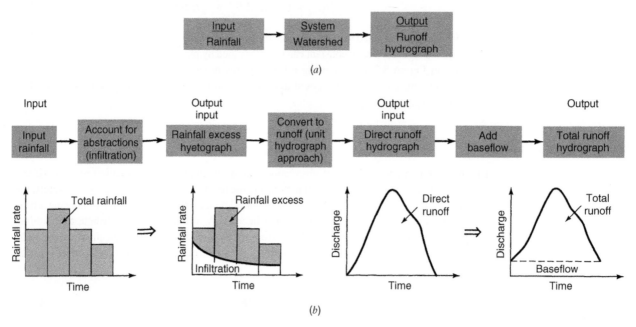

(a)

(b)

Figure 8.2.4 Storm runoff hydrographs. (*a*) Rainfall-runoff modeling; (*b*) Steps to define storm runoff.

because the direct rainfall volume is only 0.3 mm, then Φ is above the 120 mm/hr level. The direct runoff is 0.3 mm, so applying equation (8.2.2) to the two largest pulses, $r_d = 0.3$ mm $= 2[(15$ mm $- \Phi(5$ min)(1 hr/60 min)]. Solving, $\Phi = 178.2$ mm/hr.

(c) The excess rainfall hyetograph has an intensity of 180 mm/hr $-$ 178.2 mm/hr $= 1.8$ mm/hr for the two rainfall pulses.

8.2.3   Rainfall-Runoff Analysis

The objective of many hydrologic design and analysis problems is to determine the surface runoff from a watershed due to a particular storm. This process is commonly referred to as *rainfall-runoff analysis*. The processes (steps) are illustrated in Figure 8.2.4 to determine the *storm runoff hydrographs* (or streamflow or discharge hydrograph) using the unit hydrograph approach.

8.3   RAINFALL-RUNOFF ANALYSIS USING UNIT HYDROGRAPH APPROACH

The objective of *rainfall-runoff analysis* is to develop the runoff hydrograph as illustrated in Figure 8.2.4a, where the system is a watershed or river catchment, the input is the rainfall hyetograph, and the output is the runoff or discharge hydrograph. Figure 8.2.4b defines the processes (steps) to determine the runoff hydrograph from the rainfall input using the *unit hydrograph approach*.

A *unit hydrograph* is the direct runoff hydrograph resulting from 1 in (or 1 cm in SI units) of excess rainfall generated uniformly over a drainage area at a constant rate for an effective duration. The unit hydrograph is a simple linear model that can be used to derive the hydrograph resulting from any amount of excess rainfall. The following basic assumptions are inherent in the unit hydrograph approach:

1. The excess rainfall has a constant intensity within the effective duration.
2. The excess rainfall is uniformly distributed throughout the entire drainage area.
3. The base time of the direct runoff hydrograph (i.e., the duration of direct runoff) resulting from an excess rainfall of given duration is constant.
4. The ordinates of all direct runoff hydrographs of a common base time are directly proportional to the total amount of direct runoff represented by each hydrograph.
5. For a given watershed, the hydrograph resulting from a given excess rainfall reflects the unchanging characteristics of the watershed.

The following *discrete convolution equation* is used to compute direct runoff hydrograph ordinates Q_n, given the rainfall excess values P_m and given the unit hydrograph ordinates U_{n-m+1} (Chow et al., 1988):

$$Q_n = \sum_{m=1}^{n \leq M} P_m U_{n-m+1} \quad \text{for } n = 1, 2, \ldots, N \tag{8.3.1}$$

where n represents the direct runoff hydrograph time interval and m represents the precipitation time interval $(m = 1, \ldots, n)$.

The reverse process, called *deconvolution*, is used to derive a unit hydrograph given data on P_m and Q_n. Suppose that there are M pulses of excess rainfall and N pulses of direct runoff in the storm considered; then N equations can be written for Q_n, $n = 1, 2 \ldots, N$, in terms of $N - M + 1$ unknown values of the unit hydrograph, as shown in Table 8.3.1. Figure 8.3.1 diagramatically illustrates the calculation and the runoff contribution by each rainfall input pulse.

Once the unit hydrograph has been determined, it may be applied to find the direct runoff and streamflow hydrographs for given storm inputs. When a rainfall hyetograph is selected, the abstractions are subtracted to define the excess rainfall hyetograph. The time interval used in

Table 8.3.1   The Set of Equations for Discrete Time Convolution

$$Q_1 = P_1 U_1$$
$$Q_2 = P_2 U_1 + P_1 U_2$$
$$Q_3 = P_3 U_1 + P_2 U_2 + P_1 U_3$$
$$\cdots$$
$$Q_M = P_M U_1 + P_{M-1} U_2 + \cdots + P_1 U_M$$
$$Q_{M+1} = 0 + P_M U_2 + \cdots + P_2 U_M + P_1 U_{M+1}$$
$$\cdots$$
$$Q_{N-1} = 0 + 0 + \cdots + 0 + 0 + \cdots + P_M U_{N-M} + P_{M-1} U_{N-M+1}$$
$$Q_N = 0 + 0 + \cdots + 0 + 0 + \cdots + 0 + P_M U_{N-M+1}$$

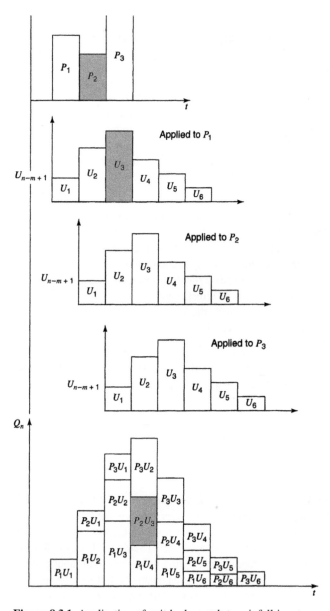

Figure 8.3.1 Application of unit hydrograph to rainfall input.

defining the excess rainfall hyetograph ordinates must be the same as that for which the unit hydrograph is specified.

EXAMPLE 8.3.1

The 1-hr unit hydrograph for a watershed is given below. Determine the runoff from this watershed for the storm pattern given. The abstractions have a constant rate of 0.3 in/h.

Time (h)	1	2	3	4	5	6
Precipitation (in)	0.5	1.0	1.5	0.5		
Unit hydrograph (cfs)	10	100	200	150	100	50

SOLUTION

The calculations are shown in Table 8.3.2. The 1-hr unit hydrograph ordinates are listed in column 2 of the table; there are $L = 6$ unit hydrograph ordinates, where $L = N - M + 1$. The number of excess rainfall intervals is $M = 4$. The excess precipitation 1-hr pulses are $P_1 = 0.2$ in, $P_2 = 0.7$ in, $P_3 = 1.2$ in, and $P_4 = 0.2$ in, as shown at the top of the table. For the first time interval $n = 1$, the discharge is computed using equation (8.3.1):

$$Q_1 = P_1 U_1 = 0.2 \times 10 = 2 \text{ cfs}$$

For the second time interval, $n = 2$,

$$Q_2 = P_1 U_2 + P_2 U_1 = 0.2 \times 100 + 0.7 \times 10 = 27 \text{ cfs}$$

and similarly for the remaining direct runoff hydrograph ordinates. The number of direct runoff ordinates is $N = L + M - 1 = 6 + 4 - 1 = 9$; i.e., there are nine nonzero ordinates, as shown in Table 8.3.2. Column 3 of Table 8.3.2 contains the direct runoff corresponding to the first rainfall pulse, $P_1 = 0.2$ in, and column 4 contains the direct runoff from the second rainfall pulse, $P_2 = 0.2$ in, etc. The direct runoff hydrograph, shown in column 7 of the table, is obtained, from the principle of superposition, by adding the values in columns 3–6.

Table 8.3.2 Calculation of the Direct Runoff Hydrograph

(1)	(2)	(3)	(4)	(5)	(6)	(7)
				Total precipitation (in)		
		0.5	1	1.5	0.5	
Time (hr)	Unit hydrograph (cfs/in)	Excess precipitation (in)				Direct runoff (cfs)
		0.2	0.7	1.2	0.2	
0	0	0	0			0
1	10	2	0	0		2
2	100	20	7	0	0	27
3	200	40	70	12	0	122
4	150	30	140	120	2	292
5	100	20	105	240	20	385
6	50	10	70	180	40	300
7	0	0	35	120	30	185
8			0	60	20	80
9				0	10	10
10					0	0

EXAMPLE 8.3.2

Determine the 1-hr unit hydrograph for a watershed using the precipitation pattern and runoff hydrograph below. The abstractions have a constant rate of 0.3 in/hr, and the baseflow of the stream is 0 cfs.

Time (h)	1	2	3	4	5	6	7	8	9	10
Precipitation (in)	0.5	1.0	1.5	0.5						
Runoff (cfs)	2	27	122	292	385	300	185	80	10	0

SOLUTION

Using the deconvolution process, we get $Q_1 = P_1 U_1$

so that for $P_1 = 0.5 - 0.3 = 0.2$ in and $Q_1 = 2$ cfs,

$U_1 = Q_1/P_1 = 2/0.2 = 10$ cfs.

$Q_2 = P_1 U_2 + P_2 U_1$, so that

$U_2 = (Q_2 - P_2 U_1)/P_1$

where

$P_2 = 1.0 - 0.3 = 0.7$ in and $Q_2 = 27$ cfs.

$U_2 = (27 - 0.7(10))/0.2 = 100$ cfs and

$Q_3 = P_1 U_3 + P_2 U_2 + P_3 U_1$

then

$U_3 = (Q_3 - P_2 U_2 - P_3 U_1)/P_1$, so that

$U_3 = (122 - 0.7(100) - 1.2(10))/0.2 = 200$ cfs.

The rest of the unit hydrograph ordinates can be calculated in a similar manner.

8.4 SYNTHETIC UNIT HYDROGRAPHS

8.4.1 Snyder's Synthetic Unit Hydrograph

When observed rainfall-runoff data are not available for unit hydrograph determination, a *synthetic unit hydrograph* can be developed. A unit hydrograph developed from rainfall and streamflow data in a watershed applies only to that watershed and to the point on the storm where the streamflow data were measured. Synthetic unit hydrograph procedures are used to develop unit hydrographs for other locations on the stream in the same watershed or other watersheds that are of similar character.

One of the most commonly used synthetic unit hydrograph procedures is Snyder's synthetic unit hydrograph. This method relates the time from the centroid of the rainfall to the peak of the unit hydrograph to geometrical characteristics of the watershed. To determine the regional parameters C_t and C_p, one can use values of these parameters determined from similar watersheds. C_t can be determined from the relationship for the *basin lag*:

$$t_p = C_1 C_t (L \cdot L_c)^{0.3} \qquad (8.4.1)$$

where C_1, L, and L_c are defined in Table 8.4.1. Solving equation (8.4.1) for C_t gives

$$C_t = \frac{t_p}{C_1 (L \cdot L_c)^{0.3}} \qquad (8.4.2)$$

To compute C_t for a gauged basin, L and L_c are determined for the gauged watershed and t_p from the derived unit hydrograph for the gauged basin.

To compute the other required parameter C_p, the expression for peak discharge of the standard unit hydrograph can be used:

$$Q_p = \frac{C_2 C_p A}{t_p} \tag{8.4.3}$$

or for a unit discharge (discharge per unit area)

$$q_p = \frac{C_2 C_p}{t_p} \tag{8.4.4}$$

Solving equation (8.4.4) for C_p gives

$$C_p = \frac{q_p t_p}{C_2} \tag{8.4.5}$$

This relationship can be used to solve for C_p for the ungauged watershed, knowing the terms in the right-hand side. Table 8.4.1 defines the steps for this procedure.

Section 8.8 discusses the SCS-unit hydrograph procedure.

EXAMPLE 8.4.1   A watershed has a drainage area of 5.42 mi^2; the length of the main stream is 4.45 mi, and the main channel length from the watershed outlet to the point opposite the center of gravity of the watershed is 2.0 mi. Using $C_t = 2.0$ and $C_p = 0.625$, determine the standard synthetic unit hydrograph for this basin. What is the standard duration? Use Snyder's method to determine the 30-min unit hydrograph parameter.

SOLUTION   For the standard unit hydrograph, equation (8.4.1) gives

$$t_p = C_1 C_t (LL_c)^{0.3} = 1 \times 2 \times (4.45 \times 2)^{0.3} = 3.85 \text{ hr}$$

The standard rainfall duration $t_r = 3.85/5.5 = 0.7$ hr. For a 30-min unit hydrograph, $t_R = 30$ min $= 0.5$ hr. The basin lag $t_{pR} = t_p - (t_r - t_R)/4 = 3.85 - (0.7 - 0.5)/4 = 3.80$ hr. The peak flow for the required unit hydrograph is $= q_p t_p / t_{pR}$, and substituting equation (8.4.4) in the previous equation, $q_{pR} = q_p t_p / t_{pR} = (C_2 C_p / t_p) t_p / t_{pR} = C_2 C_p / t_{pR}$, so that $q_{pR} = 640 \times 0.625/3.80 = 105.26$ cfs $(\text{in} \cdot \text{mi}^2)$, and the peak discharge is $Q_{pR} = q_{pR} A = 105.26 \times 5.42 = 570$ cfs/in.

The widths of the unit hydrograph are computed next. At 75 percent of the peak discharge, $W_{75} = C_{W_{75}} q_{pR}^{-1.08} = 440 \times 105.26^{-1.08} = 2.88$ hr. At 50 percent of the peak discharge, $W_{50} = C_{W_{50}} q_{pR}^{-1.08} = 770 \times 105.26^{-1.08} = 5.04$ hr.

The base time t_b may be computed assuming a triangular shape. This, however, does not guarantee that the volume under the unit hydrograph corresponds to 1 in (or 1 cm, for SI units) of excess rainfall. To overcome this, the value of t_b may be exactly computed taking into account the values of W_{50} and W_{75} by solving the equation in step 5 of Table 8.4.1 for t_b:

$$t_b = 2581 A/Q_{pR} - 1.5 W_{50} - W_{75}$$

so that, with $A = 5.42$ mi^2, $W_{50} = 5.04$ hr, $W_{75} = 2.88$ hr, and $Q_{pR} = 570$ cfs/in, $T_b = 2581(5.42)/570 - 1.5 \times 5.04 - 2.88 = 14.1$ hr.

8.4.2   Clark Unit Hydrograph

The Clark unit hydrograph procedure (Clark, 1945) is based upon using a time-area relationship of the watershed that defines the cumulative area of the watershed contributing runoff to the watershed outlet as a function of time. Ordinates of the time-area relationship are converted to a volume of runoff per second for an excess (1 cm or 1 in) and interpolated to the given time interval to define a

Table 8.4.1   Steps to Compute Snyder's Synthetic Unit Hydrograph

Step 0	Measured information from topography map of watershed

Step 0   Measured information from topography map of watershed
- L = main channel length in mi (km)
- L_c = length of the main stream channel from outflow point of watershed to a point opposite the centroid of the watershed in mi (km)
- A = watershed area in mi^2 (km^2)

Regional parameters C_t and C_p determined from similar watersheds.

Step 1   Determine time to peak (t_p) and duration (t_r) of the standard unit hydrograph:

$$t_p = C_1 C_t (L \cdot L_c)^{0.3} \quad \text{(hours)}$$
$$t_r = t_p/5.5 \quad\quad\quad \text{(hours)}$$

where $C_1 = 1.0$ (0.75 for SI units)

Step 2   Determine the time to peak t_{pR} for the desired duration t_R:

$$t_{pR} = t_p + 0.25(t_R - t_r) \quad \text{(hours)}$$

Step 3   Determine the peak discharge, Q_{pR}, in cfs/in ((m^3/s)/cm in SI units)

$$Q_{pR} = \frac{C_2 C_p A}{t_{pR}}$$

where $C_2 = 640$ (2.75 for SI units)

Step 4   Determine the width of the unit hydrograph at $0.5 Q_{pR}$ and $0.75 Q_{pR}$. W_{50} is the width at 50% of the peak given as

$$W_{50} = \frac{C_{50}}{(Q_{pR}/A)^{1.08}}$$

where $C_{50} = 770$ (2.14 for SI units). W_{75} is the width at 75% of the peak given as

$$W_{75} = \frac{C_{75}}{(Q_{pR}/A)^{1.08}}$$

where $C_{75} = 440$ (1.22 for SI units)

Step 5   Determine the base, T_b, such that the unit hydrograph represents 1 in (1 cm in SI units) of direct runoff volume;

$$1 \text{ in} = \left[\left(\frac{W_{50} + T_b}{2} \right)(0.5 Q_{pR}) + \left(\frac{W_{75} + W_{50}}{2} \right)(0.25 Q_{pR}) + \frac{1}{2} W_{75}(0.25 Q_{pR}) \right] \left(\text{hr} \times \frac{\text{ft}^3}{\text{sec}} \right)$$

$$\left(\frac{1}{A(\text{mi})^2} \times \frac{1 \text{ mi}^2}{(5,280)^2 \text{ ft}^2} \times \frac{12 \text{ in}}{\text{ft}} \times \frac{3,600 \text{ sec}}{\text{hr}} \right)$$

Solving for T_b, we get

$$T_b = 2,581 \frac{A}{Q_{pR}} - 1.5 W_{50} - W_{75}$$

for A in mi^2, Q_{pR} in cfs, W_{50} and W_{75} in hours.

Step 6   Define known points of the unit hydrograph. $\left(T_p = t_{pR} + \dfrac{t_R}{2} \right)$

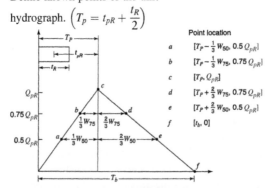

	Point location
a	$[T_P - \frac{1}{3} W_{50}, \, 0.5\, Q_{pR}]$
b	$[T_P - \frac{1}{3} W_{75}, \, 0.75\, Q_{pR}]$
c	$[T_P, \, Q_{pR}]$
d	$[T_P + \frac{2}{3} W_{75}, \, 0.75\, Q_{pR}]$
e	$[T_P + \frac{2}{3} W_{50}, \, 0.5\, Q_{pR}]$
f	$[t_b, \, 0]$

translation hydrograph. The assumption is a pure translation of the rainfall excess without storage effects of the watershed to define a translation hydrograph. This translation hydrograph is routed through a linear reservoir ($S = RQ$) in order to simulate the effects of storage of the watershed, where R is the storage coefficient. The resulting routed hydrograph for the instantaneous excess is averaged to produce the unit hydrograph for the excess (1 cm or 1 inch) occurring in the given time interval.

Synthetic time-area relationships can be expressed in the following form such as that used by the U.S. Army Corps of Engineers (1990)

$$A/A_c = 1.414(t/T_c)^{1.5} \qquad \text{for } 0 \le t/T_c \le 0.5 \qquad (8.4.6a)$$

and

$$A/A_c = 1 - 1.414(1 - t/T_c)^{1.5} \quad \text{for } 0.5 \le t/T_c \le 1.0 \qquad (8.4.6b)$$

where A is the contributing area at time t, A_c is the total watershed area, and T_c is the time of concentration of the watershed area. Some investigators such as Ford et al. (1980) indicate that a detailed time-area curve usually is not necessary for accurate synthetic unit hydrograph estimation. A comparison of the HEC (Hydrologic Engineering Center) default relation found in HEC-1 and HEC-HMS to that used in Phoenix, Arizona is given in Table 8.4.2.

The average instantaneous flow over time interval t to $t + \Delta t$, defining the translation hydrograph is denoted as $I_{ave,t}$. To compute $I_{ave,t}$, assuming a pure translation over a Δt hr time period, the flowing equations are used. For 1 cm (0.01 m), the $I_{ave,t}$ in m^3/s is expressed as

$$I_{ave,t} = (0.01 \text{ m})(\Delta A \text{ km}^2)(10^6 \text{ m}^2/\text{km}^2)(1/\Delta t \text{ hr})(1 \text{ hr}/3600 \text{ sec}) \qquad (8.4.7)$$

where ΔA is the incremental area in km^2 between runoff isochrones (lines of equal runoff at a certain time) and Δt is the time increment in hours. For 1 in in the $I_{ave,t}$ in ft^3/s is expressed as

$$I_{ave,t} = (1 \text{ in})(1 \text{ ft}/12 \text{ in})(\Delta A \text{ mi}^2)(5280^2 \text{ ft}^2/\text{mi}^2)(1/\Delta t \text{ hr})(1 \text{ hr}/3600 \text{ sec}) \qquad (8.4.8)$$

Storage effects in the watershed are incorporated by routing the translation hydrograph through a linear reservoir using the continuity equation

$$I_{ave,t} - 0.5(Q_t + Q_{t+\Delta t}) = (S_t + S_{t+\Delta t})/\Delta t \qquad (8.4.9)$$

Table 8.4.2 Synthetic Dimensionless Time-Area Relations

Time as a percent of T_c	Contributing area, as a percent of total area		
	Urban* watersheds	Natural* watersheds	HEC default
0	0	0	0.0
10	5	3	4.5
20	16	5	12.6
30	30	8	23.2
40	65	12	35.8
50	77	20	50.0
60	84	43	64.2
70	90	75	76.8
80	94	90	87.4
90	97	96	95.5
100	100	100	100

*Flood Control District of Maricopa County, Phoenix, AZ (1995)

$I_{ave,t}$ is the average instantaneous inflow over time interval t to $t + \Delta t$, defining the translation hydrograph, Q is the outflow from the linear reservoir, and S is the storage in the linear reservoir. In the linear reservoir assumption, storage S_t is assumed to be linearly proportional to Q_t

$$S_t = RQ_t \qquad (8.4.10)$$

in which R is the proportionality constant (watershed storage coefficient) with units of time. Combining equations (8.4.9) and (8.4.10) the routing equation is

$$Q_{t+\Delta t} = CI_{ave,t} + (1 - C)Q_t \qquad (8.4.11)$$

where $I_{ave,t}$ is the average translated runoff (inflow rate to the linear reservoir) during time increment, and

$$C = 2\,\Delta t/(2R + \Delta t) \qquad (8.4.12)$$

The discharge, Q_t, from the linear reservoir now includes the effects of the storage of the watershed. This hydrograph is an instantaneous (duration = 0 hr) unit hydrograph, which is converted to the desired unit hydrograph of duration τ by averaging the ordinates over the time interval

$$U_\tau(t) = 0.5[Q_t + Q_{t-\tau}] \qquad (8.4.13)$$

The above equations are the basis for the Clark unit hydrograph procedure.

EXAMPLE 8.4.2

A small watershed has an area of 10 km^2 and a time of concentration of 1.5 hr. The watershed storage coefficient is 0.75 hr. The time-area relationship is the HEC default values in Table 8.4.2. Compute the 1-hr unit hydrograph for this small watershed. Use a time interval of 0.5-hr for the computations.

SOLUTION

First the incremental areas of the watershed are determined using the HEC time-area relationship in Table 8.4.2. The translation hydrograph is then computed by applying equation (8.4.7) to each ΔA. Next the translation hydrograph is routed through a linear reservoir using the given watershed storage coefficient. Compute the routing coefficient $C = 2(0.5)/[2(0.75) + 0.5] = 0.5$, so the linear reservoir routing equation is $Q_{t+\Delta t} = 0.5I_{ave,t} + (1 - 0.5)Q_t = 0.5I_{ave,t} + 0.5Q_t$. The unit hydrograph ordinates are computed using equation (8.4.13) with $\tau = 1$ hr. For example, the unit hydrograph ordinate for time 0.5 hr is $(0 + 7.56)/2 = 3.78$ m^3/s, for time 1.0 hr is $(0 + 16.4)/2 = 8.20$ m^3/s and for 1.5 hr is $(7.56 + 15.8)/2 = 11.7$ m^3/s.

t (hr)	t/T_c	A/A_c	A (km^2)	ΔA (km^2)	$I_{ave,t}$ (m^3/s)	$Q_{t+\Delta t}$ (m^3/s)	$U_\tau(t)$ (m^3/s)
0.0							0.0
0.5	0.333	0.272	2.72	2.72	15.1	7.56	3.78
1.0	0.667	0.728	7.28	4.56	25.3	16.4	8.20
1.5	1.0	1.0	10.0	2.72	15.1	15.8	11.7
2.0					0.0	7.89	12.2
2.5						3.94	9.86
3.0						1.97	4.93
3.5						0.986	2.46
4.0						0.493	1.23
4.5						0.247	0.616
5.0						0.123	0.308

The use of the model HEC-HMS (HEC-1) requires the time of concentration, T_c, and the storage coefficient R. Various locations have developed relationships for these parameters to make the methods more accurate and easier to use. Straub et al. (2000) developed the following equations for small rural watersheds (0.02-2.3 mi^2) in Illinois

$$T_c = 1.54L^{0.875} S_o^{-0.181} \tag{8.4.14}$$

and

$$R = 16.4L^{0.342} S_o^{-0.790} \tag{8.4.15}$$

where L is the stream length measured along the main channel from the watershed outlet to the watershed divide in miles, and S_o is the main-channel slope determined from elevations at points that represent 10 and 85 percent of the distance along the channel from the watershed outlet to the watershed divide in ft/mi.

Others have used time of concentration equations that have included additional parameters. For example Phoenix, Arizona (Flood Control District of Maricopa County) uses the following time of concentration equations developed by Papadakis and Kazan (1987) for urban areas

$$T_c = 11.4L^{0.50} K_b^{0.52} S_o^{-0.31} i^{-0.38} \tag{8.4.16}$$

where T_c is the time of concentration in hours, L is the length of the longest flow path in miles, K_b is a watershed resistance coefficient ($K_b = -0.00625 \log A + 0.04$) for commercial and residential areas, A is the watershed area in acres, S is the slope of the flow path in ft/mi, and i is the rainfall intensity in in/hr. The storage coefficient is

$$R = 0.37T_c^{1.11} A^{-0.57} L^{0.80} \tag{8.4.17}$$

where A is the watershed area in mi^2.

8.5   S-HYDROGRAPHS

In order to change a unit hydrograph from one duration to another, the *S-hydrograph method*, which is based on the principle of superposition, can be used. An S-hydrograph results theoretically from a continuous rainfall excess at a constant rate for an indefinite period. This curve (see Figure 8.5.1) has an S-shape with the ordinates approaching the rate of rainfall excess at the time of equilibrium.

Basically the S-curve (hydrograph) is the summation of an infinite number of t_R duration unit hydrographs, each lagged from the preceding one by the duration of the rainfall excess, as illustrated in Figure 8.5.2.

A unit hydrograph for a new duration t_R' is obtained by: (1) lagging the S-hydrograph (derived with the t_R duration unit hydrographs) by the new (desired) duration t_R', (2) *subtracting* the two S-hydrographs from one another, and (3) *multiplying* the resulting hydrograph ordinates by the ratio t_R/t_R'. Theoretically the S-hydrograph is a smooth curve because the input rainfall excess is assumed to be a constant, continuous rate. However, the numerical processes of the procedures may result in an undulatory form that may require smoothing or adjustment of the S-hydrograph.

EXAMPLE 8.5.1   Using the 2-hr unit hydrograph in Table 8.5.1, construct a 4-hr unit hydrograph (adapted from Sanders (1980)).

SOLUTION   See the computations in Table 8.5.1.

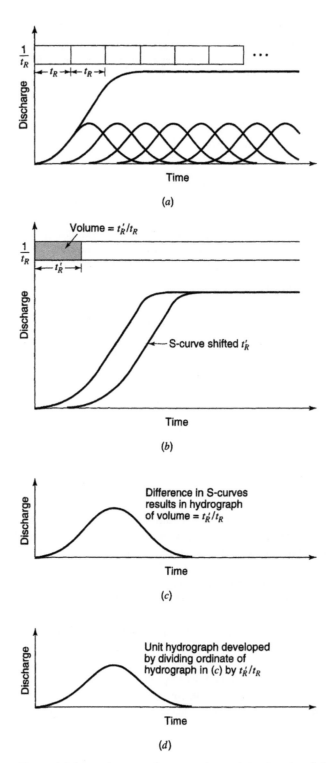

Figure 8.5.1 Development of a unit hydrograph for duration t_R' from a unit hydrograph for duration t_R.

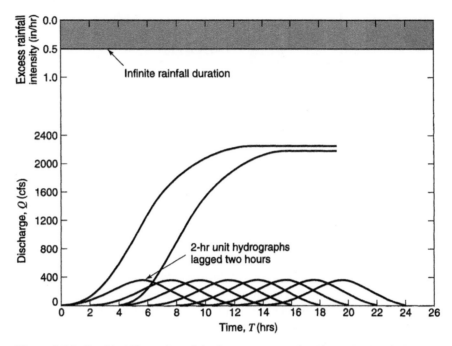

Figure 8.5.2 Graphical illustration of the S-curve construction (from Masch (1984)).

Table 8.5.1   S-Curve Determined from a 2-hr Unit Hydrograph to Estimate a 4-hr Unit Hydrograph

Time (hr)	2-hr unit hydrograph (cfs/in)	Lagged 2-hr unit hydrograph (cfs/in)		S-curve	Lagged S-curve	4-hr hydrograph	4-hr-unit hydrograph (cfs/in)
0	0			0	—	0	0
2	69	0		69	—	69	34
4	143	69	0 ...	212	0	212	106
6	328	143	69 ...	540	69	471	235
8	389	328	143 ...	929	212	717	358
10	352	389	328	1281	540	741	375
12	266	352	389	1547	929	618	309
14	192	266	352	1739	1281	458	229
16	123	192	.	1862	1547	315	158
18	84	123	.	1946	1739	207	103
20	49	84	.	1995	1862	133	66
22	20	49	.	2015	1946	69	34
24	0	20	.	*2015	1995	20	10
26	0	0	. ...	*2015	2015	0	0

*Adjusted values

Source: Sanders (1980).

8.6   NRCS (SCS) RAINFALL-RUNOFF RELATION

The U.S. Department of Agriculture Soil Conservation Service (SCS) (1972), now the National Resources Conservation Service (NRCS), developed a rainfall-runoff relation for watershed. For the storm as a whole, the depth of excess precipitation or direct runoff P_e is always less than or equal to

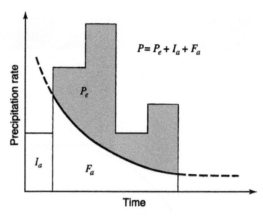

Figure 8.6.1 Variables in the SCS method of rainfall abstractions: I_a = initial abstraction, P_e = rainfall excess, F_a = continuing abstraction, and P = total rainfall.

the depth of precipitation P; likewise, after runoff begins, the additional depth of water retained in the watershed F_a is less than or equal to some *potential maximum retention S* (see Figure 8.6.1). There is some amount of rainfall I_a (initial abstraction before ponding) for which no runoff will occur, so the potential runoff is $P - I_a$. The SCS method assumes that the ratios of the two actual to the two potential quantities are equal, that is,

$$\frac{F_a}{S} = \frac{P_e}{P - I_a} \quad \frac{\text{Actual}}{\text{Potential}} \tag{8.6.1}$$

From continuity,

$$P = P_e + I_a + F_a \tag{8.6.2}$$

so that combining equations (8.6.1) and (8.6.2) and solving for P_e gives

$$P_e = \frac{(P - I_a)^2}{P - I_a + S} \tag{8.6.3}$$

which is the basic equation for computing the depth of excess rainfall or direct runoff from a storm by the SCS method.

From the study of many small experimental watersheds, an empirical relation was developed for I_a:

$$I_a = 0.2S \tag{8.6.4}$$

so that equation (8.6.3) is now expressed as

$$P_e = \frac{(P - 0.2S)^2}{P + 0.8S} \tag{8.6.5}$$

Empirical studies by the SCS indicate that the potential maximum retention can be estimated as

$$S = \frac{1000}{CN} - 10 \tag{8.6.6}$$

where CN is a runoff curve number that is a function of land use, antecedent soil moisture, and other factors affecting runoff and retention in a watershed. The curve number is a dimensionless number defined such that $0 \leq CN \leq 100$. For impervious and water surfaces $CN = 100$; for natural surfaces $CN < 100$.

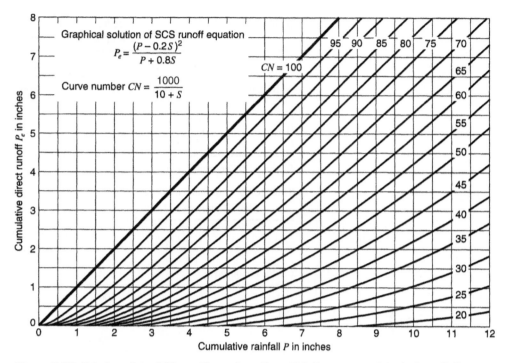

Figure 8.6.2 Solution of the SCS runoff equations (from U.S. Department of Agriculture Soil Conservation Service (1972)).

The SCS rainfall-runoff relation (equation (8.6.5)) can be expressed in graphical form using the curve numbers as illustrated in Figure 8.6.2. Equation (8.6.5) or Figure 8.6.2 can be used to estimate the volume of runoff when the precipitation volume P and the curve number CN are known.

8.7 CURVE NUMBER ESTIMATION AND ABSTRACTIONS

8.7.1 Antecedent Moisture Conditions

The curve numbers shown in Figure 8.6.2 apply for normal *antecedent moisture conditions* (AMC II). Antecedent moisture conditions are grouped into three categories:

AMC I—Low moisture

AMC II—Average moisture condition, normally used for annual flood estimates

AMC III—High moisture, heavy rainfall over the preceding few days

For dry conditions (AMC I) or wet conditions (AMC III), equivalent curve numbers can be computed using

$$CN(\text{I}) = \frac{4.2CN(\text{II})}{10 - 0.058CN(\text{II})} \tag{8.7.1}$$

and

$$CN(\text{III}) = \frac{23CN(\text{II})}{10 + 0.13CN(\text{II})} \tag{8.7.2}$$

The range of antecedent moisture conditions for each class is shown in Table 8.7.1. Table 8.7.2 lists the adjustment of curve numbers to conditions I and III for known II conditions.

Table 8.7.1   Classification of Antecedent Moisture Classes (AMC) for the SCS Method of Rainfall Abstractions

	Total 5-day antecedent rainfall (in)	
AMC group	Dormant season	Growing season
I	Less than 0.5	Less than 1.4
II	0.5 to 1.1	1.4 to 2.1
III	Over 1.1	over 2.1

Source: U.S. Department of Agriculture Soil Conservation Service (1972).

Table 8.7.2   Adjustment of Curve Numbers for Dry (Condition I) and Wet (Condition III) Antecedent Moisture Conditions

	Corresponding CN for condition	
CN for condition II	I	III
100	100	100
95	87	99
90	78	98
85	70	97
80	63	94
75	57	91
70	51	87
65	45	83
60	40	79
55	35	75
50	31	70
45	27	65
40	23	60
35	19	55
30	15	50
25	12	45
20	9	39
15	7	33
10	4	26
5	2	17
0	0	0

Source: U.S. Department of Agriculture Soil Conservation Service (1972).

8.7.2   Soil Group Classification

Curve numbers have been tabulated by the Soil Conservation Service on the basis of soil type and land use in Table 8.7.3. The four soil groups in Table 8.7.3 are described as:

Group A: Deep sand, deep loess, aggregated silts
Group B: Shallow loess, sandy loam
Group C: Clay loams, shallow sandy loam, soils low in organic content, and soils usually high in clay
Group D: Soils that swell significantly when wet, heavy plastic clays, and certain saline soils

The values of CN for various land uses on these soil types are given in Table 8.7.3. For a watershed made up of several soil types and land uses, a composite CN can be calculated.

Minimum infiltration rates for the various soil groups are:

Group	Minimum infiltration rate (in/hr)
A	0.30 – 0.45
B	0.15 – 0.30
C	0 – 0.05

Table 8.7.3   Runoff Curve Numbers (Average Watershed Condition, $I_a = 0.2S$)

Land use description		Curve numbers for hydrologic soil group			
		A	B	C	D
Fully developed urban areas[a] (vegetation established)					
Lawns, open spaces, parks, golf courses, cemeteries, etc.					
Good condition; grass cover on 75% or more of the area		39	61	74	80
Fair condition; grass cover on 50% to 75% of the area		49	69	79	84
Poor condition; grass cover on 50% or less of the area		68	79	86	89
Paved parking lots, roofs, driveways, etc.		98	98	98	98
Streets and roads					
Paved with curbs and storm sewers		98	98	98	98
Gravel		76	85	89	91
Dirt		72	82	87	89
Paved with open ditches		83	89	92	93
	Average % impervious[b]				
Commercial and business areas	85	89	92	94	95
Industrial districts	72	81	88	91	93
Row houses, town houses, and residential with lot sizes 1/8 acre or less	65	77	85	90	92
Residential: average lot size					
1/4 acre	38	61	75	83	87
1/3 acre	30	57	72	81	86
1/2 acre	25	54	70	80	85
1 acre	20	51	68	79	84
2 acre	12	46	65	77	82
Developing urban areas[c] (no vegetation established)					
Newly graded area		77	86	91	94

		Cover				
Land use	Treatment of practice	Hydrologic condition[d]				
Cultivated agricultural land						
Fallow	Straight row		77	86	91	94
	Conservation tillage	Poor	76	85	90	93
	Conservation tillage	Good	74	83	88	90
Row crops	Straight row	Poor	72	81	88	91
	Straight row	Good	67	78	85	89
	Conservation tillage	Poor	71	80	87	90

(Continued)

Table 8.7.3 (*Continued*)

Cover			Curve numbers for hydrologic soil group			
Land use	Treatment of practice	Hydrologic condition[d]	A	B	C	D
	Conservation tillage	Good	64	75	82	85
	Contoured	Poor	70	79	84	88
	Contoured	Good	65	75	82	86
	Contoured and conservation tillage	Poor	69	78	83	87
		Good	64	74	81	85
	Contoured and terraces	Poor	66	74	80	82
	Contoured and terraces	Good	62	71	78	81
	Contoured and terraces and conservation tillage	Poor	65	73	79	81
		Good	61	70	77	80
Small grain	Straight row	Poor	65	76	84	88
	Straight row	Good	63	75	83	87
	Conservation tillage	Poor	64	75	83	86
	Conservation tillage	Good	60	72	80	84
	Contoured	Poor	63	74	82	85
	Contoured	Good	61	73	81	84
	Contoured and conservation tillage	Poor	62	73	81	84
		Good	60	72	80	83
	Contoured and terraces	Poor	61	72	79	82
	Contoured and terraces	Good	59	70	78	81
	Contoured and terraces and conservation tillage	Poor	60	71	78	81
		Good	58	69	77	80
Close-seeded legumes or rotation meadow	Straight row	Poor	66	77	85	89
	Straight row	Good	58	72	81	85
	Contoured	Poor	64	75	83	85
	Contoured	Good	55	69	78	83
	Contoured and terraces	Poor	63	73	80	83
	Contoured and terraces	Good	51	67	76	80
Noncultivated agricultural land, pasture or range	No mechanical treatment	Poor	68	79	86	89
	No mechanical treatment	Fair	49	69	79	84
	No mechanical treatment	Good	39	61	74	80
	Contoured	Poor	47	67	81	88
	Contoured	Fair	25	59	75	83
	Contoured	Good	6	35	70	79
Meadow		—	30	58	71	78
Forested—grass or orchards—evergreen or deciduous		Poor	55	73	82	86
		Fair	44	65	76	82
		Good	32	58	72	79
Brush		Poor	48	67	77	83
		Good	20	48	65	73
Woods		Poor	45	66	77	83
		Fair	36	60	73	79
		Good	25	55	70	77
Farmsteads		—	59	74	82	86
Forest-range						
Herbaceous		Poor		79	86	92
		Fair		71	80	89
		Good		61	74	84

Table 8.7.3 (*Continued*)

Cover		Hydrologic condition[d]	Curve numbers for hydrologic soil group			
Land use	Treatment of practice		A	B	C	D
Oak–aspen		Poor		65	74	
		Fair		47	57	
		Good		30	41	
Juniper–grass		Poor		72	83	
		Fair		58	73	
		Good		41	61	
Sage–grass		Poor		67	80	
		Fair		50	63	
		Good		35	48	

[a]For land uses with impervious areas, curve numbers are computed assuming that 100% of runoff from impervious areas is directly connected to the drainage system. Pervious areas (lawn) are considered to be equivalent to lawns in good condition and the impervious areas have a CN of 98.

[b]Includes paved streets.

[c]Use for the design of temporary measures during grading and construction. Impervious area percent for urban areas under development vary considerably. The user will determine the percent impervious. Then using the newly graded area CN and Figure 8.7.1a or b, the composite CN can be computed for any degree of development.

[d]For conservation tillage in poor hydrologic condition, 5 percent to 20 percent of the surface is covered with residue (less than 750-lb/acre row crops or 300-lb/acre small grain).

For conservation tillage in good hydrologic condition, more than 20 percent of the surface is covered with residue (greater than 750-lb/acre row crops or 300-lb/acre small grain).

[e]Close-drilled or broadcast.

For noncultivated agricultural land:
Poor hydrologic condition has less than 25 percent ground cover density.
Fair hydrologic condition has between 25 percent and 50 percent ground cover density.
Good hydrologic condition has more than 50 percent ground cover density.

For forest–range:
Poor hydrologic condition has less than 30 percent ground cover density.
Fair hydrologic condition has between 30 percent and 70 percent ground cover density.
Good hydrologic condition has more than 70 percent ground cover density.

Source: U.S. Department of Agriculture Soil Conservation Service (1986).

8.7.3 Curve Numbers

Table 8.7.3 gives the curve numbers for average watershed conditions, $I_a = 0.2S$, and antecedent moisture condition II. For watersheds consisting of several subcatchments with different CNs, the area-averaged composite CN can be computed for the entire watershed. This analysis assumes that the impervious areas are directly connected to the watershed drainage system (Figure 8.7.1a). If the percent imperviousness is different from the value listed in Table 8.7.3 or if the impervious areas are not directly connected, then Figures 8.7.1a or b, respectively, can be used. The pervious CN used in these figures is equivalent to the open-space CN in Table 8.7.3. If the total impervious area is less than 30 percent, Figure 8.7.1b is used to obtain a composite CN. For natural desert landscaping and newly graded areas, Table 8.7.3 gives only the CNs for pervious areas.

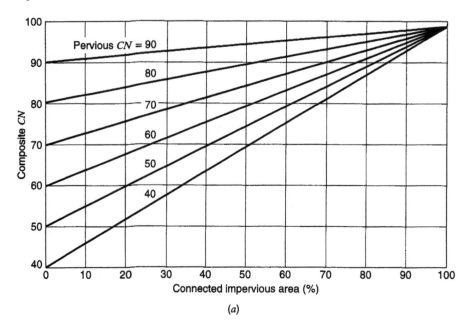

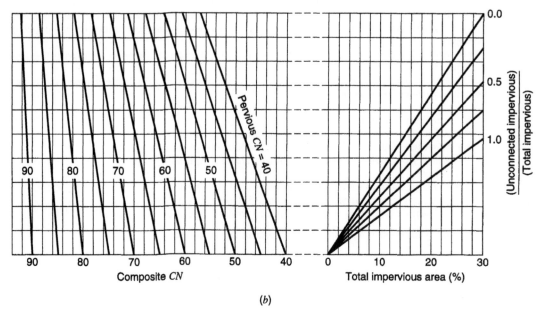

Figure 8.7.1 Relationships for determining composite *CN*. (*a*) Connected impervious area; (*b*) Unconnected impervious area (from U.S. Department of Agriculture Soil Conservation Service (1986)).

EXAMPLE 8.7.1

Determine the weighted curve numbers for a watershed with 40 percent residential (1/4-acre lots), 25 percent open space, good condition, 20 percent commercial and business (85 percent impervious), and 15 percent industrial (72 percent impervious), with corresponding soil groups of C, D, C, and D.

SOLUTION

The corresponding curve numbers are obtained from Table 8.7.3:

Land use (%)	Soil group	Curve number
40	C	83
25	D	80
20	C	94
15	D	93

The weighted curve number is

$$CN = 0.40(83) + 0.25(80) + 0.20(94) + 0.15(93)$$
$$= 33.2 + 20 + 18.8 + 13.95$$
$$= 85.95 \text{(use 86)}$$

EXAMPLE 8.7.2

The watershed in example 8.7.1 experienced a rainfall of 6 in. What is the runoff volume?

SOLUTION

Using equation (8.6.5), P_e = runoff volume is

$$P_e = \frac{(P - 0.2S)^2}{P + 0.8S}$$

where S is computed with the weighted curve number of 86 from example 8.7.1:

$$S = \frac{1000}{86} - 10 = 1.63$$

So

$$P_e = \frac{[6 - 0.2(1.63)]^2}{6 + 0.8(1.63)} = \frac{32.19}{7.3}$$
$$= 4.41 \text{ in of runoff}$$

EXAMPLE 8.7.3

For the watershed in examples 8.7.1 and 8.7.2, the 6-in rainfall pattern was 2 in the first hour, 3 in the second hour, and 1 in the third hour. Determine the cumulative rainfall and cumulative rainfall excess as functions of time.

SOLUTION

The initial abstractions are computed as $I_a = 0.2S$ with $S = 1.63$ from example 8.7.2, so $I_a = 0.2(1.63) = 033$ in. The remaining losses for time period (the first hour) are computed using the following equation, derived by combining equations (8.6.1) and (8.6.2):

$$F_{a,t} = \frac{S(P_t - I_a)}{P_t - I_a + S} = \frac{1.63(P_t - 0.33)}{P_t - 0.33 + 1.63} = \frac{1.63(P_t - 0.33)}{P_t + 1.3}$$

$$F_{a,1} = \frac{1.63(2 - 0.33)}{2 + 1.3} = 0.82 \text{ in}$$

The total loss for the first hour is $0.33 + 0.82 = 1.15$ in, and the excess is

$$P_{e1} = P_1 - I_a - F_{a,1} = 2 - 0.33 - 0.82 = 0.85 \text{ in}$$

For the second hour, $P_t = 2 + 3 = 5$ in, so

$$F_{a,2} = \frac{1.63(5 - 0.33)}{5 + 1.3} = 1.21 \text{ in}$$

and the cumulative rainfall excess is $P_{e_2} = 5 - 0.33 - 1.21 = 3.46$ in.

For the third hour, $P_3 = 2 + 3 + 1 = 6$ in, so

$$F_{a,3} = \frac{1.63(6 - 0.33)}{6 + 1.3} = 1.27 \text{ in}$$

and $P_{e_3} = 6 - 0.33 - 1.27 = 4.40$ in (which compares well with the results of example 8.7.2). The results are summarized below, along with the rainfall excess hyetograph.

Time (h)	Cumulative rainfall P_t (in)	Cumulative abstractions		Cumulative rainfall excess P_e (in)	Rainfall excess hyetograph (in)
		I_a (in)	$F_{a,t}$ (in)		
1	2	0.33	0.82	0.85	0.85
2	5	0.33	1.21	3.46	2.61
3	6	0.33	1.27	4.40	0.94

8.8   NRCS (SCS) UNIT HYDROGRAPH PROCEDURE

The SCS dimensionless unit hydrograph and mass curve are shown in Figure 8.8.1 and tabulated in Table 8.8.1. The SCS dimensionless equivalent triangular unit hydrograph is also shown in Figure 8.8.1. The following section discusses how to develop a unit hydrograph from these dimensionless unit hydrographs.

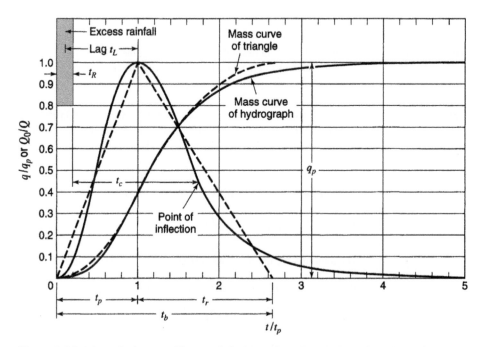

Figure 8.8.1 Dimensionless curvilinear unit hydrograph and equivalent triangular hydrograph (from U.S. Department of Agriculture Soil Conservation Service (1986)).

Table 8.8.1 Ratios for Dimensionless Unit Hydrograph and Mass Curve

Time ratios t/t_p	Discharge ratios q/q_p	Mass curve ratios Q_a/Q
0	0.000	0.000
0.1	0.030	0.001
0.2	0.100	0.006
0.3	0.190	0.012
0.4	0.310	0.035
0.5	0.470	0.065
0.6	0.660	0.107
0.7	0.820	0.163
0.8	0.930	0.228
0.9	0.990	0.300
1.0	1.000	0.375
1.1	0.990	0.450
1.2	0.930	0.522
1.3	0.860	0.589
1.4	0.780	0.650
1.5	0.680	0.700
1.6	0.560	0.751
1.7	0.460	0.790
1.8	0.390	0.822
1.9	0.330	0.849
2.0	0.280	0.871
2.2	0.207	0.908
2.4	0.147	0.934
2.6	0.107	0.953
2.8	0.077	0.967
3.0	0.055	0.977
3.2	0.040	0.984
3.4	0.029	0.989
3.6	0.021	0.993
3.8	0.015	0.995
4.0	0.011	0.997
4.5	0.005	0.999
5.0	0.000	1.000

Source: U.S. Department of Agriculture Soil Conservation Service (1972).

8.8.1 Time of Concentration

The *time of concentration* for a watershed is the time for a particle of water to travel from the hydrologically most distant point in the watershed to a point of interest, such as the outlet of the watershed. SCS has recommended two methods for time of concentration, the *lag method* and the *upland*, or *velocity method.*

The lag method relates the *time lag* (t_L), defined as the time in hours from the center of mass of the rainfall excess to the peak discharge, to the slope (Y) in percent, the hydraulic length (L) in feet, and the potential maximum retention (S), expressed as

$$t_L = \frac{L^{0.8}(S+1)^{0.7}}{1900Y^{0.5}}$$

(8.8.1)

The SCS uses the following relationship between the time of concentration (t_c) and the lag (t_L):

$$t_c = \frac{5}{3} t_L \qquad (8.8.2)$$

or

$$t_c = \frac{L^{0.8}(S+1)^{0.7}}{1140 Y^{0.5}} \qquad (8.8.3)$$

where t_c is in hours. Refer to Figure 8.8.1 to see the SCS definition of t_c and t_L.

The velocity (upland) method is based upon defining the time of concentration as the ratio of the hydraulic flow length (L) to the velocity (V):

$$t_c = \frac{L}{3600V} \qquad (8.8.4)$$

where t_c is in hours, L is in feet, and V is in ft/s. The velocity can be estimated knowing the land use and the slope in Figure 8.8.2. Alternatively, we can think of the concentration as being the sum of

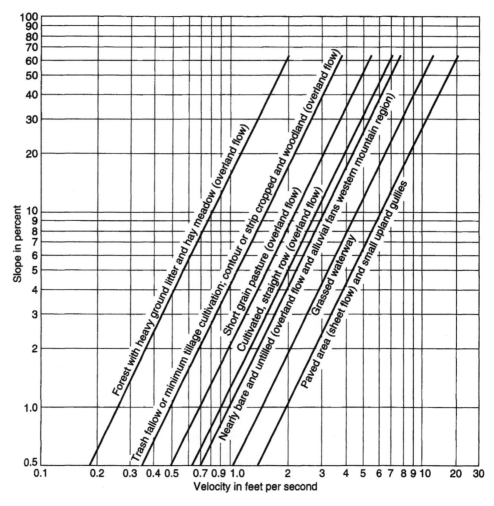

Figure 8.8.2 Velocities for velocity upland method of estimating t_c (from U.S. Department of Agriculture Soil Conservation Service (1986)).

travel times for different segments

$$t_c = \frac{1}{3600} \sum_{i=1}^{k} \frac{L_i}{V_i} \qquad (8.8.5)$$

for k segments, each with different land uses.

8.8.2   Time to Peak

Time to peak (t_p) is the time from the beginning of rainfall to the time of the peak discharge (Figure 8.8.1)

$$t_p = \frac{t_R}{2} + t_L \qquad (8.8.6)$$

where t_p is in hours, t_R is the duration of the rainfall excess in hours, and t_L is the lag time in hours. The SCS recommends that t_R be 0.133 of the time of concentration of the watershed, t_c:

$$t_R = 0.133t_c \qquad (8.8.7)$$

and because $t_L = 0.6t_c$ by equation (8.8.2), then by equation (8.8.6) we get

$$t_p = \frac{0.133t_c}{2} + 0.6t_c$$
$$t_p = 0.67t_c \qquad (8.8.8)$$

8.8.3   Peak Discharge

The area of the unit hydrograph equals the volume of direct runoff Q, which was estimated by equation (8.6.5). With the equivalent triangular dimensionless unit hydrograph of the curvilinear dimensionless unit hydrograph in Figure 8.8.1, the time base of the dimensionless triangular unit hydrograph is 8/3 of the time to peak t_p, as compared to $5t_p$ for the curvilinear. The areas under the rising limb of the two dimensionless unit hydrographs are the same (37 percent).

Based upon geometry (Figure 8.8.1), we see that

$$Q = \frac{1}{2} q_p \left(t_p + t_r \right) \qquad (8.8.9)$$

for the direct runoff Q, which is 1 in where t_r is the recession time of the dimensionless triangular unit hydrograph and q_p is the peak discharge. Solving equation (8.8.9) for q_p gives

$$q_p = \frac{Q}{t_p} \left[\frac{2}{1 + t_r/t_p} \right] \qquad (8.8.10)$$

Letting $K = \left[\dfrac{2}{1 + t_r/t_p} \right]$, then

$$q_p = \frac{KQ}{t_p} \qquad (8.8.11)$$

where Q is the volume, equals to 1 in for a unit hydrograph.

The above equation can be modified to express q_p in ft³/sec, t_p in hours, and Q in inches:

$$q_p = 645.33K\frac{AQ}{t_p} \qquad (8.8.12)$$

The factor 645.33 is the rate necessary to discharge 1 in of runoff from 1 mi^2 in 1 hr. Using $t_r = 1.67t_p$ gives $K = [2/(1 + 1.67)] = 0.75$; then equation (8.8.12) becomes

$$q_p = \frac{484AQ}{t_p} \tag{8.8.13}$$

For SI units,

$$q_p = \frac{2.08AQ}{t_p} \tag{8.8.14}$$

where A is in square kilometers.

The steps in developing a unit hydrograph are:

Step 1   Compute the time of concentration using the lag method (equation (8.8.3)) or the velocity method (equation (8.8.4) or (8.8.5)).

Step 2   Compute the time to peak $t_p = 0.67t_c$ (equation (8.8.8)) and then the peak discharge q_p using equation (8.8.13) or (8.8.14).

Step 3   Compute time base t_b and the recession time t_r:
Triangular hydrograph: $t_b = 2.67t_p$
Curvilinear hydrograph: $t_b = 5t_p$
$t_r = t_b - t_p$

Step 4   Compute the duration $t_R = 0.133\ t_c$ and the lag $t_L = 0.6\ t_c$ by using equations (8.8.7) and (8.8.2), respectively.

Step 5   Compute the unit hydrograph ordinates and plot. For the triangular only t_p, q_p, and t_r are needed. For the curvilinear, use the dimensionless ratios in Table 8.8.1.

EXAMPLE 8.8.1

For the watershed in example 8.7.1, determine the triangular SCS unit hydrograph. The average slope of the watershed is 3 percent and the area is 3.0 mi^2. The hydraulic length is 1.2 mi.

SOLUTION

Step 1   The time of concentration is computed using equation (8.8.1), with S = 1.63 from example 8.7.2:

$$t_L = \frac{(6336)^{0.8}(1.63 + 1)^{0.7}}{1900\sqrt{3}} = 0.66 \text{ hr}$$

and $t_c = \dfrac{5}{3}t_L = 1.1$ hr

Step 2   The time to peak $t_p = 0.67t_c = 0.67(1.1) = 0.74$ hr.

Step 3   The time base is $t_b = 2.67t_p = 1.97$ hr.

Step 4   The duration is $t_R = 0.133t_c = 0.133(1.1) = 0.15$ hr, and t_L is 0.66 hr.

Step 5   The peak is (for $Q = 1$ in)

$$q_p = \frac{484AQ}{t_p} = \frac{484(3)(1)}{0.74} = 1962 \text{ cfs.}$$

In summary, the triangular unit hydrograph has a peak of 1962 cfs at the time to peak of 0.74 hr with a time base of 1.97 hr. This is a 0.15-hr duration unit hydrograph.

8.9   KINEMATIC-WAVE OVERLAND FLOW RUNOFF MODEL

Hortonian overland flow occurs when the rainfall rate exceeds the infiltration capacity and sufficient water ponds on the surface to overcome surface tension effects and fill small depressions. Overland flow is surface runoff that occurs in the form of sheet flow on the land surface without concentrating

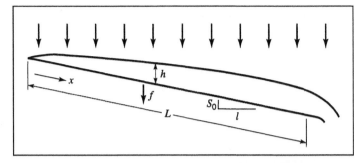

Figure 8.9.1 Definition sketch of overland flow on a plane as a one-dimensional flow (from Woolhiser et al. (1990)).

in clearly defined channels (Ponce, 1989). For the purposes of rainfall-runoff analysis, this flow can be viewed as a one-dimensional flow process (Figure 8.9.1) in which the flux is proportional to some power of the storage per unit area, expressed as (Woolhiser et al., 1990):

$$Q = \alpha h^m \tag{8.9.1}$$

where Q is the discharge per unit width, h is the storage of water per unit area (or depth if the surface is a plane), and α and m are parameters related to slope, surface roughness, and whether the flow is laminar or turbulent.

The mathematical description of overland flow can be accomplished through the continuity equation in one-dimensional form and a simplified form of the momentum equation. This model is referred to as the *kinematic wave model. Kinematics* refers to the study of motion exclusive of the influence of mass and force. A *wave* is a variation in flow, such as a change in flow rate or water surface elevation. *Wave celerity* is the velocity with which this variation travels. *Kinematic waves* govern flow when inertial and pressure forces are negligible.

The *kinematic wave equations* (also see Chapter 9) for a one-dimensional flow are expressed as follows:
Continuity:

$$\frac{\partial A}{\partial t} + \frac{\partial Q}{\partial x} = q(x, t) \tag{8.9.2}$$

Momentum:

$$S_0 - S_f = 0 \tag{8.9.3}$$

where A is the cross-sectional area of flow, Q is the discharge, t is time, x is the spatial coordinate, $q(x, t)$ is the lateral inflow rate, S_0 is the overland flow slope, and S_f is the friction slope.

Equation (8.9.3) indicates that the gravity and friction forces are balanced, so that flow does not accelerate appreciably. The inertial (local and convective acceleration) term and pressure term are neglected in the kinematic wave model (refer to Section 9.4). Eliminating these terms eliminates the mechanism to describe backwater effects and flood wave peak attenuation.

Considering that h is the storage per unit area or depth, then $A = h$, so that equation (8.9.2) becomes

$$\frac{\partial h}{\partial t} + \frac{\partial Q}{\partial x} = q(x, t) \tag{8.9.4}$$

Substituting equation (8.9.1) into (8.9.4) gives

$$\frac{\partial h}{\partial t} + \frac{\partial(\alpha h^m)}{\partial x} = q(x, t) \tag{8.9.5}$$

The one-dimensional overland flow on a plane surface (illustrated in Figure 8.9.1) is not the type of flow found in most watershed situations (Woolhiser et al., 1990). The kinematic assumption does not require sheet flow as shown; it requires only that the discharge be some unique function of the amount of water stored per unit of area.

Woolhiser and Liggett (1967) and Morris and Woolhiser (1980) showed that the kinematic-wave formulation is an excellent approximation for most overland flow conditions. Keep in mind that these equations are a simplification of the Saint-Venant equations (see Chapter 9).

The kinematic-wave equation (8.9.5) for overland flow can be solved numerically using a four-point implicit method where the finite-difference approximations for the spatial and temporal derivatives are, respectively,

$$\frac{\partial h}{\partial x} = \theta \frac{h_{i+1}^{j+1} - h_i^{j+1}}{\Delta x} + (1 - \theta) \frac{h_{i+1}^j - h_i^j}{\Delta x} \tag{8.9.6}$$

and

$$\frac{\partial h}{\partial t} = \frac{1}{2} \left[\frac{h_i^{j+1} - h_i^j}{\Delta t} + \frac{h_{i+1}^{j+1} - h_{i+1}^j}{\Delta t} \right]$$

or

$$\frac{\partial h}{\partial t} = \frac{h_i^{j+1} + h_{i+1}^{j+1} - h_i^j - h_{i+1}^j}{2\Delta t} \tag{8.9.7}$$

and

$$q = \frac{1}{2} \left(\bar{q}_{i+1} + \bar{q}_i \right) \tag{8.9.8}$$

where θ is a weighting parameter for spatial derivative, $\theta = \Delta't/\Delta t$ (see Figure 8.9.2). The derivative $\partial h/\partial t$ is the average of the temporal derivatives at locations i and $i + 1$ or for the midway locations between i and $i + 1$, and $\bar{q}_i$ and $\bar{q}_{i+1}$ are the average lateral inflows at i and $i + 1$, respectively. Notation for the finite-difference grid is shown in Figure 8.9.2. Substituting these finite difference expressions (8.9.6), (8.9.7), and (8.9.8) into (8.9.5) and simplifying results in the following finite-difference equation:

$$h_{i+1}^{j+1} - h_{i+1}^j + h_i^{j+1} - h_i^j$$

$$+ \frac{2\Delta t}{\Delta x} \left\{ \theta \left[\alpha_{i+1}^{j+1} \left(h_{i+1}^{j+1} \right)^m - \alpha_i^{j+1} \left(h_i^{j+1} \right)^m \right] + (1 - \theta) \left[\alpha_{i+1}^j \left(h_{i+1}^j \right)^m - \alpha_i^j \left(h_i^j \right)^m \right] \right\} \tag{8.9.9}$$

$$- \Delta t \left(\bar{q}_{i+1} + \bar{q}_i \right) = 0$$

The only unknown in the above equation is h_{i+1}^{j+1}, which must be solved by using Newton's method (see Appendix A). Using Manning's equation to express equation (8.9.1), $Q = \alpha h^m$, we

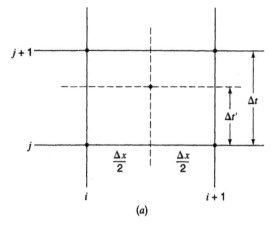

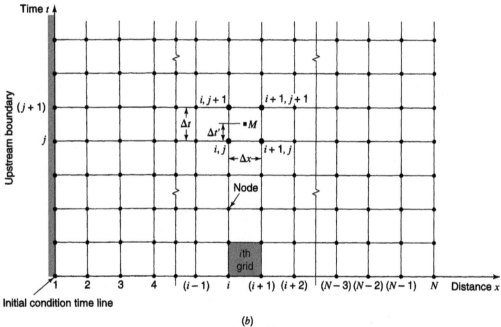

Figure 8.9.2 The *x-t* solution plane: The finite-difference forms of the Saint-Venant equations are solved at a discrete number of points (values of the independent variables *x* and *t*) arranged to the time axis represent locations along the plane, and those parallel to the distance axis represent times. (*a*) Four points of finite-difference grid; (*b*) Finite difference grid.

find

$$Q = \left[\frac{1.49S^{1/2}}{n}\right] h^{5/3} \tag{8.9.10}$$

where

$$\alpha = \frac{1.49S_0^{1/2}}{n}, \; m = 5/3 \tag{8.9.11}$$

Table 8.9.1 Recommended Manning's Roughness Coefficients for Overland Flow

Cover or treatment	Residue rate, Tons/Acre	Value recommended	Range
Concrete or asphalt		0.011	0.010–0.013
Bare sand		0.01	0.010–0.016
Graveled surface		0.02	0.012–0.03
Bare clay loam (eroded)		0.02	0.012–0.033
Fallow, no residue		0.05	0.006–0.16
Chisel plow	<1/4	0.07	0.006–0.17
	<1/4–1	0.18	0.07–0.34
	1–3	0.30	0.19–0.47
	>3	0.40	0.34–0.46
Disk/harrow	<1/4	0.08	0.008–0.41
	1/4–1	0.16	0.10–0.25
	1–3	0.25	0.14–0.53
	>3	0.30	—
No till	<1/4	0.04	0.03–0.07
	1/4–1	0.07	0.01–0.13
	1–3	0.30	0.16–0.47
Moldboard plow (fall)		0.06	0.02–0.10
Colter		0.10	0.05–0.13
Range (natural)		0.13	0.01–0.32
Range (clipped)		0.10	0.02–0.24
Grass (bluegrass sod)		0.45	0.39–0.63
Short grass prairie		0.15	0.10–0.20
Dense grass[1]		0.24	0.17–0.30
Bermuda grass[1]		0.41	0.30–0.48

[1]Weeping lovegrass, bluegrass, buffalo grass, blue gamma grass, native grass mix (OK), alfalfa, lespedeza (from Palmer, 1946).

Sources: Woolhiser (1975), Engman (1986), Woolhiser et al. (1990).

where n is Manning's roughness coefficient and S_0 is the slope of the overland flow plane. Recommended values of Manning's roughness coefficients for overland flow are given in Table 8.9.1. The *time to equilibrium* of a plane of length L and slope S_0 can be derived using Manning's equation as

$$t_c = \frac{nL}{1.49 S_0^{1/2} h^{2/3}} \tag{8.9.12}$$

The U.S. Department of Agriculture Agricultural Research Service (Woolhiser et al., 1990) has developed a KINematic runoff and EROSion model referred to as KINEROS. This model is event-oriented, i.e., it is a physically based model describing the processes of interception, infiltration, surface runoff, and erosion from small agricultural and urban watersheds. The model is distributed because flows are modeled for both the watershed and the channel elements, as illustrated in Figures 8.9.3 and 8.9.4. The model is *event-oriented* because it does not have components describing evapotranspiration and soil water movement between storms. In other words, there is no hydrologic balance between storms.

Figures 8.9.3 and 8.9.4 illustrate that the approach to describing a watershed is to divide it into a branching system of channels with plane elements contributing lateral flow to channels. The

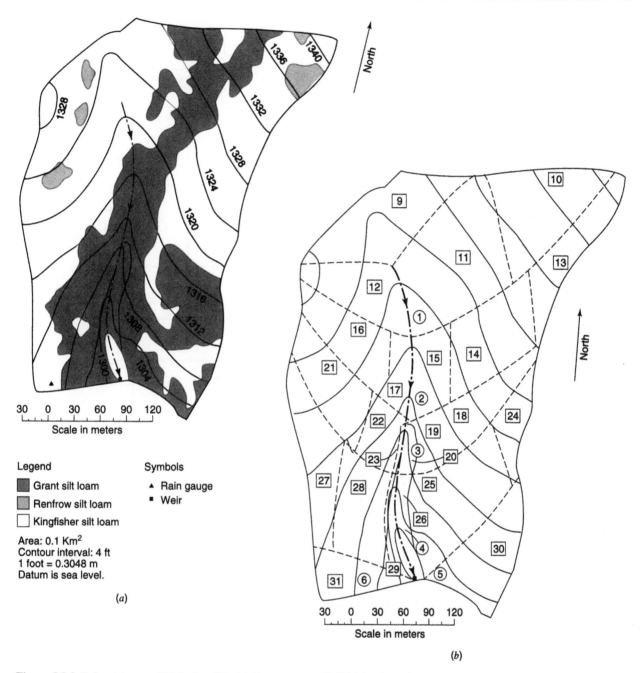

Figure 8.9.3 R-5 catchment, Chickasha, OK. (*a*) Contour map; (*b*) Division into plane and channel elements (from Woolhiser et al. (1990)).

KINEROS model takes into account interception, infiltration, overland flow routing, channel routing, reservoir routing, erosion, and sediment transport. Overland flow routing has been described in this section. Channel routing is performed using the kinematic-wave approximation described in Chapter 9. The reservoir routing in KINEROS is basically a level-pool routing procedure, as described in Chapter 9.

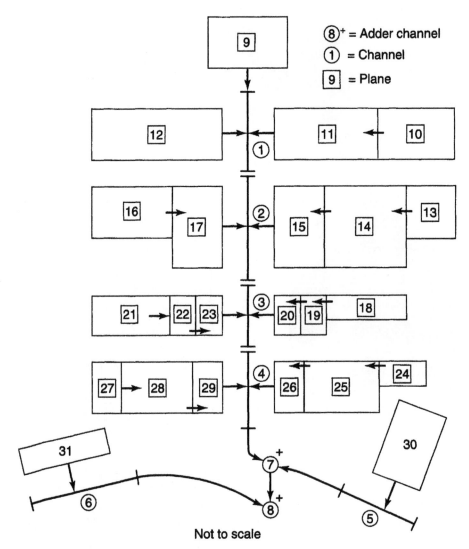

Figure 8.9.4 Schematic of R-5 plane and channel configuration (from Woolhiser et al. (1990)).

8.10 COMPUTER MODELS FOR RAINFALL-RUNOFF ANALYSIS

Computer models for runoff analysis can be classified as event-based models or continuous simulation models. *Event-based models* are used to simulate the discrete rainfall-runoff events using, for example, the unit hydrograph approach. These models emphasize infiltration and surface-runoff components with the objective of determining direct runoff, and are applicable to excess-water flow calculations in cases where direct runoff is the major contributor to streamflow. The HEC-HMS (HEC-1) and the TR-20 and TR-55 models are single-event models. The KINEROS model (Woolhiser et al., 1990) described in the previous section is an overland flow model based on the kinematic-wave routing. This model is a *distributed event-based model*. Other examples include the kinematic-wave model for overland flow routing in the HEC-HMS (HEC-1) model.

Continuous-simulation models account for the overall moisture balance of the basin, including moisture accounting between storm events. These models explicitly account for all runoff components including surface flow and indirect runoff such as interflow and baseflow, and are well suited for *long-term runoff forecasting*.

PROBLEMS

8.1.1 Suppose that the runoff peak discharge (Q_p) from an urban watershed is related to rainfall intensity (I) and drainage area (A) in the following form, $Q_p = a \times I^b \times A^b$, in which a and b are unknown parameters to be determined from the data given in the following table.

Q_p (ft³/sec)	23	45	68	62
I (in/hr)	3.2	4.6	6.1	7.4
A (acres)	12	21	24	16

(a) Use the least-squares method to determine the values of unknown parameters a and b.
(b) Assess how good the resulting equation is.
(c) Estimate the expected peak discharge for watershed with an area of 20 acres resulting from a 4-in/hr rain.

8.1.2 The following data were obtained to study the relationship between peak discharge (Q_p) of a watershed with drainage area (A) and average rainfall intensity (I).

Peak discharge (Q_p in ft³/sec)	23	45	44	64	68
Rainfall intensity (I in in/hr)	3.2	4.2	5.1	3.8	6.1
Drainage area (A in acres)	12	21	18	32	24

Suppose that the relationship between Q_p, I, and A can be expressed as $Q_p = I^a \times A^b$ in which a and b are constants.

(a) Determine the values of the two constants, i.e., a and b, by the least-squares method.
(b) How good is the least-squares fit?
(c) Based on the result in part (a), estimate the peak discharge per unit area for a watershed with basin area of 15 acres under a storm with rainfall depth of 15 in in 3 hr.

8.1.3 The following table contains measurements of peak discharge, average rainfall intensity, and drainage area.

Data	Peak discharge Q (m³/s)	Rainfall intensity I (mm/hr)	Drainage area A (km²)
1	0.651	81.3	0.0486
2	1.246	129.5	0.0728
3	1.812	96.5	0.1295
4	1.926	154.9	0.0971

(a) Determine the coefficients a and b in the equation $Q = a \times (I \times A)^b$ using the least-squares method.
(b) Determine the corresponding coefficient of determination and standard error of estimate for part (a).
(c) Physically, when a storm with constant rainfall intensity I continues indefinitely, the term $I \times A$ represents the steady-state peak discharge. Under the steady-state condition, let $b = 1$ in the above equation. Determine the least-squares estimate for the constant a and the corresponding coeffi-

cient of determination. How do you interpret the meaning of constant a? Does your estimated value of a make sense from the physical viewpoint? Please explain.

8.1.4 For a particular watershed, the observed runoff volume and the corresponding peak discharges, Q_p, are shown below.

Q_p (m³/s)	0.42	0.12	0.58	0.28	0.66
Runoff (cm)	1.28	0.40	2.29	1.42	3.19

(a) Determine the coefficients (a and b) in the equation $QP = a \times RO^b$ using the least-squares method where $Q_p =$ peak discharge, $RO =$ runoff volume, and $a, b =$ coefficients.
(b) Estimate the peak discharge for the runoff volume of 4 cm.

8.2.1 Consider a 10-km² catchment receiving rainfall with intensity that varies from 0 to 30 mm/hr in a linear fashion for the first 6 hr. Then, rainfall intensity stays constant at 30 mm/hr over the next 6 hr before it stops. During the rainfall period of 12 hr, the rate of surface runoff from the catchment increases linearly from 0 at $t = 0$ hr to 30 mm/hr at $t = 12$ hr.

(a) Assume that hydrologic losses, such as infiltration, evaporation, etc. are negligible in the catchment. Determine the time when the surface runoff ends, if surface runoff decreases linearly to zero from 30 mm/hr at $t = 12$hr.
(b) At what time does the total storage in the catchment reach its maximum and what is the corresponding storage (in km³)?
(c) Identify the times at which the rate of increase and decrease in storage are the largest.

8.2.2 The following table contains the records of rainfall and surface runoff data from a catchment with the drainage area of 2.25 km². Column (2) is the rainfall hyetograph for the averaged intensity whereas column (3) is the runoff hydrograph for instantaneous flow rate.

Time	Avg. intensity (mm/h)	Instantaneous discharge (m³/s)
1:00		0
1:10	60	12
1:20	150	24
1:30	30	48
1:40	18	30
1:50	60	18
2:00	72	24
2:10	24	30
2:20	12	18
2:30		12
2:40		6
2:50		0

(a) Determine the rolling-time and clock-time maximum rainfall depth (in mm) for durations of 30 min and 50 min.

(b) Determine the time and the corresponding volume (in m^3) when the maximum storage occurs. Assume that the initial storage volume is zero.

(c) Determine the percentage of rainfall volume that becomes the surface runoff.

8.2.3 Based on the table given below showing the hydrograph ordinates at 24-hour intervals, use graph paper to (a) identify the time instant at which the direct runoff ends and (b) determine the recession constant, k, in the following equation:

$$Q(t_2) = Q(t_1) \times e^{-k(t_2 - t_1)}$$

in which $t_2 > t_1$ are two time points on the baseflow recession of a hydrograph.

Time (day)	Flow (m^3/s)	Time (day)	Flow (m^3/s)
1	6	8	91
2	970	9	79
3	707	10	68
4	400	11	58
5	254	12	50
6	162	13	43
7	121	14	37

8.2.4 The following table contains the records of rainfall and surface runoff data of a particular storm event for a watershed with the drainage area of 8.4 km^2.

(a) Determine the maximum rainfall intensity for durations of 30 min, 60 min, and 120 min.

(b) Determine the percentage of total rainfall that is lost from appearing as the surface runoff.

(c) Determine the time period that the storage volume of the water is increasing in the watershed.

Time of day	Cumulative rainfall (mm)	Instantaneous runoff (m^3/s)
4:00	0	0.0
4:30	30	0.3
5:00	80	4.2
5:30	90	11.3
6:00	95	13.0
6:30	127	34.0
7:00	130	45.3
7:30	132	21.2
8:00		9.3
8:30		4.8
9:00		2.5
9:30		1.4
10:00		0.8
10:30		0.0

8.2.5 The following table contains the records of rainfall and surface runoff data of a storm event for a watershed having the drainage area of 8.0 km^2.

(a) Determine the maximum rainfall intensity for durations of 15 min, 30 min, and 60 min.

(b) Determine the percentage of total rainfall that is lost from appearing as the surface runoff.

(c) Determine the time period that the storage volume of the water is increasing in the watershed and the maximum storage volume (in cm).

(d) Determine the Φ index and the corresponding rainfall excess hyetograph.

Time	Cumul. rainfall (mm)	Instant. flow rate (m^3/s)
4:00	0	0
4:15	15	1
4:30	40	4
4:45	45	11
5:00	48	13
5:15	64	35
5:30	65	45
5:45	66	21
6:00	66	10
6:15	66	5
6:30		3
6:45		1
7:00		0

8.2.6 Consider a watershed with drainage area of 1 km^2, For a given storm event, the recorded average rainfall intensity and total instantaneous discharge at the outlet of the watershed are listed in the following table. Assume that the baseflow is 1 m^3/s and rainfall amount in the first 30 min is the initial loss.

(a) Determine the total amount of rainfall.

(b) Determine the volume of direct runoff.

(c) Determine the average infiltration *rate*, i.e., ϕ index

(d) Determine the rainfall excess *intensity* hyetograph corresponding to the ϕ index obtained in (c). Also, what is the duration of rainfall excess hyetograph?

Time (min)	Intensity (mm/h)	Discharge (m^3/s)
0	10	1
30	20	1
60	38	4
90	26	8
120	10	6
150	20	4
180	24	3
210		1

8.2.7 On a particular day in 1990, there was a rainstorm event that produced the following rainfall and runoff on a 36-km^2 drainage basin.

Time (min)	0	15	30	45	60	75	90	105	120
Cumulative rainfall (mm)	0	10	50	75	90	100			
Instantaneous runoff (m^3/s)	10	30	160	360	405	305	125	35	10

For the above rainstorm event, assume that the basefiow is 10 m^3/s.

(a) Find the volume of direct runoff (in mm).
(b) Assuming the initial loss is 10 mm, determine the value of ϕ index (in mm/hr) and the corresponding effective rainfall intensity hyetograph (in mm/hr).

8.3.1 Determine the 4-hr unit hydrograph using the following data for a watershed having a drainage area of 200 km^2, assuming a constant rainfall abstraction rate and a constant baseflow of 20 m^3/s.

Four-hour period	1	2	3	4	5	6	7	8	9	10	11
Rainfall (cm)	1.0	2.5	4.0	2.0							
Storm flow (m^3/s)	20	30	60	95	130	170	195	175	70	25	20

8.3.2 Using the unit hydrograph developed in problem 8.3.1, determine the direct runoff from the 200 km^2 watershed using the following rainfall excess pattern.

Four-hour period	1	2	3	4
Rainfall excess (cm)	2.0	3.0	0.0	1.5

8.3.3 The ordinates at 1-hr intervals of a 1-hr unit hydrograph are 100, 300, 500, 700, 400, 200, and 100 cfs. Determine the direct runoff hydrograph from a 3-hr storm in which 1 in of excess rainfall occurred in the first hour, 2 in the second hour, and 0.5 in the third hour. What is the area of the watershed in mi^2?

8.3.4 You are responsible for developing the runoff hydrograph for a watershed that fortunately has gauged information at a very nearby location on the stream. You have obtained information on an actual rainfall-runoff event for this watershed. This information includes the actual rainfall hyetograph and the resulting runoff hydrograph. You have been asked to develop the runoff hydrograph for a design rainfall event. The peak discharge from this new developed storm hydrograph will be used to design a hydraulic structure. You will assume a constant baseflow and a constant rainfall abstraction. Explain the hydrologic analysis procedure that you will use to solve this problem.

8.3.5 Consider a drainage basin with an area of 226.8 km^2. From a storm event, the observed cumulative rainfall depth and the corresponding runoff hydrograph are given in the following table. Assume that the baseflow is 10 m^3/s.

Time (hr)	0	3	6	9	12	15	18	21	24	27	30	33	36
Cumulative rainfall (cm)	0.0	0.5	1.5	2.5	4.5	6.5	8.0	9.5					
Discharge (m^3/s)	10	10	30	50	130	175	260	240	220	145	70	40	10

(a) Determine the value of the ϕ index and the corresponding effective rainfall hyetograph.
(b) Set up the system of equations, $Pu = Q$, and solve for the unit hydrograph.

8.3.6 Suppose the 4-hr unit hydrograph for a watershed is

Time (hr)	0	2	4	6	8	10	12
UH (m^3/s/cm)	0	20	50	75	90	40	0

(a) Determine the 2-hr unit hydrograph.
(b) Suppose that a 10-year design storm has a total effective rainfall of 50 mm and the corresponding hyetograph has a rainfall of 30 mm in the first 2 hr period, no rain in the second period, and 20 mm in the third period. Assuming the baseflow is 10 m^3/s, calculate the total runoff hydrograph from this 10-year design storm.

8.3.7 Consider a drainage basin with an area of 151.2 km^2. From a storm event, the observed cumulative rainfall depth and the corresponding runoff hydrograph are given in the following table. Assume that the baseflow is 10 m^3/s.

Time (hr)	0	2	4	6	10	12	14	16	18	20	22	24	26
Cumulative rainfall (cm)	0.0	0.5	1.5	2.5	4.5	6.5	8.0	9.5					
Instantaneous discharge (m^3/s)	10	10	30	50	130	175	260	240	220	145	70	40	10

(a) Determine the value of the ϕ index and the corresponding effective rainfall hyetograph.
(b) Set up the system of equations, $Pu = Q$, for deriving the 6-hr unit hydrograph and define the elements in P, u, and Q. Solve for the unit hydrograph.

8.3.8 Suppose the 30-min unit hydrograph for a watershed is

Time (hr)	0	3	6	9	12	15	18	21
UH (m^3/s/cm)	0	15	45	65	50	25	10	0

(a) Determine the 1-hr unit hydrograph.
(b) A storm has a total depth of 4 cm with an effective rainfall hyetograph of 2 cm in the first 2 hr, 1 cm in the second 2 hr. Assuming the baseflow is 10 m^3/s, calculate the total runoff hydrograph from the storm.

8.3.9 The following table lists a 15-min unit hydrograph.

Time (min)	0	5	10	15	20	25	30	35	40	45
15 min UH (m^3/s/cm)	0.0	1.5	3.5	6.0	5.5	5.0	3.0	2.0	0.5	0.0

(a) Determine the corresponding area of the drainage basin.
(b) Determine the 10-min unit hydrograph.
(c) Determine the total runoff hydrograph resulting from the following effective rainfall, assuming the base flow is $2\,\text{m}^3/\text{s}$.

Time (min)	0	10	20	30
Cumul. rainfall excess (mm)	0	10	30	35

8.3.10 Consider there are two rain gauges (A and B) in a watershed with an area of $21.6\,\text{km}^2$. The cumulative rainfall depths over time for a particular storm event at both rain gauge stations are given in the table below. The storm also produces a runoff hydrograph at the outlet of the watershed, shown in the last row of the table. Assume that the baseflow is $10\,\text{m}^3/\text{s}$. It is known, by the Thiessen polygon, that the contributing areas for the rain gauges are identical.

Time (hr)	0	1	2	3	4	5	6	7	8	9	10
Cumul. Rainfall at Station A (mm)	0	25	80	105	100	140	180	230	230		
Cumul. rainfall at Station B (mm)	0	15	40	85	130	180	210	230	230		
Instantaneous discharge (m³/s)	10	10	30	60	30	40	80	40	30	10	10

(a) Determine the basin-wide representative rainfall hyetograph (in mm/hr) for the storm event. Furthermore, determine the value of ϕ index (in mm/hr) and the corresponding effective rainfall hyetograph (in mm/hr).
(b) Set up the system of equations $Pu = Q$.
(c) Solve part (b) *by the least-squares method* for the unit hydrograph. What is the duration of the unit hydrograph obtained?

8.3.11 Consider a drainage basin with an area of $167.95\,\text{km}^2$. From a storm event, the observed cumulative rainfall depth and the corresponding runoff hydrograph are given in the following table. Assume that the baseflow is $10\,\text{m}^3/\text{s}$.

Time (hr)	0	2	4	6	8	10	12	14	16	18	20	22	24
Cumulative rainfall (cm)	0.0	0.5	1.5	2.5	4.5	6.5	8.0	9.5					
Instantaneous discharge (m³/s)	10	10	30	50	130	175	260	240	220	145	70	40	10

(a) Determine the value of the ϕ index and the corresponding effective rainfall hyetograph.
(b) Set up the system of equations, $Pu = Q$, for deriving the 4-hr unit hydrograph and define the elements in P, u, and Q.
(c) Solve for the unit hydrograph ordinates.

8.3.12 Consider a drainage basin with an area of $216\,\text{km}^2$. From a storm event, the observed cumulative rainfall depth and the corresponding runoff hydrograph are given in the following table. Assume that the baseflow is $20\,\text{m}^3/\text{s}$.

Time (hr)	0	6	12	18	24	30
Cumulative rainfall (cm)	0.0	1.5	4.5	9.5		
Discharge (m³/s)	20	50	160	340	180	20

(a) Determine the value of the ϕ index and the corresponding effective rainfall hyetograph.
(b) Determine the 6-hr unit hydrograph by the least-squares method.

8.3.13 On a particular date, the measured rainfall and runoff from a drainage basin of 54 ha ($54 \times 10^4\,\text{m}^2$) are listed in the table below.

(a) Assuming that the base flow is $2\,\text{m}^3/\text{s}$, determine the effective rainfall hyetograph by the ϕ index method.
(b) Based on the effective rainfall hyetograph and direct runoff hydrograph obtained in Part (a), use the least-squares method to determine the 15-min unit hydrograph.

Time (min)	0	15	30	45	60	75
Cumul. rain (mm)	0	5	20	40	50	
Instant. flow (m³/s)	2	6	10	7	3	2

8.3.14 You have been given information on an actual rainfall-runoff event and asked to develop the runoff hydrograph for another storm hydrograph for the watershed. The following is known:

Time (hr)	Rainfall (in)	Discharge (cfs)
0		10
1	1.1	20
2	2.1	130
3	0	410
4	1.1	570
5		510
6		460
7		260
8		110
9		60
10		10

A constant base flow of 10 cfs and a uniform rainfall loss of 0.1 in/hr are applicable. Determine the size of the drainage basin. Then determine the direct runoff hydrograph for a 2.0-in excess precipitation for the first hour, no rainfall for the second hour, followed by a 2.0-in excess precipitation for the third hour.

8.4.1 A watershed has a drainage area of $14\,\text{km}^2$; the length of the main stream is 7.16 km, and the main channel length from the watershed outlet to the point opposite the center of gravity of the watershed is 3.22 km. Use $C_t = 2.0$ and $C_p = 0.625$ to determine the standard synthetic unit hydrograph for the watershed. What is the standard duration? Use Snyder's method to determine the 30-min unit hydrograph for the watershed.

8.4.2 Watershed A has a 2-hr unit hydrograph with $Q_{pR} = 276\,\text{m}^3/\text{s}$, $t_{pR} = 6$ hr, $W_{50} = 4.0$ hr, and $W_{75} = 2$ hr. The watershed area = $259\,\text{km}^2$, $L_c = 16.1$ km, and $L = 38.6$ km. Watershed B is assumed to be hydrologically similar with an area of $181\,\text{km}^2$, $L = 25.1$ km, and $L_c = 15.1$ km. Determine the 1-hr synthetic unit hydrograph for watershed B. Determine the direct runoff hydrograph for a 2-hr storm that has 1.5 cm of excess rainfall the first hour and 2.5 cm of excess rainfall the second hour.

8.4.3 A watershed has an area of 39.3 mi^2 and a main channel length of 18.1 mi, and the main channel length from the watershed outlet to the point opposite the centroid of the watershed is 6.0 mi. The regional parameters are $C_t = 2.0$ and $C_p = 0.6$. Compute the T_P, Q_P, W_{50}, W_{75}, and T_B for Snyder's standard synthetic unit hydrograph and the same information for a 3-hr Snyder's synthetic unit hydrograph. Also, what is the duration of the standard synthetic unit hydrograph?

8.4.4 Compute the 3-hr Snyder's synthetic unit hydrograph for the watershed in problem 8.4.3.

8.4.5 You are performing a hydrologic study (rainfall-runoff analysis) for a watershed Z. Unfortunately, there is no gauged data or other hydrologic studies so you do not have a unit hydrograph. However, you do have data for another nearby watershed, known as watershed X, which has gauged information (including the discharge hydrograph for a known rainfall event). What procedure would you use to develop a design runoff hydrograph for a design storm for watershed Z? What are you assuming about the watersheds in this procedure?

8.4.6 You have determined the following from the basin map of a given watershed: $L = 100$ km, $L_c = 50$ km, and drainage area = 2000 km^2. From the unit hydrograph developed for the watershed, the following were determined: $t_R = 6$ hr, $t_{pR} = 15$ hr, and the peak discharge = 80 m^3/s/cm. Determine the regional parameters used in the Snyder's synthetic unit hydrograph procedure.

8.4.7 Derive a 3-hr unit hydrograph by the Snyder method for a watershed of 54 km^2 area. It has a main stream that is 10-km long. The distance measured from the watershed outlet to a point on the stream nearest to the centroid of the watershed is 3.75 km. Take $C_t = 2.0$ and $C_p = 0.65$. Sketch the 3-hr unit hydrograph for the watershed.

8.4.8 Derive a 2-hr unit hydrograph by the Snyder method for a watershed of 50 km^2 area. It has a main stream that is 8 km long. The distance measured from the watershed outlet to a point on the stream nearest to the centroid of the watershed is 4 km. Take $C_t = 2.0$ and $C_p = 0.65$. Graph the 2-hr unit hydrograph for the watershed.

8.4.9 Use the Clark unit hydrograph procedure to compute the 1-hr unit hydrograph for a watershed that has an area of 5.0 km^2 and a time of concentration of 5.5 hr. The Clark storage coefficient is estimated to be 2.5 hr. Use a 1-hr time interval for the computations. Use the HEC U.S. Army Corps of Engineer synthetic time-area relationship.

8.4.10 Compute the Clark unit hydrograph parameters (T_c and R) for a 2.17-mi^2 (1389 acre) urban watershed in Phoenix, Arizona that has a flow path of 1.85 mi, a slope of 30.5 ft/mi, and an imperviousness of 21 percent. A rainfall intensity of 2.56 in/hr is to be used. The time of concentration for the watershed is computed using $T_c = 11.4L^{0.50} K_b^{0.52} S^{-0.31} i^{-0.38}$, where T_c is the time of concentration in hours, L is the length of the longest flow path in miles, K_b is a watershed resistance coefficient ($K_b = -0.00625\log A + 0.04$) for commercial and residential areas, A is the watershed area in acres, S is the slope of the flow path

in ft/mi, and i is the rainfall intensity in in/hr. The storage coefficient is $R = 0.37T_c^{1.11} A^{-0.57} L^{0.80}$, where A is the area in mi^2.

8.4.11 Compute the 15-min Clark unit hydrograph for the Phoenix watershed in problem 8.4.10.

8.4.12 For the situation in problem 8.4.10, compute the time of concentration. However, now the rainfall intensity is not given, instead use the rainfall intensity duration frequency relation in Figure 7.2.15, with a 25-year return period.

8.4.13 Develop the 15-min Clark unit hydrograph for a 2.17-mi^2 (1389 acre) rural watershed that has a flow path of 1.85 mi, a slope of 30.5 ft/mi, and an imperviousness of 21 percent. A rainfall intensity of 2.56 in/hr is to be used. The time of concentration for the watershed is computed using $T_c = 11.4L^{0.50} K_b^{0.52} S^{-0.31} i^{-0.38}$, where T_c is the time of concentration in hours, L is the length of the longest flow path in miles, K_b is a watershed resistance coefficient ($K_b = -0.01375\log A + 0.08$), A is the watershed area in acres, and S is the slope of the flow path in ft/mi, and i is the rainfall intensity in in/hr.

8.5.1 Using the 4-hr unit hydrograph developed in problem 8.3.1, use the S-curve method to develop the 8-hr unit hydrograph for this 200 km^2 watershed.

8.5.2 Using the one-hour unit hydrograph given in problem 8.3.3, develop the 3-hr unit hydrograph using the S-curve method.

8.5.3 Suppose the 4-hr unit hydrograph for a watershed is

Time (hr)	0	2	4	6	8	10	12	14
UH (m^3/s/10 mm)	0	15	45	65	50	25	10	0

(a) Determine the 2-hr unit hydrograph.
(b) From the frequency analysis, the 10-year 4-hr storm has a total depth of 3 cm with an effective rainfall intensity of 1 cm/hr in the first 2 hr and 0.5 cm/hr in the second 2 hr. Assuming the baseflow is 10 m^3/s, calculate the total runoff hydrograph from the 10-year 4-hr storm.

8.5.4 Suppose the 4-hr unit hydrograph for the watershed is

Time (hr)	0	2	4	6	8	10	12	14	16
UH (m^3/s/cm)	0	19	38	32	22	13	6	2	0

(a) What is the area of the watershed?
(b) Determine the 2-hr unit hydrograph.
(c) Suppose that a 25-year design storm has a total effective rainfall of 7 cm and the corresponding hyetograph has 5 cm in the first 2 hr and 2 cm in the second 2 hr. Assuming the baseflow is 10 m^3/s, calculate the total runoff hydrograph from this 25-year design storm.

8.5.5 Suppose the 6-hr unit hydrograph for the watershed is

Time (hr)	0	3	6	9	12	15	18	21
UH (m^3/s/cm)	0	15	45	65	50	25	10	0

(a) What is the area of the watershed?
(b) Determine the 3-hr unit hydrograph.
(c) Suppose that a 10-year design storm has a total effective rainfall of 6 cm and the corresponding hyetograph has 4 cm in the first 3 hr and 2 cm in the second 3 hr. Assuming the baseflow is 10 m^3/s, calculate the total runoff hydrograph from this 10-year design storm.

8.5.6 Suppose the 6-hr unit hydrograph for a watershed is

Time (hr)	0	3	6	9	12	15	18
UH ($m^3/s/cm$)	0	60	90	50	30	10	0

(a) Determine the 3-hr unit hydrograph.
(b) Suppose that a 50-year design storm has a total effective rainfall of 9 cm and the corresponding hyetograph has 2 cm in the first 3 hr, 5 cm in the second 3 hr, and 2 cm in the third 3 hr. Assuming the baseflow is 20 m^3/s, determine the total runoff hydrograph from this 50-year design storm.

8.5.7 Suppose the 6-hr unit hydrograph for a watershed is

Time (hr)	0	3	6	9	12	15	18	21
UH ($m^3/s/cm$)	0	15	45	65	50	25	10	0

(a) Determine the 2-hr unit hydrograph.
(b) Assuming the baseflow is 10 m^3/s, calculate the total runoff hydrograph from an effective rainfall hyetograph of 2 cm in the first 2 hr and 1 cm in the second 2 hr.

8.7.1 Determine the weighted curve numbers for a watershed with 60 percent residential (1/4-acre lots), 20 percent open space, good condition, and 20 percent commercial and business (85 percent impervious) with corresponding soil groups of C, D, and C.

8.7.2 Rework example 8.7.1 with corresponding soil groups of B, C, D, and B.

8.7.3 The watershed in problem 8.7.1 experienced a rainfall of 5 in; what is the runoff volume per unit area?

8.7.4 Rework example 8.7.2 with a 7-in rainfall.

8.7.5 Calculate the cumulative abstractions and the excess rainfall hyetograph for the situation in problems 8.7.1 and 8.7.3. The rainfall pattern is 1.5 in during the first hour, 2.5 in during the second hour, and 1.0 in during the third hour.

8.7.6 Calculate the cumulative abstraction and the excess rainfall hyetograph for the situation in problems 8.7.2 and 8.7.4. The rainfall pattern is 2.0 in during the first hour, 3.0 in during the second hour, and 2.0 in during the third hour.

8.7.7 Consider an urban drainage basin having 60 percent soil group B and 40 percent soil group C. The land use pattern is 1/2 commercial area and 1/2 industrial district. Determine the rainfall excess *intensity* hyetograph under the dry antecedent moisture condition (AMC I) from the recorded storm given in problem 8.2.6.

8.7.8 Consider a drainage basin having 60 percent soil group B and 40 percent soil group C. Five years ago, the watershed land use pattern was 1/2 wooded area with good cover and 1/2 pasture with good condition. Now, the land use has been changed to 1/3 wooded area, 1/3 pasture land, and 1/3 residential area (1/4-acre lot). Estimate the volume of increased runoff due to the land use change over the past 5-year period for a storm with 6 in of rainfall under the dry antecedent moisture condition (AMC I).

8.7.9 Refer to problem 8.7.8 for the present watershed land use pattern. Determine the effective rainfall hyetograph for the following storm event using the SCS method under the dry antecedent moisture condition (AMC I). Next determine the value of the ϕ index corresponding to the effective rainfall hyetograph.

Time (h)	0–0.5	0.5–1.0	1.0–1.5	1.5–2.0
Rainfall intensity (in/h)	6.0	3.0	2.0	1.0

8.7.10 Consider a drainage basin having 60 percent soil group A and 40 percent soil group B. Five years ago, the land use pattern in the basin was 1/2 wooded area with poor cover and 1/2 cultivated land with good conservation treatment. Now, the land use has been changed to 1/3 wooded area, 1/3 cultivated land, and 1/3 commercial and business area. Estimate the increased runoff volume during the dormant season due to the land use change over the past 5-year period for a storm of 350 mm in total depth. This storm depth corresponds to a duration of 6-hr and 100-year return period. The total 5-day antecedent rainfall amount is 30 mm. Note: 1 inch = 25.4 mm.

8.7.11 Consider a drainage basin of 500-ha having hydrologic soil group D. In 1970, the watershed land use pattern was 50% wooded area with good cover, 25% range land with good condition, and 25% residential area (1/4-acre lot). Ten years later, the land use has been changed to 30% wooded area, 20% pasture land, 40% residential area (1/4-acre lot), and 10% commercial and business area.

(a) Compute the weighted curve numbers in 1970 and 1980.
(b) Using the SCS method, estimate the percentage of change (increase or decrease) in runoff volume with respect to year 1970 due to the land use change over the 10-year period for a 2-hr, 50-mm rainfall event under the wet antecedent moisture condition.
(c) Suppose that the 2-hr, 50-mm rainfall event has the following hyetograph. Determine the rainfall excess intensity hyetograph (in mm/hr) and the corresponding incremental infiltration (in mm) over the storm duration under 1980 conditions.

Time (hr)	0–0.5	0.5–1.0	1.0–1.5	1.5–2.0
Rainfall intensity (mm/hr)	4.0	50.0	30.0	16.0

8.7.12 Consider a drainage basin of 36 km^2 having hydrologic soil group B. In 1970, the watershed land use pattern was 60 percent wooded area with good cover, 25 percent range land with good condition, and 15 percent residential area (1/4-acre lot). Twenty years later (i.e., in 1990), the land use has been changed to 30 percent wooded area with good cover, 20 percent pasture land with good condition, 40 percent residential area (1/4-acre lot), and 10 percent commercial and business area.

On a particular day in 1990, there was a rainstorm event that produced rainfall and runoff recorded in the following table.

Time (min)	0	15	30	45	60	75	90	105	120
Cumulative rainfall (mm)	0	10	50	75	90	100			
Instantaneous runoff (m³/s)	10	30	160	360	405	305	125	35	10

(a) Using the SCS method, estimate the percentage of change (increase or decrease) in runoff volume in 1990 with respect to that of 1970 for the above rainstorm event under the normal antecedent moisture condition.

(b) Under the wet antecedent moisture condition, determine the rainfall excess intensity hyetograph (in mm/hr) and the corresponding incremental infiltration (in mm) over the storm duration for the above rainstorm event under the 1990 conditions.

(c) For a baseflow of 10 m³/s, determine the volume of direct runoff.

(d) Assuming an initial loss of 10 mm, determine the Φ index (in mm/hr) and the corresponding excess rainfall hyetograph.

8.7.13 Consider a drainage basin with an area of 200 km². From a storm event, the observed cumulative rainfall depth and the corresponding runoff hydrograph are given in the following table. Assume that the baseflow is 20 m³/s.

Time (hr)	0	4	8	12	16	20
Cum. rainfall (cm)	0.0	1.6	5.5	7.5	7.5	7.5
Discharge (m³/s)	20	40	130	300	155	20

(a) Determine the effective rainfall hyetograph by the SCS method with a curve number $CN = 85$.

(b) Determine the 4-hr unit hydrograph by the least-squares method.

8.7.14 During a rain storm event, rain gauge at location X broke down and rainfall record was not available. Fortunately, three rain gauges nearby did not have a technical problem and their rainfall readings, along with other information, are provided in the table below.

Rain gauge	A	B	C	X
Event depth (mm)	40	60	50	Missing
Distance to gauge X (km)	10	8	12	0
Elevation (m)	20	50	40	60
Mean annual rainfall (mm)	1500	2000	1800	2200
Polygon area (km²)	10	30	20	40

The four rain gauges are located either within or in the neighborhood of a watershed having a drainage area of 100 km². According to the Thiessen polygon method, the contributing area of each rain gauge is shown in the above table.

For this particular storm event, the duration of the storm is 3 hr and the percentage of rainfall depth in the first hour is 50 percent, in the second hour 30 percent, and in the third hour 20 percent. The watershed largely consists of woods and meadow of good hydrologic condition with soil group B. Among the two land cover types, woods occupy 60 percent of the total area while the meadow makes up the remaining 40 percent.

For this watershed, the 1-hr unit hydrograph resulting from 1 cm of effective rainfall has been derived and is given below:

Time (hr)	0	1	2	3	4	5
Flow rate (m³/s/cm)	0	10	30	20	10	0

(a) Select a method to estimate missing rainfall depth at station X by an appropriate method. Explain the reasons why the method is selected.

(b) Determine the basin-wide equivalent uniform rainfall depth and the corresponding hyetograph for this particular storm.

(c) According to part (b), determine the total effective rainfall depth and the corresponding rainfall excess hyetograph for the storm. It is known that a storm occurred 2 days before this particular storm and, therefore, the ground is quite wet.

(d) What is the magnitude of peak runoff discharge produced by this particular storm? It is reasonable to assume that the baseflow is 5 m³/s.

8.7.15 Consider a drainage basin having 60 percent soil group A and 40 percent soil group B. Five years ago, the land use pattern in the basin was 1/2 wooded area with poor cover and 1/2 cultivated land with good conservation treatment. Now the land use has been changed to 1/3 wooded area, 1/3 cultivated land, and 1/3 commercial and business area.

(a) Estimate the increased runoff volume during the dormant season due to the land use change over the past 5-year period for a storm of 35 cm total depth under the dry antecedent moisture condition (AMC I). This storm depth corresponds to a duration of 6-hr and 100-year return period. The total 5-day antecedent rainfall amount is 30 mm. (Note: 1 in = 25.4 mm.)

(b) Under the present watershed land use pattern, find the effective rainfall hyetograph (in cm/hr) for the following storm event using SCS method under the dry antecedent moisture condition (AMC I).

Time (hr)	0–0.5	0.5–1.0	1.0–1.5	1.5–2.0
Avg. rainfall intensity (cm/hr)	16.0	9.0	5.0	3.0

8.7.16 Consider a drainage basin having 60 percent soil group B and 40 percent soil group C. Five years ago, the watershed land use pattern was 1/2 wooded area with good cover and 1/2 pasture with good condition. Today, the wooded area and pasture each have been reduced down to 1/3 and the remaining 1/3 of the drainage basin has become a residential area (1/4-acre lot). Consider the rainstorm event having the observed rainfall mass data given below.

Time (min)	0	30	60	90	120
Cumulative Rainfall (mm)	0	152	228	278	304

(a) Estimate the increased total runoff volume due to the land use change over the past 5-year period for the above rainstorm event under the dry antecedent moisture condition (AMC I).
(b) Referring to the present watershed land use condition, find the effective rainfall hyetograph (in mm/hr) for the rainstorm event given the above using the SCS method under the dry antecedent moisture condition (AMC I).

8.7.17 Consider a drainage basin of 1,200 hectares having 60 percent soil group B and 40 percent soil group C. Five years ago, the watershed land use pattern was 1/2 wooded area with good cover and 1/2 pasture with good condition. Now the land use has been changed to 1/3 wooded area, 1/3 pasture land, and 1/3 residential area (1/4-acre lot). For a storm event, the cumulative rainfall and the corresponding runoff hydrograph are shown in the following table and figure, respectively. (Note: 1 hectare = 10,000 m^2; 1 in = 25.4 mm).

Time (hr)	0	1	2	3	4
Cumulative rainfall (mm)	0	60	120	160	180

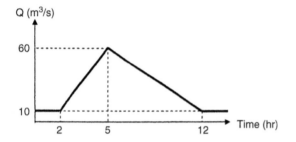

(a) Estimate the increased total runoff volume by the SCS method due to the land use change over the past 5-year period for this storm under the wet antecedent moisture condition.
(b) Find the effective rainfall intensity hyetograph for the storm event using the SCS method under the wet antecedent moisture condition and current land use condition.

(c) Find the value of the ϕ index corresponding to the above direct runoff hydrograph assuming the baseflow is 10 m^3/s.

8.7.18 Consider a drainage basin having soil group D. The land use pattern in the basin is 1/3 wooded area with good cover, 1/3 cultivated land with conservation treatment, and 1/3 commercial and business area. For the rainstorm event given below, use the SCS method under the dry antecedent moisture condition (AMC I).

(a) Estimate the total volume of effective rainfall (in cm).
(b) Find the effective rainfall hyetograph (in cm/hr) for the following storm event.

Time (hr)	0–0.5	0.5–1.0	1.0–1.5	1.5–2.0
Avg. rainfall intensity (cm/hr)	16.0	9.0	5.0	3.0

8.8.1 Develop the SCS triangular unit hydrograph for a 400-acre watershed that has been commercially developed. The flow length is 1,500 ft, the slope is 3 percent, and the soil group is group B.

8.8.2 Prior to development of the 400-acre watershed in problem 8.8.1, the land use was contoured pasture land with fair condition. Compute the SCS triangular unit hydrograph and compare with the one for commercially developed conditions.

8.8.3 Using the watershed defined in problems 8.8.1 and 8.8.2, determine the SCS triangular unit hydrograph assuming residential lot size. Compare with the results in problem 8.8.2.

8.8.4 A 20.7-km^2 watershed has a time of concentration of 1.0 hr. Calculate the 10-min unit hydrograph for the watershed using the SCS triangular unit hydrograph method. Determine the direct runoff hydrograph for a 30-min storm having 1.5 cm of excess rainfall in the first 10 min, 0.5 cm in the second 10 min, and 1.0 cm in the third 10 min.

8.9.1 Develop a flowchart of the kinematic overland flow runoff model described in Section 8.9.

8.9.2 Develop the appropriate equations to solve equation (8.9.9) by Newton's method.

8.9.3 Derive equation (8.9.12).

REFERENCES

Chow, V. T. (editor), *Handbook of Applied Hydrology*, McGraw-Hill, New York, 1964.

Chow, V. T., D. R. Maidment, and L. W. Mays, *Applied Hydrology*, McGraw-Hill, New York, 1988.

Clark, C. O., "Storage and the Unit Hydrograph," *Trans. American Society of Civil Engineers*, Vol. 110, pp. 1419–1488, 1945.

Engman, E. T., "Roughness Coefficients for Routing Surface Runoff", *Journal of Irrigation and Drainage Engineering; American Society of Civil Engineers*, 112(1), pp. 39–53, 1986.

Flood Control District of Maricopa County, *Drainage Design Manual for Maricopa County, Arizona*, Phoenix, AZ, 1995.

Ford, D., E. C. Morris, and A. D. Feldman, "Corps of Engineers' Experience with Automatic Calibration Precipitation-Runoff Model," in *Water and Related Land Resource Systems* edited by Y. Haimes and J. Kindler, p. 467–476, Pergamon Press, New York, 1980.

Hewlett, J. D., and W. L. Nutter, *An Outline of Forest Hydrology*, University of Georgia Press, Athens, GA, 1969.

Horton, R. E., "Erosional Development of Streams and Their Drainage Basins; Hydrological Approach to Quantitative Morphology," *Bull. Geol. Soc. Am.*, vol. 56, pp. 275–370, 1945.

Masch, F. D., *Hydrology*, Hydraulic Engineering Circular No. 19, FHWA-10-84-15, Federal Highway Administration, U.S. Department of the Interior, McLean, VA, 1984.

Morris, E. M., and D. A. Woolhiser, "Unsteady One-Dimensional Flow over a Plane: Partial Equilibrium and Recession Hydrographs," *Water Resources Research*, vol. 16, no, 2, pp. 355–360, 1980.

Mosley, M. P., and A. I. McKerchar, "Streamflow," in *Handbook of Hydrology* (edited by D. R. Maidment), McGraw-Hill, New York, 1993.

Palmer, V. J., "Retardance Coefficients for Low Flow in Channels Lined with Vegetation," *Transactions of the American Geophysical Union*, 27(11), pp. 187–197, 1946.

Papadakis, C. N., and M. N. Kazan, "Time of Concentration in Small, Rural Watersheds," *Proceedings of the Engineering Hydrology Symposium*, ASCE, Williamsburg, Virginia, pp. 633–638, 1987.

Ponce, V. M., *Engineering Hydrology; Principles and Practices*, Prentice-Hall, Englewood Cliffs, NJ, 1989.

Sanders, T. G. (editor), *Hydrology for Transportation Engineers*, U.S. Dept. of Transportation, Federal Highway Administration, 1980.

Strahler, A. N., "Quantitative Geomorphology of Drainage Basins and Channel Networks," section 4-II in *Handbook of Applied Hydrology* (edited by V. T. Chow), McGraw-Hill, New York, 1964.

Straub, T. D., C. S. Melching, and K. E. Kocher, *Equations for Estimating Clark Unit-Hydrograph Parameters for Small Rural Watersheds in Illinois*, U.S. Geological Survey Water-Resources Investigations Report 00-4184, Urbana, IL, 2000.

U.S. Army Corps of Engineers, Hydrologic Engineering Center, HEC-1 Flood Hydrograph Package, User's Manual, Davis, CA, 1990.

U.S. Department of Agriculture Soil Conservation Service, National Engineering Handbook, Section 4, *Hydrology*, available from U.S. Government Printing Office, Washington, DC, 1972.

U.S. Department of Agriculture Soil Conservation Service, "Urban Hydrology for Small Watersheds," Tech. Release No., 55, Washington, DC, June, 1986.

Woolhiser, D. A., and J. A. Liggett, "Unsteady, One-Dimensional Flow over a Plane—the Rising Hydrograph," *Water Resources Research*, vol. 3(3), pp. 753–771, 1967.

Woolhiser, D. A., R. E. Smith, and D. C. Goodrich, *KINEROS, A. Kinematic Runoff and Erosion Model: Documentation and User Manual*, U. S. Department of Agricultural Research Service, ARS-77, Tucson, AZ, 1990.

Chapter 9

Reservoir and Stream Flow Routing

9.1 ROUTING

Figure 9.1.1 illustrates how stream flow increases as the *variable source area* extends into the drainage basin. The variable source area is the area of the watershed that is actually contributing flow to the stream at any point. The variable source area expands during rainfall and contracts thereafter.

Flow routing is the procedure to determine the time and magnitude of flow (i.e., the flow hydrograph) at a point on a watercourse from known or assumed hydrographs at one or more points upstream. If the flow is a flood, the procedure is specifically known as flood routing. Routing by lumped system methods is called *hydrologic (lumped) routing*, and routing by distributed systems methods is called *hydraulic (distributed) routing*.

For hydrologic routing, input $I(t)$, output $Q(t)$, and storage $S(t)$ as functions of time are related by the continuity equation (3.3.10)

$$\frac{dS}{dt} = I(t) - Q(t) \tag{9.1.1}$$

Even if an inflow hydrograph $I(t)$ is known, equation (9.1.1) cannot be solved directly to obtain the outflow hydrograph $Q(t)$, because, both Q and S are unknown. A second relationship, or storage function, is required to relate S, I, and Q; coupling the storage function with the continuity equations provides a solvable combination of two equations and two unknowns.

The specific form of the storage function depends on the nature of the system being analyzed. In reservoir routing by the level pool method (Section 9.2), storage is a nonlinear function of Q, $S = f(Q)$, and the function $f(Q)$ is determined by relating reservoir storage and outflow to reservoir water level. In the Muskingum method (Section 9.3) for flow routing in channels, storage is linearly related to I and Q.

The effect of storage is to redistribute the hydrograph by shifting the centroid of the inflow hydrograph to the position of the outflow hydrograph in a *time of redistribution*. In very long channels, the entire flood wave also travels a considerable distance, and the centroid of its hydrograph may then be shifted by a time period longer than the time of redistribution. This additional time may be considered the *time of translation*. The total time of flood movement between the centroids of the inflow and outflow hydrographs is equal to the sum of the time of redistribution and the time of translation. The process of redistribution modifies the shape of the hydrograph, while translation changes its position.

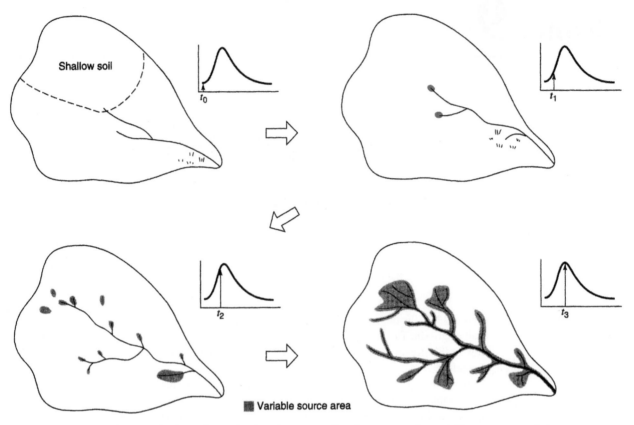

Figure 9.1.1 The small arrows in the hydrographs show how streamflow increases as the variable source extends into swamps, shallow soils, and ephemeral channels. The process reverses as streamflow declines (from Hewlett (1982)).

9.2   HYDROLOGIC RESERVOIR ROUTING

Level pool routing is a procedure for calculating the outflow hydrograph from a reservoir assuming a horizontal water surface, given its inflow hydrograph and storage-outflow characteristics. Equation (9.1.1) can be expressed in the infinite-difference form to express the change in storage over a time interval (see Figure 9.2.1) as

$$S_{j+1} - S_j = \frac{I_j + I_{j+1}}{2}\Delta t - \frac{Q_j + Q_{j+1}}{2}\Delta t \qquad (9.2.1)$$

The inflow values at the beginning and end of the jth time interval are I_j and I_{j+1}, respectively, and the corresponding values of the outflow are Q_j and Q_{j+1}. The values of I_j and I_{j+1} are prespecified. The values of Q_j and S_j are known at the jth time interval from calculations for the previous time interval. Hence, equation (9.2.1) contains two unknowns, Q_{j+1} and S_{j+1}, which are isolated by multiplying (9.2.1) through by $2/\Delta t$, and rearranging the result to produce:

$$\left[\frac{2S_{j+1}}{\Delta t} + Q_{j+1}\right] = (I_j + I_{j+1}) + \left[\frac{2S_j}{\Delta t} - Q_j\right] \qquad (9.2.2)$$

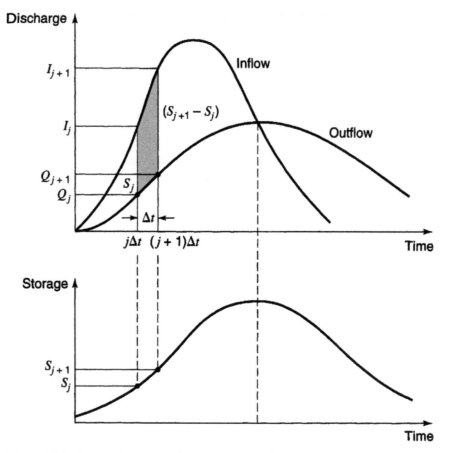

Figure 9.2.1 Change of storage during a routing period Δt.

In order to calculate the outflow Q_{j+1}, a storage-outflow function relating $2S/\Delta t + Q$ and Q is needed. The method for developing this function using elevation-storage and elevation-outflow relationships is shown in Figure 9.2.2. The relationship between water surface elevation and reservoir storage can be derived by planimetering topographic maps or from field surveys. The elevation-discharge relation is derived from hydraulic equations relating head and discharge for various types of spillways and outlet works. (See Chapter 17.) The value of Δt is taken as the time interval of the inflow hydrograph. For a given value of water surface elevation, the values of storage S and discharge Q are determined (parts (a) and (b) of Figure 9.2.2), and then the value of $2S/\Delta t + Q$ is calculated and plotted on the horizontal axis of a graph with the value of the outflow Q on the vertical axis (part (c) of Figure 9.2.2).

In routing the flow through time interval j, all terms on the right side of equation (9.2.2) are known, and so the value of $2S_{j+1}/\Delta t + Q_{j+1}$ can be computed. The corresponding value of Q_{j+1} can be determined from the storage-outflow function $2S/\Delta t + Q$ versus Q, either graphically or by linear interpolation of tabular values. To set up the data required for the next time interval, the value of $(2S_{j+1}/\Delta t - Q_{j+1})$ is calculated using

$$\left[\frac{2S_{j+1}}{\Delta t} - Q_{j+1}\right] = \left[\frac{2S_{j+1}}{\Delta t} + Q_{j+1}\right] - 2Q_{j+1} \tag{9.2.3}$$

The computation is then repeated for subsequent routing periods.

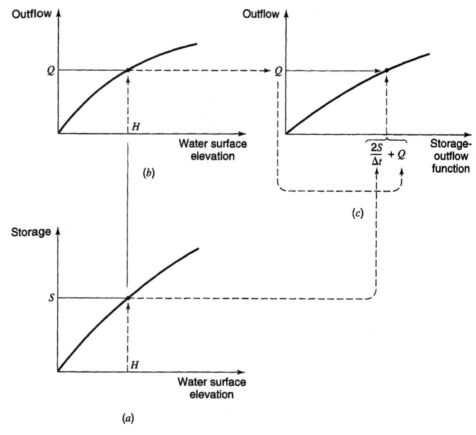

Figure 9.2.2 Development of the storage-outflow function for level pool routing on the basis of storage-elevation-outflow curves (from Chow et al. (1988)).

EXAMPLE 9.2.1

Consider a 2-acre stormwater detention basin with vertical walls. The triangular inflow hydrograph increases linearly from zero to a peak of 60 cfs at 60 min and then decreases linearly to a zero discharge at 180 min. Route the inflow hydrograph through the detention basin using the head-discharge relationship for the 5-ft diameter pipe spillway in columns (1) and (2) of Table 9.2.1. The pipe is located at the bottom of the basin. Assuming the basin is initially empty, use the level pool routing procedure with a 10-min time interval to determine the maximum depth in the detention basin.

SOLUTION

The inflow hydrograph and the head-discharge (columns (1) and (2)) and discharge-storage (columns (2) and (3)) relationships are used to determine the routing relationship in Table 9.2.1. A routing interval of 10 min is used to determine the routing relationship $2S/\Delta t + Q$ vs. Q, which is columns (2) and (4) in Table 9.2.1. The routing computations are presented in Table 9.2.2. These computations are carried out using equation (9.2.3). For the first time interval, $S_1 = Q_1 = 0$ because the reservoir is empty at $t = 0$; then $(2S_1/\Delta t - Q_1) = 0$. The value of the storage-outflow function at the end of the time interval is

$$\left[\frac{2S_2}{\Delta t} + Q_2\right] = (I_1 + I_2) + \left[\frac{2S_1}{\Delta t} - Q_1\right] = (0 + 10) + 0 = 10$$

The value of Q_2 is determined using linear interpolation, so that

$$Q_2 = 0 + \frac{(3 - 0)}{(148.2 - 0)}(10 - 0) = 0.2 \text{ cfs}$$

Table 9.2.1   Elevation-Discharge-Storage Data for Example 9.2.1

1 Head H (ft)	2 Discharge Q (cfs)	3 Storage S (ft^3)	4 $\frac{2S}{\Delta t} + Q$ (cfs)
0.0	0	0	0.00
0.5	3	43,500	148.20
1.0	8	87,120	298.40
1.5	17	130,680	452.60
2.0	30	174,240	610.80
2.5	43	217,800	769.00
3.0	60	261,360	931.20
3.5	78	304,920	1094.40
4.0	97	348,480	1258.60
4.5	117	392,040	1423.80
5.0	137	435,600	1589.00

Table 9.2.2   Routing of Flow Through Detention Reservoir by the Level Pool Method (Example 9.2.1)

Time t (min)	Inflow I_j (cfs)	$I_j + I_{j+1}$ (cfs)	$\frac{2S_j}{\Delta t} - Q_j$ (cfs)	$\frac{2S_{j+1}}{\Delta t} + Q_{j+1}$ (cfs)	Outflow (cfs)
0.00	0.00				0.00
10.00	10.00	10.00	0.00	10.00	0.20
20.00	20.00	30.00	9.60	39.60	0.80
30.00	30.00	50.00	37.99	87.99	1.78
40.00	40.00	70.00	84.43	154.43	3.21
50.00	50.00	90.00	148.01	238.01	5.99
60.00	60.00	110.00	226.04	336.04	10.20
70.00	55.00	115.00	315.64	430.64	15.72
80.00	50.00	105.00	399.21	504.21	21.24
90.00	45.00	95.00	461.72	556.72	25.56
100.00	40.00	85.00	505.61	590.61	28.34
110.00	35.00	75.00	533.93	608.93	29.85
120.00	30.00	65.00	549.24	614.24	30.28
130.00	25.00	55.00	553.67	608.67	29.83
140.00	20.00	45.00	549.02	594.02	28.62
150.00	15.00	35.00	536.78	571.78	26.79
160.00	10.00	25.00	518.19	543.19	24.44
170.00	5.00	15.00	494.30	509.30	21.66
180.00	0.00	5.00	465.98	470.98	18.51
190.00	0.00	0.00	433.96	433.96	15.91
200.00	0.00	0.00	402.14	402.14	14.05
210.00	0.00	0.00	374.03	374.03	12.41
220.00	0.00	0.00	349.20	349.20	10.97
230.00	0.00	0.00	327.27	327.27	9.69
240.00	0.00	0.00	307.90	307.90	8.55

With $Q_1 = 0.2$, then $2S_2/\Delta t - Q_2$ for the next iteration is

$$\left[\frac{2S_2}{\Delta t} - Q_2\right] = \left[\frac{2S_2}{\Delta t} + Q_2\right] - 2Q_2 = 10 - 2(0.2) = 9.6 \text{ cfs}$$

The computation now proceeds to the next time interval. Refer to Table 9.2.2 for the remaining computations.

9.3 HYDROLOGIC RIVER ROUTING

The *Muskingum method* is a commonly used hydrologic routing method that is based upon a variable discharge-storage relationship. This method models the storage volume of flooding in a river channel by a combination of wedge and prism storage (Figure 9.3.1). During the advance of a flood wave, inflow exceeds outflow, producing a wedge of storage. During the recession, outflow exceeds inflow, resulting in a negative wedge. In addition, there is a prism of storage that is formed by a volume of constant cross-section along the length of prismatic channel.

Assuming that the cross-sectional area of the flood flow is directly proportional to the discharge at the section, the *volume of prism storage* is equal to KQ, where K is a proportionality coefficient (approximate as the travel time through the reach), and the *volume of wedge storage* is equal to $KX(I - Q)$, where X is a weighting factor having the range $0 \leq X \leq 0.5$. The total storage is defined as the sum of two components,

$$S = KQ + KX(I - Q) \tag{9.3.1}$$

which can be rearranged to give the storage function for the Muskingum method

$$S = K[XI + (I - X)Q] \tag{9.3.2}$$

and represents a linear model for routing flow in streams.

The value of X depends on the shape of the modeled wedge storage. The value of X ranges from 0 for reservoir-type storage to 0.5 for a full wedge. When $X = 0$, there is no wedge and hence no backwater; this is the case for a level-pool reservoir. In natural streams, X is between 0 and 0.3, with a mean value near 0.2. Great accuracy in determining X may not be necessary because the results of the method are relatively insensitive to the value of this parameter. The parameter K is the time of travel of the flood wave through the channel reach. For hydrologic routing, the values of K and X are assumed to be specified and constant throughout the range of flow.

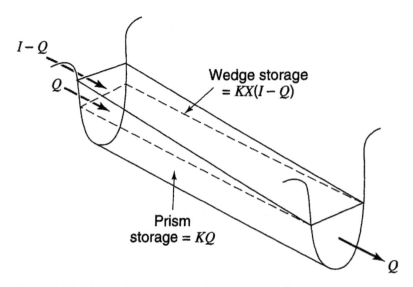

Figure 9.3.1 Prism and wedge storages in a channel reach.

The values of storage at time j and $j + 1$ can be written, respectively, as

$$S_j = K[XI_j + (1 - X)Q_j] \tag{9.3.3}$$

$$S_{j+1} = K[XI_{j+1} + (1 - X)Q_{j+1}] \tag{9.3.4}$$

Using equations (9.3.3) and (9.3.4), the change in storage over time interval Δt is

$$S_{j+1} - S_j = K\{[XI_{j+1} + (1 - X)Q_{j+1}] - [XI_j + (1 - X)Q_j]\} \tag{9.3.5}$$

The change in storage can also be expressed using equation (9.2.1). Combining equations (9.3.5) and (9.2.1) and simplifying gives

$$Q_{j+1} = C_1I_{j+1} + C_2I_j + C_3Q_j \tag{9.3.6}$$

which is the routing equation for the Muskingum method, where

$$C_1 = \frac{\Delta t - 2KX}{2K(1 - X) + \Delta t} \tag{9.3.7}$$

$$C_2 = \frac{\Delta t + 2KX}{2K(1 - X) + \Delta t} \tag{9.3.8}$$

$$C_3 = \frac{2K(1 - X) - \Delta t}{2K(1 - X) + \Delta t} \tag{9.3.9}$$

Note that $C_1 + C_2 + C_3 = 1$.

The routing procedure can be repeated for several sub-reaches (N_{steps}) so that the total travel time through the reach is K. To insure that the method is computationally stable and accurate, the U.S. Army Corps of Engineers (1990) uses the following criterion to determine the number of routing reaches:

$$\frac{1}{2(1 - X)} \le \frac{K}{N_{steps}\Delta t} \le \frac{1}{2X} \tag{9.3.10}$$

If observed inflow and outflow hydrographs are available for a river reach, the values of K and X can be determined. Assuming various values of X and using known values of the inflow and outflow, successive values of the numerator and denominator of the following expression for K, derived from equations (9.3.5) and (9.2.1), can be computed using

$$K = \frac{0.5\Delta t[(I_{j+1} + I_j) - (Q_{j+1} + Q_j)]}{X(I_{j+1} - I_j) + (1 - X)(Q_{j+1} - Q_j)} \tag{9.3.11}$$

The computed values of the numerator (storage) and denominator (weighted discharges) are plotted for each time interval, with the numerator on the vertical axis and the denominator on the horizontal axis. This usually produces a graph in the form of a loop, as shown in Figure 9.3.2. The value of X that produces a loop closest to a single line is taken to be the correct value for the reach, and K, according to equation (9.3.11), is equal to the slope of the line. Since K is the time required for the incremental flood wave to traverse the reach, its value may also be estimated as the observed time of travel of peak flow through the reach.

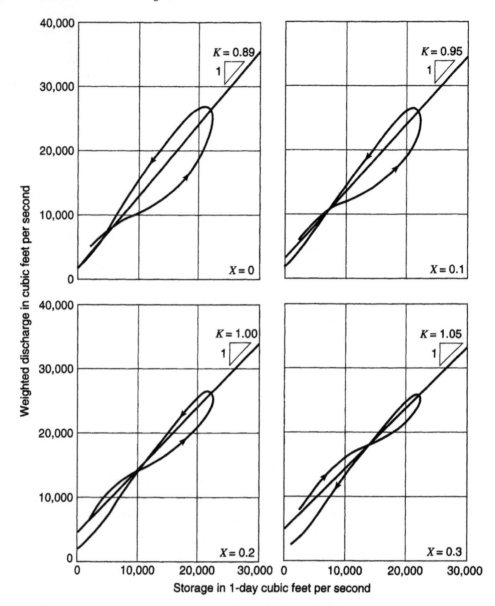

Figure 9.3.2 Typical valley storage curves (after Cudworth (1989)).

EXAMPLE 9.3.1

The objective of this example is to determine K and X for the Muskingum routing method using the February 26 to March 4, 1929 data on the Tuscarawas River from Dover to Newcomerstown. This example is taken from the U.S. Army Corps of Engineers (1960) as used in Cudworth (1989). Columns (2) and (3) in Table 9.3.1 are the inflow and outflow hydrographs for the reach. The numerator and denominator of equation (9.3.11) were computed (for each time period) using four values of $X = 0, 0.1, 0.2,$ and 0.3. The accumulated numerators are in column (9) and the accumulated denominators (weighted discharges) are in columns (11), (13), (15), and (17). In Figure 9.3.2, the accumulated numerator (storages) from column (9) are plotted against the corresponding accumulated denominator (weighted discharges) for each of the four X values. According to Figure 9.3.2, the best fit (linear relationship) appears to be for $X = 0.2$, which has a resulting $K = 1.0$. To perform a routing, K should equal Δt, so that if $\Delta t = 0.5$ day, as in this case, the reach should be subdivided into two equal reaches ($N_{steps} = 2$) and the value of K should be 0.5 day for each reach.

Table 9.3.1 Determination of Coefficients K and X for the Muskingum Routing Method. Tuscarawas River, Muskingum Basin, Ohio Reach from Dover to Newcomerstown, February 26 to March 4, 1929

(1) Date $\Delta t = 0.5$ day	(2) In-flow[1], ft³/s	(3) Out-flow[2], ft³/s	(4) I_2+I_1, ft³/s	(5) O_2+O_1, ft³/s	(6) I_2-I_1, ft³/s	(7) O_2-O_1, ft³/s	(8) [3]N	(9) ΣN	(10) [4]D X=0	(11) ΣD X=0	(12) D X=0.1	(13) ΣD X=0.1	(14) D X=0.2	(15) ΣD X=0.2	(16) D X=0.3	(17) ΣD X=0.3
2-26-29 a.m.	2200	2000	16,700	9000	12,300	5000	1900	1900	5000	5000	5700	5700	6500	6500	7200	7200
p.m.	14,500	7000	42,900	18,700	13,900	4700	6100	8000	4700	9700	5600	11,300	6500	13,000	7500	14,700
2-27-29 a.m.	28,400	11,700	60,200	28,200	3400	4800	8000	16,000	4800	14,500	4600	15,900	4500	17,500	4300	19,000
p.m.	31,800	16,500	61,500	40,500	-2100	7500	5200	21,200	7500	22,000	6700	22,600	5600	23,100	4600	23,600
2-28-29 a.m.	29,700	24,000	55,000	53,100	-4400	5100	500	21,700	5100	27,100	4100	26,700	3200	26,300	2300	25,900
p.m.	25,300	29,100	45,700	57,500	-4900	-700	-2900	18,800	-700	26,400	-1100	25,600	-1500	24,800	-2000	23,900
3-01-29 a.m.	20,400	28,400	36,700	52,200	-4100	-4600	-3900	14,900	-4600	21,800	-4600	21,000	-4500	20,300	-4400	19,500
p.m.	16,300	23,800	28,900	43,200	-3700	-4400	-3600	11,300	-4400	17,400	-4300	16,700	-4300	16,000	-4200	15,300
3-02-29 a.m.	12,600	19,400	21,900	34,700	-3300	-4100	-3200	8100	-4100	13,300	-4000	12,700	-3900	12,100	-3900	11,400
p.m.	9300	15,300	16,000	26,500	-2600	-4100	-2500	5500	-4100	9200	-4000	8700	-3800	8300	-3600	7800
3-03-29 a.m.	6700	11,200	11,700	19,400	-1700	-3000	-1900	3600	-3000	6200	-2800	5900	-2800	5500	-2600	5200
p.m.	5000	8200	9100	14,600	-900	-1800	-1400	2200	-1800	4400	-1700	4200	-1600	3900	-1600	3600
3-04-29 a.m.	4100	6400	7700	11,600	-500	-1200	-1000	1200	-1200	3200	-1200	3000	-1100	2800	-900	2700
p.m.	3600	5200	6000	9800	-1200	-600	-1000	200	-600	2600	-600	2400	-700	2100	-800	1900
3-05-29 a.m.	2400	4600	—	—	—	—	—	—	—	—	—	—	—	—	—	—

[1]Inflow to reach was adjusted to equal volume of outflow.

[2]Outflow is the hydrograph at Newcomerstown.

[3]Numerator, N, is $\Delta t/2$, column (4) – column (5).

[4]Denominator, D, is column (7) + X [column (6) - column (7)].

Note: From plottings of column (9) versus columns (11), (13), (15), and (17), the plot giving the best fit is considered to define K and X.

$$K = \frac{\text{Numerator, } N}{\text{Denominator, } D} = \frac{0.5\Delta t[(I_2+I_1)-(O_2+O_1)]}{X(I_2-I_1)+(1-X)(O_2-O_1)}$$

Source: Cudworth (1989).

EXAMPLE 9.3.2 Route the inflow hydrograph below using the Muskingum method; $\Delta t = 1$ hr, $X = 0.2$, $K = 0.7$ hr.

Time (hr)	0	1	2	3	4	5	6	7
Inflow (cfs)	0	800	2000	4200	5200	4400	3200	2500

Time (hr)	8	9	10	11	12	13
Inflow (cfs)	2000	1500	1000	700	400	0

$$C_1 = \frac{1.0 - 2(0.7)(0.2)}{2(0.7)(1-0.2)+1.0} = 0.3396$$

$$C_2 = \frac{1.0 + 2(0.7)(0.2)}{2(0.7)(1-0.2)+1.0} = 0.6038$$

$$C_3 = \frac{2(0.7)(1-0.2) - 1.0}{2(0.7)(1-0.2)+1.0} = 0.0566$$

(Adapted from Masch (1984).)

Check to see if $C_1 + C_2 + C_3 = 1$:

$$0.3396 + 0.6038 + 0.0566 = 1$$

Using equation (9.3.6) with $I_1 = 0$ cfs, $I_2 = 800$ cfs, and $Q_1 = 0$ cfs, compute Q_2 at $t = 1$ hr:

$$
\begin{aligned}
Q_2 &= C_1 I_2 + C_2 I_1 + C_3 Q_1 \\
&= (0.3396)(800) + 0.6038(0) + 0.0566(0) \\
&= 272 \text{ cfs } (7.7 \text{ m}^3/\text{s})
\end{aligned}
$$

Next compute Q_3 at $t = 2$ hr :

$$
\begin{aligned}
Q_3 &= C_1 I_3 + C_2 I_2 + C_3 Q_2 \\
&= (0.3396)(2000) + 0.6038(800) + 0.0566(272) \\
&= 1{,}178 \text{ cfs } (33 \text{ m}^3/\text{s})
\end{aligned}
$$

The remaining computations result in

Time (hr)	0	1	2	3	4	5	6	7
Q (cfs)	0	272	1178	2701	4455	4886	4020	3009

Time (hr)	8	9	10	11	12	13	14	15
Q (cfs)	2359	1851	1350	918	610	276	16	1

9.4 HYDRAULIC (DISTRIBUTED) ROUTING

Distributed routing or *hydraulic routing*, also referred to as *unsteady flow routing*, is based upon the one-dimensional unsteady flow equations referred to as the *Saint-Venant equations*. The hydrologic river routing and the hydrologic reservoir routing procedures presented previously are lumped procedures and compute flow rate as a function of time alone at a downstream location. Hydraulic (distributed) flow routings allow computation of the flow rate and water surface elevation (or depth) as a function of both space (location) and time. The Saint-Venant equations are presented in Table 9.4.1 in both the *velocity-depth (nonconservation) form* and the *discharge-area (conservation) form*.

The momentum equation contains terms for the physical processes that govern the flow momentum. These terms are: the *local acceleration term*, which describes the change in momentum due to the change in velocity over time, the *convective acceleration term*, which describes the change in momentum due to change in velocity along the channel, the *pressure force term*,

Table 9.4.1   Summary of the Saint-Venant Equations*

Continuity equation

Conservation form

$$\frac{\partial Q}{\partial x} + \frac{\partial A}{\partial t} = 0$$

Nonconservation form

$$V\frac{\partial y}{\partial x} + \frac{\partial V}{\partial x} - \frac{\partial y}{\partial t} = 0$$

Momentum equation

Conservation form

$$\frac{1}{A}\frac{\partial Q}{\partial t} \quad + \quad \frac{1}{A}\frac{\partial}{\partial x}\left(\frac{Q^2}{A}\right) \quad + \quad g\frac{\partial y}{\partial x} \quad - \quad g(S_0 \quad - \quad S_f) \quad = 0$$

Local acceleration term	Convective acceleration term	Pressure force term	Gravity force term	Friction force term

Nonconservation form (unit with element)

$$\frac{\partial V}{\partial t} \quad + \quad V\frac{\partial V}{\partial x} \quad + \quad g\frac{\partial y}{\partial x} \quad - \quad g(S_0 \quad - \quad S_f) \quad = 0$$

$\qquad\qquad\qquad\qquad\qquad\qquad\qquad\qquad\qquad\qquad$ **Kinematic wave**

$\qquad\qquad\qquad\qquad\qquad\qquad\qquad\qquad$ **Diffusion wave**

$\qquad\qquad\qquad\qquad\qquad\qquad$ **Dynamic wave**

*Neglecting lateral inflow, wind shear, and eddy losses, and assuming $\beta = 1$.

x = longitudinal distance along the channel or river, t = time, A = cross-sectional area of flow, h = water surface elevation, S_f = friction slope, S_0 = channel bottom slope, g = acceleration due to gravity, V = velocity of flow, and y = depth of flow.

proportional to the change in the water depth along the channel, the gravity force term, proportional to the bed slope S_0, and the friction force term, proportional to the friction slope S_f. The local and convective acceleration terms represent the effect of inertial forces on the flow.

Alternative distributed flow routing models are produced by using the full continuity equation while eliminating some terms of the momentum equation (refer to Table 9.4.1). The simplest distributed model is the *kinematic wave model*, which neglects the local acceleration, convective acceleration, and pressure terms in the momentum equation; that is, it assumes that $S_0 = S_f$ and the friction and gravity forces balance each other. The *diffusion wave model* neglects the local and convective acceleration terms but incorporates the pressure term. The *dynamic wave model* considers all the acceleration and pressure terms in the momentum equation.

The momentum equation can also be written in forms that take into account whether the flow is steady or unsteady, and uniform or nonuniform, as illustrated in Table 9.4.1. In the continuity equation, $\partial A/\partial t = 0$ for a steady flow, and the lateral inflow q is zero for a uniform flow.

9.4.1 Unsteady Flow Equations: Continuity Equation

The *continuity equation* for an unsteady variable-density flow through a control volume can be written as in equation (3.3.1):

$$0 = \frac{d}{dt}\int_{CV} \rho\,d\forall + \int_{CS} \rho\mathbf{V}\cdot d\mathbf{A} \qquad (9.4.1)$$

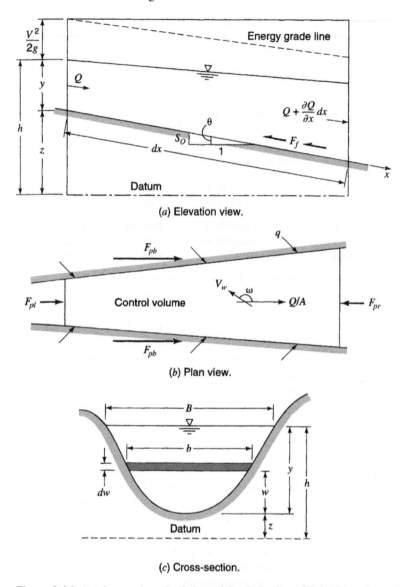

Figure 9.4.1 An elemental reach of channel for derivation of Saint-Venant equations.

Consider an elemental control volume of length dx in a channel. Figure 9.4.1 shows three views of the control volume: (a) an elevation view from the side, (b) a plan view from above, and (c) a channel cross-section. The inflow to the control volume is the sum of the flow Q entering the control volume at the upstream end of the channel and the lateral inflow q entering the control volume as a distributed flow along the side of the channel. The dimensions of q are those of flow per unit length of channel, so the rate of lateral inflow is qdx and the mass inflow rate is

$$\int_{inlet} \rho \mathbf{V} \cdot \mathbf{dA} = -\rho(Q + qdx) \tag{9.4.2}$$

This is negative because inflows are considered negative in the control volume approach (Reynolds transport theorem). The mass outflow from the control volume is

$$\int_{outlet} \rho \mathbf{V} \cdot \mathbf{dA} = \rho\left(Q + \frac{\partial Q}{\partial x}dx\right) \tag{9.4.3}$$

where $\partial Q/\partial x$ is the rate of change of channel flow with distance. The volume of the channel element is Adx, where A is the average cross-sectional area, so the rate of change of mass stored within the control volume is

$$\frac{d}{dt}\int_{CV}\rho\, d\forall = \frac{\partial(\rho A dx)}{\partial t} \tag{9.4.4}$$

where the partial derivative is used because the control volume is defined to be fixed in size (though the water level may vary within it). The net outflow of mass from the control volume is found by substituting equations (9.4.2)–(9.4.4) into (9.4.1):

$$\frac{\partial(\rho A dx)}{dt} - \rho(Q + qdx) + \rho\left(Q + \frac{\partial Q}{\partial x}dx\right) = 0 \tag{9.4.5}$$

Assuming the fluid density ρ is constant, equation (9.4.5) is simplified by dividing through by ρdx and rearranging to produce the *conservation form* of the continuity equation,

$$\frac{\partial Q}{\partial x} + \frac{\partial A}{\partial t} - q = 0 \tag{9.4.6}$$

which is applicable at a channel cross-section. This equation is valid for a *prismatic* or a *nonprismatic* channel; a prismatic channel is one in which the cross-sectional shape does not vary along the channel and the bed slope is constant.

For some methods of solving the Saint-Venant equations, the *nonconservation form* of the continuity equation is used, in which the average flow velocity V is a dependent variable, instead of Q. This form of the continuity equation can be derived for a unit width of flow within the channel, neglecting lateral inflow, as follows. For a unit width of flow, $A = y \times 1 = y$ and $Q = VA = Vy$. Substituting into equation (9.4.6) yields

$$\frac{\partial(Vy)}{\partial x} + \frac{\partial y}{\partial t} = 0 \tag{9.4.7}$$

or

$$V\frac{\partial y}{\partial x} + y\frac{\partial V}{\partial x} + \frac{\partial y}{\partial t} = 0 \tag{9.4.8}$$

9.4.2 Momentum Equation

Newton's second law is written in the form of Reynolds transport theorem as in equation (3.5.5):

$$\sum \mathbf{F} = \frac{d}{dt}\int_{CV}\mathbf{V}\rho\, d\forall + \sum_{CS}\mathbf{V}\rho\mathbf{V}\cdot d\mathbf{A} \tag{9.4.9}$$

This states that the sum of the forces applied is equal to the rate of change of momentum stored within the control volume plus the net outflow of momentum across the control surface. This equation, in the form $\sum F = 0$, was applied to steady uniform flow in an open channel in Chapter 5. Here, unsteady nonuniform flow is considered.

Forces. There are five forces acting on the control volume:

$$\sum F = F_g + F_f + F_e + F_p \tag{9.4.10}$$

where F_g is the *gravity force* along the channel due to the weight of the water in the control volume, F_f is the *friction force* along the bottom and sides of the control volume, F_e is the *contraction/*

expansion force produced by abrupt changes in the channel cross-section, and F_p is the *unbalanced pressure force* (see Figure 9.4.1). Each of these four forces is evaluated in the following paragraphs.

Gravity. The volume of fluid in the control volume is Adx and its weight is $\rho gAdx$. For a small angle of channel inclination θ, $S_0 \approx \sin\theta$ and the gravity force is given by

$$F_g = \rho gAdx \sin\theta \approx \rho gAS_0 dx \tag{9.4.11}$$

where the channel bottom slope S_0 equals $-\partial z/\partial x$.

Friction. Frictional forces created by the shear stress along the bottom and sides of the control volume are given by $-\tau_0 Pdx$, where $\tau_0 = \gamma RS_f = \rho g(A/P)S_f$ is the bed shear stress and P is the wetted perimeter. Hence the friction force is written as

$$F_f = -\rho gAS_f dx \tag{9.4.12}$$

where the friction slope S_f is derived from resistance equations such as Manning's equation.

Contraction/expansion. Abrupt contractions or expansions of the channel cause energy losses through eddy motion. Such losses are similar to minor losses in a pipe system. The magnitude of eddy losses is related to the change in velocity head $V^2/2g = (Q/A)^2/2g$ through the length of channel causing the losses. The drag forces creating these eddy losses are given by

$$F_e = -\rho gAS_e dx \tag{9.4.13}$$

where S_e is the eddy loss slope

$$S_e = \frac{K_e}{2g}\frac{\partial(Q/A)^2}{\partial x} \tag{9.4.14}$$

in which K_e is the nondimensional expansion or contraction coefficient, negative for channel expansion (where $\partial(Q/A)^2/\partial x$ is negative) and positive for channel contractions.

Pressure. Referring to Figure 9.4.1, the unbalanced pressure force is the resultant of the hydrostatic force on the each side of the control volume. Chow et al. (1988) provide a detailed derivation of the pressure force F_p as simply

$$F_p = \rho gA\frac{\partial y}{\partial x}dx \tag{9.4.15}$$

The sum of the forces in equation (9.4.10) can be expressed, after substituting equations (9.4.11), (9.4.12), (9.4.13), and (9.4.15), as

$$\sum F = \rho AS_0 dx - \rho gAS_f dx - \rho gAS_e dx - \rho gA\frac{\partial y}{\partial x}dx \tag{9.4.16}$$

Momentum. The two momentum terms on the right-hand side of equation (9.4.9) represent the rate of change of storage of momentum in the control volume, and the net outflow of momentum across the control surface, respectively.

Net momentum outflow. The mass inflow rate to the control volume (equation (9.4.2)) is $-\rho(Q+qdx)$, representing both stream inflow and lateral inflow. The corresponding momentum is computed by multiplying the two mass inflow rates by their respective velocity and a *momentum correction factor* β:

$$\int_{\text{inlet}} \mathbf{V}\rho\mathbf{V}\,d\mathbf{A} = -\rho(\beta VQ + \beta v_x qdx) \tag{9.4.17}$$

where $-\rho\beta VQ$ is the momentum entering from the upstream end of the channel, and $-\rho\beta v_x qdx$ is the momentum entering the main channel with the lateral inflow, which has a velocity v_x in the x

direction. The term β is known as the *momentum coefficient* or *Boussinesq coefficient*; it accounts for the nonuniform distribution of velocity at a channel cross-section in computing the momentum. The value of β is given by

$$\beta = \frac{1}{V^2 A} \int v^2 dA \tag{9.4.18}$$

where v is the velocity through a small element of area dA in the channel cross-section. The value of β ranges from 1.01 for straight prismatic channels to 1.33 for river valleys with floodplains (Chow, 1959; Henderson, 1966).

The momentum leaving the control volume is

$$\int_{\text{outlet}} V\rho V \, dA = \rho \left[\beta VQ + \frac{\partial(\beta VQ)}{\partial x} dx \right] \tag{9.4.19}$$

The net outflow of momentum across the control surface is the sum of equations (9.4.17) and (9.4.19):

$$\int_{\text{CS}} V\rho V \, dA = -\rho(\beta VQ + \beta v_x q dx) + \rho \left[\beta VQ + \frac{\partial(\beta VQ)}{\partial x} dx \right] = -\rho \left[\beta v_x q - \frac{\partial(\beta VQ)}{\partial x} \right] dx \tag{9.4.20}$$

Momentum storage. The time rate of change of momentum stored in the control volume is found by using the fact that the volume of the elemental channel is $A dx$, so its momentum is $\rho A dx V$, or $\rho Q dx$, and then

$$\frac{d}{dt} \int_{\text{CV}} V\rho d\forall = \rho \frac{\partial Q}{\partial x} dx \tag{9.4.21}$$

After substituting the force terms from equation (9.4.16) and the momentum terms from equations (9.4.20) and (9.4.21) into the momentum equation (9.4.9), it reads

$$\rho g A S_0 dx - \rho g A S_f dx - \rho g A \frac{\partial y}{\partial x} dx = -\rho \left[\beta v_x q - \frac{\partial(\beta VQ)}{\partial x} \right] dx + \rho \frac{\partial Q}{\partial t} dx \tag{9.4.22}$$

Dividing through by ρdx, replacing V with Q/A, and rearranging produces the conservation form of the momentum equation:

$$\frac{\partial Q}{\partial t} + \frac{\partial(\beta Q^2/A)}{\partial t} + gA\left(\frac{\partial y}{\partial x} - S_0 + S_f + S_e \right) - \beta q v_x = 0 \tag{9.4.23}$$

The depth y in equation (9.4.23) can be replaced by the water surface elevation h, using

$$h = y + z \tag{9.4.24}$$

where z is the elevation of the channel bottom above a datum such as mean sea level. The derivative of equation (9.4.24) with respect to the longitudinal distance x along the channel is

$$\frac{\partial h}{\partial x} = \frac{\partial y}{\partial x} + \frac{\partial z}{\partial x} \tag{9.4.25}$$

but $\partial z/\partial x = -S_0$, so

$$\frac{\partial h}{\partial x} = \frac{\partial y}{\partial x} - S_0 \tag{9.4.26}$$

The momentum equation can now be expressed in terms of h by using equation (9.4.26) in (9.4.23):

$$\frac{\partial Q}{\partial t} + \frac{\partial(\beta Q^2/A)}{\partial x} + gA\left(\frac{\partial h}{\partial x} + S_f + S_e\right) - \beta q v_x = 0 \qquad (9.4.27)$$

The Saint-Venant equations, (9.4.6) for continuity and (9.4.27) for momentum, are the governing equations for one-dimensional, unsteady flow in an open channel. The use of the terms S_f and S_e in equation (9.4.27), which represent the rate of energy loss as the flow passes through the channel, illustrates the close relationship between energy and momentum considerations in describing the flow. Strelkoff (1969) showed that the momentum equation for the Saint-Venant equations can also be derived from energy principles, rather than by using Newton's second law as presented here.

The nonconservation form of the momentum equation can be derived in a similar manner to the nonconservation form of the continuity equation. Neglecting eddy losses, wind shear effect, and lateral inflow, the nonconservation form of the momentum equation for a unit width in the flow is

$$\frac{\partial V}{\partial t} + V\frac{\partial V}{\partial x} + g\left(\frac{\partial y}{\partial x} - S_0 + S_f\right) = 0 \qquad (9.4.28)$$

9.5 KINEMATIC WAVE MODEL FOR CHANNELS

In Section 8.9, a kinematic wave overland flow runoff model was presented. This is an implicit nonlinear kinematic model that is used in the KINEROS model. This section presents a general discussion of the kinematic wave followed by a brief description of the very simplest linear models, such as those found in the U.S. Army Corps of Engineers HEC-HMS and HEC-1, and the more complicated models such as the KINEROS model (Woolhiser et al., 1990).

Kinematic waves govern flow when inertial and pressure forces are not important. Dynamic waves govern flow when these forces are important, as in the movement of a large flood wave in a wide river. In a kinematic wave, the gravity and friction forces are balanced, so the flow does not accelerate appreciably.

For a kinematic wave, the energy grade line is parallel to the channel bottom and the flow is steady and uniform ($S_0 = S_f$) within the differential length, while for a dynamic wave the energy grade line and water surface elevation are not parallel to the bed, even within a differential element.

9.5.1 Kinematic Wave Equations

A *wave* is a variation in a flow, such as a change in flow rate or water surface elevation, and the *wave celerity* is the velocity with which this variation travels along the channel. The celerity depends on the type of wave being considered and may be quite different from the water velocity. For a kinematic wave, the acceleration and pressure terms in the momentum equation are negligible, so the wave motion is described principally by the equation of continuity. The name kinematic is thus applicable, as *kinematics* refers to the study of motion exclusive of the influence of mass and force; in *dynamics* these quantities are included.

The kinematic wave model is defined by the following equations.

Continuity:

$$\frac{\partial Q}{\partial x} + \frac{\partial A}{\partial t} = q(x, t) \qquad (9.5.1)$$

Momentum:

$$S_0 = S_f \qquad (9.5.2)$$

where $q(x, t)$ is the net lateral inflow per unit length of channel.

The momentum equation can also be expressed in the form

$$A = \alpha Q^\beta \tag{9.5.3}$$

For example, Manning's equation written with $S_0 = S_f$ and $R = A/P$ is

$$Q = \frac{1.49 S_0^{1/2}}{n P^{2/3}} A^{5/3} \tag{9.5.4}$$

which can be solved for A as

$$A = \left(\frac{n P^{2/3}}{1.49 \sqrt{S_0}}\right)^{3/5} Q^{3/5} \tag{9.5.5}$$

so $\alpha = [nP^{2/3}/(1.49\sqrt{S_0})]^{0.6}$ and $\beta = 0.6$ in this case.

Equation (9.5.1) contains two dependent variables, A and Q, but A can be eliminated by differentiating equation (9.5.3):

$$\frac{\partial A}{\partial t} = \alpha\beta Q^{\beta-1}\left(\frac{\partial Q}{\partial t}\right) \tag{9.5.6}$$

and substituting for $\partial A/\partial t$ in equation (9.5.1) to give

$$\frac{\partial Q}{\partial x} + \alpha\beta Q^{\beta-1}\left(\frac{\partial Q}{\partial t}\right) = q \tag{9.5.7}$$

Alternatively, the momentum equation could be expressed as

$$Q = aA^B \tag{9.5.8}$$

where a and B are defined using Manning's equation. Using

$$\frac{\partial Q}{\partial x} = \frac{dQ}{dA}\frac{\partial A}{\partial x} \tag{9.5.9}$$

the governing equation is

$$\frac{\partial A}{\partial t} + \frac{dQ}{dA}\frac{\partial A}{\partial x} = q \tag{9.5.10}$$

where dQ/dA is determined by differentiating equation (9.5.8):

$$\frac{dQ}{dA} = aBA^{B-1} \tag{9.5.11}$$

and substituting in equation (9.5.10):

$$\frac{\partial A}{\partial t} + aBA^{B-1}\frac{\partial A}{\partial x} = q \tag{9.5.12}$$

The kinematic wave equation (9.5.7) has Q as the dependent variable and the kinematic wave equation (9.5.12) has A as the dependent variable. First consider equation (9.5.7), by taking the logarithm of (9.5.3):

$$\ln A = \ln\alpha + \beta\ln Q \tag{9.5.13}$$

and differentiating

$$\frac{dQ}{Q} = \frac{1}{\beta}\left(\frac{dA}{A}\right) \tag{9.5.14}$$

This defines the relationship between relative errors dA/A and dQ/Q. For Manning's equation $\beta < 1$, so that the discharge estimation error would be magnified by the ratio $1/\beta$ if A were the dependent variable instead of Q.

Next consider equation (9.5.12); by taking the logarithm of (9.5.8):

$$\ln Q = \ln a + B \ln A$$

$$\frac{dA}{A} = \frac{1}{B}\left(\frac{dQ}{Q}\right) \tag{9.5.15}$$

or

$$\frac{dQ}{Q} = B\left(\frac{dA}{A}\right) \tag{9.5.16}$$

In this case $\beta > 1$, so that the discharge estimation error would be decreased by B if A were the dependent variable instead of Q. In summary, if we use equation (9.5.3) as the form of the momentum equation, then Q is the dependent variable with equation (9.5.7) being the governing equation; if we use equation (9.5.8) as the form of the momentum equation, then A is the dependent variable with equation (9.5.12) being the governing equation.

9.5.2 U.S. Army Corps of Engineers Kinematic Wave Model for Overland Flow and Channel Routing

The HEC-1 (HEC-HMS) computer program actually has two forms of the kinematic wave. The first is based upon equation (9.5.12) where an explicit finite difference form is used (refer to Figures 9.5.1 and 8.9.2):

$$\frac{\partial A}{\partial t} = \frac{A_{i+1}^{j+1} - A_{i+1}^{j}}{\Delta t} \tag{9.5.17}$$

$$\frac{\partial A}{\partial x} = \frac{A_{i+1}^{j} - A_{i}^{j}}{\Delta x} \tag{9.5.18}$$

$$A = \frac{A_{i+1}^{j} + A_{i}^{j}}{2} \tag{9.5.19}$$

and

$$q = \frac{q_{i+1}^{j+1} + q_{i+1}^{j}}{2} \tag{9.5.20}$$

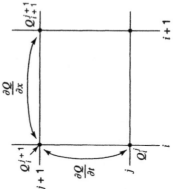

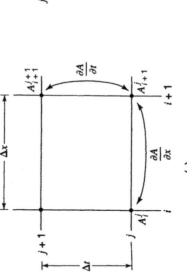

Figure 9.5.1 Finite difference forms. (a) HEC-1 "standard form;" (b) HEC-1 "conservation form."

Substituting these finite-difference approximations into equation (9.5.12) gives

$$\frac{1}{\Delta t}(A_{i+1}^{j+1} - A_{i+1}^j) + aB\left[\frac{A_{i+1}^{j+1} + A_i^j}{2}\right]^{B-1}\left[\frac{A_{i+1}^{j+1} - A_i^j}{\Delta x}\right] = \frac{q_{i+1}^{j+1} + q_{i+1}^j}{2} \tag{9.5.21}$$

The only unknown in equation (9.5.21) is A_{i+1}^{j+1}, so

$$A_{i+1}^{j+1} = A_{i+1}^j - aB\left(\frac{\Delta t}{\Delta x}\right)\left[\frac{A_{i+1}^{j+1} + A_i^j}{2}\right]^{B-1}(A_{i+1}^j - A_{i,j}) + (q_{i+1}^{j+1} + q_{i+1}^j)\frac{\Delta t}{2} \tag{9.5.22}$$

After computing A_{i+1}^{j+1} at each grid along a time line going from upstream to downstream (see Figure 8.9.2), compute the flow using equation (9.5.8):

$$Q_{i+1}^{j+1} = a(A_{i+1}^{j+1})^B \tag{9.5.23}$$

The HEC-1 model uses the above kinematic wave model as long as a stability factor $R < 1$ (Alley and Smith, 1987), defined by

$$R = \frac{a}{q\Delta x}\left[\left((q\Delta t + A_i^j)\right)^B - \left(A_i^j\right)^B\right]^{B-1} \text{ for } q > 0 \tag{9.5.24a}$$

$$R = aB(A_i^j)^{B-1}\frac{\Delta t}{\Delta x} \text{ for } q = 0 \tag{9.5.24b}$$

Otherwise the form of equation (9.5.1) is used, where (see Figure 9.5.1)

$$\frac{\partial Q}{\partial x} = \frac{Q_{i+1}^{j+1} - Q_i^{j+1}}{\Delta x} \tag{9.5.25}$$

$$\frac{\partial A}{\partial t} = \frac{A_i^{j+1} - A_i^j}{\Delta t} \tag{9.5.26}$$

so

$$\frac{Q_{i+1}^{j+1} - Q_i^{j+1}}{\Delta x} + \frac{A_i^{j+1} - A_i^j}{\Delta t} = q \tag{9.5.27}$$

Solving for the only unknown Q_{i+1}^{j+1} yields

$$Q_{i+1}^{j+1} = Q_i^{j+1} + q\Delta x - \frac{\Delta x}{\Delta t}(A_i^{j+1} - A_i^j) \tag{9.5.28}$$

Then solve for A_{i+1}^{j+1} using equation (9.5.23):

$$A_{i+1}^{j+1} = \left(\frac{1}{a}Q_{i+1}^{j+1}\right)^{1/B} \tag{9.5.29}$$

The *initial condition* (values of A and Q at time 0 along the grid, referring to Figure 8.9.2) are computed assuming uniform flow or nonuniform flow for an initial discharge. The *upstream boundary* is the inflow hydrograph from which Q is obtained.

The kinematic wave schemes used in the HEC-1 (HEC-HMS) model are very simplified. Chow, et al. (1988) presented both linear and nonlinear kinematic wave schemes based upon the equation (9.5.7) formulation. An example of a more desirable kinematic wave formulation is that by Woolhiser et al. (1990) presented in the next subsection.

9.5.3   KINEROS Channel Flow Routing Model

The KINEROS channel routing model uses the equation (9.5.10) form of the kinematic wave equation (Woolhiser et al., 1990):

$$\frac{\partial A}{\partial t} + \frac{dQ}{dA}\frac{\partial A}{\partial x} = q(x, t) \tag{9.5.10}$$

where $q(x, t)$ is the net lateral inflow per unit length of channel. The derivatives are approximated using an implicit scheme in which the spatial and temporal derivatives are, respectively,

$$\frac{\partial A}{\partial x} = \theta\left(\frac{A_{i+1}^{j+1} - A_i^{j+1}}{\Delta x}\right) + (1-\theta)\frac{A_{i+1}^j - A_i^j}{\Delta x} \tag{9.5.30}$$

$$\frac{dQ}{dA}\frac{\partial A}{\partial x} = \theta\left(\frac{dQ}{dA}\right)^{j+1}\left(\frac{A_{i+1}^{j+1} - A_i^{j+1}}{\Delta x}\right) + (1-\theta)\left(\frac{dQ}{dA}\right)^{j+1}\left(\frac{A_{i+1}^j - A_i^j}{\Delta x}\right) \tag{9.5.31}$$

and

$$\frac{\partial A}{\partial t} = \frac{1}{2}\left[\frac{A_i^{j+1} - A_i^j}{\Delta t} + \frac{A_{i+1}^{j+1} - A_{i+1}^j}{\Delta t}\right] \tag{9.5.32}$$

or

$$\frac{\partial A}{\partial t} = \frac{A_i^{j+1} + A_{i+1}^{j+1} - A_i^j - A_{i+1}^j}{2\Delta t} \tag{9.5.33}$$

Substituting equations (9.5.31) and (9.5.33) into (9.5.10), we have

$$\frac{A_{i+1}^{j+1} - A_i^{j+1} + A_{i+1}^j - A_i^j}{2\Delta t} + \left\{\theta\left(\frac{dQ}{dA}\right)^{j+1}\left(\frac{A_{i+1}^{j+1} - A_i^{j+1}}{\Delta x}\right) + (1-\theta)\left[\left(\frac{dQ}{dA}\right)^{j+1}\left(\frac{A_{i+1}^j - A_i^j}{\Delta x}\right)\right]\right\}$$
$$= \frac{1}{2}\left(q_{i+1}^{j+1} + q_i^{j+1} + q_{i+1}^j + q_i^j\right) \tag{9.5.34}$$

The only unknown in this equation is A_{i+1}^{j+1}, which must be solved for numerically by use of an iterative scheme such as the Newton-Raphson method (see Appendix A). Woolhiser et al. (1990) use the following relationship between channel discharge and cross-sectional area, which embodies the kinematic wave assumption:

$$Q = \alpha R^{m-1} A \tag{9.5.35}$$

where R is the hydraulic radius and $\alpha = 1.49 S^{1/2}/n$ and $m = 5/3$ for Manning's equation.

9.5.4   Kinematic Wave Celerity

Kinematic waves result from changes in Q. An increment in flow dQ can be written as

$$dQ = \frac{\partial Q}{\partial x}dx + \frac{\partial Q}{\partial t}dt \tag{9.5.36}$$

Dividing through by dx and rearranging produces:

$$\frac{\partial Q}{\partial x} + \frac{dt}{dx}\frac{\partial Q}{\partial t} = \frac{dQ}{dx} \tag{9.5.37}$$

Equations (9.5.7) and (9.5.37) are identical if

$$\frac{dQ}{dx} = q \tag{9.5.38}$$

and

$$\frac{dx}{dt} = \frac{1}{\alpha\beta Q^{\beta-1}} \tag{9.5.39}$$

Differentiating equation (9.5.3) and rearranging gives

$$\frac{dQ}{dA} = \frac{1}{\alpha\beta Q^{\beta-1}} \tag{9.5.40}$$

and by comparing equations (9.5.39) and (9.5.40), it can be seen that

$$\frac{dx}{dt} = \frac{dQ}{dA} \tag{9.5.41}$$

or

$$c_k = \frac{dx}{dt} = \frac{dQ}{dA} \tag{9.5.42}$$

where c_k is the kinematic wave celerity. This implies that an observer moving at a velocity $dx/dt = c_k$ with the flow would see the flow rate increasing at a rate of $dQ/dx = q$. If $q = 0$, the observer would see a constant discharge. Equations (9.5.38) and (9.5.42) are the *characteristic equations* for a kinematic wave, two ordinary differential equations that are mathematically equivalent to the governing continuity and momentum equations.

The kinematic wave celerity can also be expressed in terms of the depth y as

$$c_k = \frac{1}{B}\frac{dQ}{dy} \tag{9.5.43}$$

9.6 MUSKINGUM–CUNGE MODEL

Cunge (1969) proposed a variation of the kinematic wave method based upon the Muskingum method (see Chapter 8). With the grid shown in Figure 9.6.1, the unknown discharge Q_{i+1}^{j+1} can be expressed using the Muskingum equation ($Q_{i+1} = C_1 I_{i+1} + C_2 I_i + C_3 Q_i$):

$$Q_{i+1}^{j+1} = C_1 Q_i^{j+1} + C_2 Q_i^j + C_3 Q_{i+1}^j \tag{9.6.1}$$

where $Q_{i+1}^{j+1} = Q_{i+1}; Q_i^{j+1} = Q_i^j; Q_i^j = I_{j+1} = I_{j;}; Q_i^j = I_{j;}$ and $Q_{i+1}^j = Q_j$. The Muskingum coefficients are

$$C_1 = \frac{\Delta t - 2KX}{2K(1-X)+\Delta t} \tag{9.6.2}$$

where $dA = Bdy$.

Both kinematic and dynamic wave motion are present in natural flood waves. In many cases the channel slope dominates in the momentum equation; therefore, most of a flood wave moves as a kinematic wave. Lighthill and Whitham (1955) proved that the velocity of the main part of a natural flood wave approximates that of a kinematic wave. If the other momentum terms $(\partial V/\partial t, V(\partial V/\partial x)$ and $(1/g)\partial v/\partial x)$ are not negligible, then a dynamic wave front exists that can propagate both upstream and downstream from the main body of the flood wave.

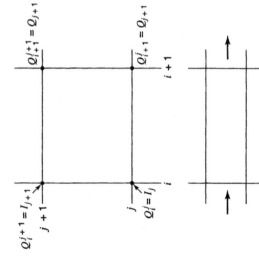

Figure 9.6.1 Finite-difference grid for the Muskingum–Cunge method.

$$C_2 = \frac{\Delta t + 2KX}{2K(1-X) + \Delta t} \tag{9.6.3}$$

$$C_3 = \frac{2K(1-X) - \Delta t}{2K(1-X) + \Delta t} \tag{9.6.4}$$

Cunge (1969) showed that when K and Δt are considered constant, equation (9.6.1) is an approximate solution of the kinematic wave. He further demonstrated that equation (9.6.1) can be considered an approximation of a modified diffusion equation if

$$K = \frac{\Delta x}{c_k} = \frac{\Delta x}{dQ/dA} \tag{9.6.5}$$

and

$$X = \frac{1}{2}\left(1 - \frac{Q}{Bc_k S_0 \Delta x}\right) \tag{9.6.6}$$

where c_k is the celerity corresponding to Q and B, and B is the width of the water surface. The value of $\Delta x/(dQ/dA)$ in equation (9.6.5) represents the time propagation of a given discharge along a channel reach of length Δx. Numerical stability requires $0 \leq X \leq 1/2$. The solution procedure is basically the same as the kinematic wave.

9.7   IMPLICIT DYNAMIC WAVE MODEL

The conservation form of the Saint-Venant equations is used because this form provides the versatility required to simulate a wide range of flows from gradual long-duration flood waves in rivers to abrupt waves similar to those caused by a dam failure. The equations are developed from equations (9.4.6) and (9.4.25) as follows.

Weighted four-point finite-difference approximations given by equations (9.7.1)–(9.7.3) are used for dynamic routing with the Saint-Venant equations. The spatial derivatives $\partial Q/\partial x$ and $\partial h/\partial x$ are estimated between adjacent time lines:

$$\frac{\partial Q}{\partial x} = \theta \frac{Q_{i+1}^{j+1} - Q_i^{j+1}}{\Delta x_i} + (1-\theta)\frac{Q_{i+1}^j - Q_i^j}{\Delta x_i} \tag{9.7.1}$$

and the time derivatives are:

$$\frac{\partial(A+A_0)}{\partial t} = \frac{(A+A_0)_i^{j+1} + (A+A_0)_{i+1}^{j+1} - (A+A_0)_i^j - (A+A_0)_{i+1}^j}{2\Delta t_j} \tag{9.7.2}$$

$$\frac{\partial h}{\partial x} = \theta \frac{h_{i+1}^{j+1} - h_i^{j+1}}{\Delta x_i} + (1-\theta)\frac{h_{i+1}^j - h_i^j}{\Delta x_i} \tag{9.7.3}$$

The nonderivative terms, such as q and A, are estimated between adjacent time lines, using:

$$q = \theta \frac{q_i^{j+1} + q_{i+1}^{j+1}}{2} + (1-\theta)\frac{q_i^j + q_{i+1}^j}{2} \tag{9.7.4}$$

$$A = \theta\left[\frac{A_i^{j+1} + A_{i+1}^{j+1}}{2}\right] + (1-\theta)\left[\frac{A_i^j + A_{i+1}^j}{2}\right] = \theta \overline{A}_i^{\,j+1} + (1-\theta)\overline{A}_i^{\,j} \tag{9.7.5}$$

$$\frac{\partial Q}{\partial t} = \frac{Q_i^{j+1} + Q_{i+1}^{j+1} - Q_i^j - Q_{i+1}^j}{2\Delta t_j} \tag{9.7.6}$$

where $\overline{q}_i$ and $\overline{A}_i$ indicate the lateral flow and cross-sectional area averaged over the reach Δx_i. The finite-difference form of the continuity equation is produced by substituting equations (9.7.1), (9.7.3), and (9.7.5) into (9.4.6):

$$\theta\left(\frac{Q_{i+1}^{j+1} - Q_i^{j+1}}{\Delta x_i} - \overline{q}_i^{\,j+1}\right) + (1-\theta)\left(\frac{Q_{i+1}^j - Q_i^j}{\Delta x_i} - \overline{q}_i^{\,j}\right)$$
$$+ \frac{(A+A_0)_i^{j+1} + (A+A_0)_{i+1}^{j+1} - (A+A_0)_i^j - (A+A_0)_{i+1}^j}{2\Delta t_j} = 0 \tag{9.7.7}$$

Similarly, the momentum equation (9.4.27) is written in finite-difference form as:

$$\frac{Q_i^{j+1} + Q_{i+1}^{j+1} - Q_i^j - Q_{i+1}^j}{2\Delta t_j}$$
$$+ \theta\left[\frac{(\beta Q^2/A)_{i+1}^{j+1} - (\beta Q^2/A)_i^{j+1}}{\Delta x_i} + g\overline{A}_i^{\,j+1}\left(\frac{h_{i+1}^{j+1} - h_i^{j+1}}{\Delta x_i} + (\overline{S}_f)_i^{j+1} + (\overline{S}_e)_i^{j+1}\right) - (\overline{\beta q v_x})_i^{j+1}\right]$$
$$+ (1-\theta)\left[\frac{(\beta Q^2/A)_{i+1}^j - (\beta Q^2/A)_i^j}{\Delta x_i} + g\overline{A}_i^{\,j}\left(\frac{h_{i+1}^j - h_i^j}{\Delta x_i} + (\overline{S}_f)_i^j + (\overline{S}_e)_i^j\right) - (\overline{\beta q v_x})_i^j\right] = 0 \tag{9.7.8}$$

The four-point finite-difference form of the continuity equation can be further modified by multiplying equation (9.7.7) by Δx_i to obtain

$$\theta\left(Q_{i+1}^{j+1} - Q_i^{j+1} - \overline{q}_i^{\,j+1}\Delta x_i\right) + (1-\theta)\left(Q_{i+1}^j - Q_i^j - \overline{q}_i^{\,j}\Delta x_i\right)$$
$$+ \frac{\Delta x_i}{2\Delta t_j}\left[(A+A_0)_i^{j+1} + (A+A_0)_{i+1}^{j+1} - (A+A_0)_i^j - (A+A_0)_{i+1}^j\right] = 0 \tag{9.7.9}$$

Similarly, the momentum equation can be modified by multiplying by Δx_i to obtain

$$\frac{\Delta x_i}{2\Delta t_j}\left(Q_i^{j+1} + Q_{i+1}^{j+1} - Q_i^j - Q_{i+1}^j\right)$$

$$+\theta\left\{\left(\frac{\beta Q^2}{A}\right)_{i+1}^{j+1} - \left(\frac{\beta Q^2}{A}\right)_i^{j+1} + g\overline{A}_i^{j+1}\left[h_{i+1}^{j+1} - h_i^{j+1} + (\overline{S}_f)_i^{j+1} + (\overline{S}_e)_i^{j+1}\,\Delta x_i\right] - (\overline{\beta q v_x})_i^{j+1}\,\Delta x_i\right\}$$

$$+(1-\theta)\left\{\left(\frac{\beta Q^2}{A}\right)_{i+1}^{j} - \left(\frac{\beta Q^2}{A}\right)_i^{j} + g\overline{A}_i^j\left[h_{i+1}^j - h_i^j + (\overline{S}_f)_i^j\Delta x_i + (\overline{S}_e)_i^j\Delta x_i\right] - (\overline{\beta q v_x})_i^j\Delta x_i\right\}=0$$

$$(9.7.10)$$

where the average values (marked with an overbar) over a reach are defined as

$$\overline{\beta}_i = \frac{\beta_i + \beta_{i+1}}{2} \tag{9.7.11}$$

$$\overline{A}_i = \frac{A_i + A_{i+1}}{2} \tag{9.7.12}$$

$$\overline{B}_i = \frac{B_i + B_{i+1}}{2} \tag{9.7.13}$$

$$\overline{Q}_i = \frac{Q_i + Q_{i+1}}{2} \tag{9.7.14}$$

Also,

$$\overline{R}_i = \overline{A}_i / \overline{B}_i \tag{9.7.15}$$

for use in Manning's equation. Manning's equation may be solved for S_f and written in the form shown below, where the $|Q|Q$ has magnitude Q^2 and sign positive or negative depending on whether the flow is downstream or upstream, respectively:

$$(\overline{S}_f)_i = \frac{\overline{n}_i^2|\overline{Q}_i|\overline{Q}_i}{2.208\overline{A}_i^2\overline{R}_i^{4/3}} \tag{9.7.16}$$

The minor headlosses arising from contraction and expansion of the channel are proportional to the difference between the squares of the downstream and upstream velocities, with a contraction/expansion loss coefficient K_e:

$$(\overline{S}_e)_i = \frac{(K_e)_i}{2g\Delta x_i}\left[\left(\frac{Q}{A}\right)_{i+1}^2 - \left(\frac{Q}{A}\right)_i^2\right] \tag{9.7.17}$$

The terms having superscript j in equations (9.7.9) and (9.7.10) are known either from initial conditions or from a solution of the Saint-Venant equations for a previous time line. The terms g, Δx_i, β_i, K_e, C_w, and V_w are known and must be specified independently of the solution. The unknown terms are Q_i^{j+1}, Q_{i+1}^{j+1}, h_i^{j+1}, A_i^{j+1}, A_{i+1}^{j+1}, B_i^{j+1}, and B_{i+1}^{j+1}. However, all the terms can be expressed as functions of the unknowns Q_i^{j+1}, Q_{i+1}^{j+1}, h_i^{j+1}, and h_{i+1}^{j+1}, so there are actually four unknowns. The unknowns are raised to powers other than unity, so equations (9.7.9) and (9.7.10) are nonlinear equations.

The continuity and momentum equations are considered at each of the $N-1$ rectangular grids shown in Figure 9.7.1 between the upstream boundary at $i=1$ and the downstream boundary at

Initial condition time line

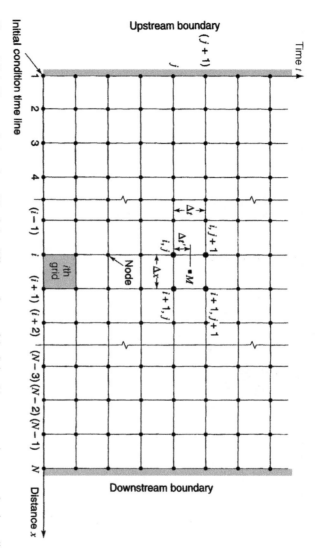

Figure 9.7.1 The *x-t* solution plane. The finite-difference forms of the Saint–Venant equations are solved at a discrete number of points (values of the independent variables *x* and *t*) arranged to form the rectangular grid shown. Lines parallel to the time axis represent locations along the channel, and those parallel to the distance axis represent times (from Fread (1974)).

$i = N$. This yields $2N$-2 equations. There are two unknowns at each of the N grid points (Q and h), so there are $2N$ unknowns in all. The two additional equations required to complete the solution are supplied by the upstream and downstream boundary conditions. The upstream boundary condition is usually specified as a known inflow hydrograph, while the downstream boundary condition can be specified as a known stage hydrograph, a known discharge hydrograph, or a known relationship between stage and discharge, such as a rating curve. The U.S. National Weather Service FLDWAV model (hsp.nws.noaa.gov/oh/hrl/rvmech) uses the above to describe the implicit dynamic wave model formulation.

PROBLEMS

9.1.1 Consider a river segment with the surface area of 10 km². For a given flood event, the measured time variation of inflow rate (called inflow hydrograph) at the upstream section of the river segment and the outflow hydrograph at the downstream section are shown in Figure P9.1.1. Assume that the initial storage of water in the river segment is 15 mm in depth.

(a) Determine the time at which the change in storage of the river segment is increasing, decreasing, and at its maximum.

(b) Calculate the storage change (in mm) in the river segment during the time periods of [0, 4 hr], and [6, 8 hr].

(c) Determine the amount of water (in mm) that is 'lost' or 'gained' in the river segment over the time period of 12 hours.

(d) What is the storage volume (in mm) at the end of the twelfth hour?

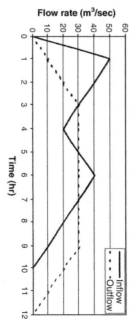

Figure P9.1.1

9.1.2 Consider a river segment with the surface area of 5 km². For a given flood event, the measured time variation of inflow rate (called inflow hydrograph, in m³/sec) at the upstream section of the river segment and the outflow hydrograph at the downstream section are shown in Figure P9.1.1. Assume that the initial storage of water in the river segment is 10 mm in depth.

(a) Determine the time at which the change in storage of the river segment is increasing or decreasing, at its maximum.

(b) Calculate the storage change (in mm) in the river segment during the time periods of [0, 4 hr], and [6, 8 hr].

(c) Determine the amount of water (in mm) that is "lost" or "gained" in the river segment over the time period of 12 hr.

(d) What is the storage volume (in mm) at the end of the 12th hour?

9.1.3 There are two reservoirs, A and B, connected in series. Reservoir A is located upstream and releases its water to reservoir B. The surface area of reservoir A is half of the surface area of B. In the past one year, the two reservoirs received the same amount of rainfall of 100 cm and evaporated 30 cm of water. However, during the same period, reservoir A experienced a change in storage of 20 cm whereas storage in reservoir B remained constant. Please clearly state any assumption used in the calculation.

(a) Determine the outflow volume from reservoir A to B during the past year.

(b) Determine the outflow volume from reservoir B during the past year.

(c) How big is the flow rate from reservoir B as compared with that of reservoir A?

9.2.1 The storage-outflow characteristics for a reservoir are given below. Determine the storage-outflow function $2S/\Delta t + Q$ versus Q for each of the tabulated values using $\Delta t = 1.0$ hr. Plot a graph of the storage-outflow function.

Storage (10^6 m^3)	70	80	85	100	115
Outflow (m^3/s)	0	50	150	350	700

9.2.2 Route the inflow hydrograph given below through the reservoir with the storage-outflow characteristics given in problem 9.2.1 using the level pool method. Assume the reservoir has an initial storage of 70×10^6 m^3.

Time (h)	0	1	2	3	4	5	6	7	8
Inflow (m^3/s)	0	40	60	150	200	300	250	200	180

Time (h)	9	10	11	12	13	14	15	16	
Inflow (m^3/s)	220	320	400	280	190	150	50	0	

9.2.3 Rework problem 9.2.2 assuming the reservoir storage is initially 80×10^6 m^3.

9.2.4 Write a computer program to solve problems 9.2.2 and 9.2.3.

9.2.5 Rework example 9.2.2 using a 1.5-acre detention basin.

9.2.6 Rework example 9.2.2 using a triangular inflow hydrograph that increases linearly from zero to a peak of 90 cfs at 120 min and then decreases linearly to a zero discharge at 240 min. Use a 30-min routing interval.

9.2.7 Consider a reservoir with surface area of 1 km^2. Initially, the reservoir has a storage volume of 500,000 m^3 with no flow coming

out of it. Suppose it receives, from the beginning, a uniform inflow of 40 cm/h continuously for 5 hr. During the time instant when the reservoir receives the inflow, it starts to release water simultaneously. After 10 hr, the reservoir is empty and outflow becomes zero thereafter. It is also known that the outflow discharge reaches its peak value at the same time instant when the inflow stops. For simplicity, assume that the variation in outflow is linear during its rise as well as during its recession.

(a) Determine the peak outflow discharge in m^3/s.

(b) Determine the time when the total storage volume (in m^3) in the reservoir is maximum and its corresponding total storage volume.

(c) What is the total storage volume (in m^3) in the reservoir at the end of the eighth hour?

9.2.8 A rectangular reservoir equipped with an outflow-control weir has the following characteristics: $S = 5 \times h$ and $Q = 2 \times h$, in which S is the storage (m^3/s-day), Q is the outflow discharge (in m^3/s), and h is the water elevation (in m). With an initial water elevation being 0.25 m, route the following inflow hydrograph to determine:

(a) the percentage of reduction in peak discharge by the reservoir; and

(b) the peak water surface elevation.

Time	6:00 am	9:00 am	12:00 nn	3:00 pm	6:00 pm
Inflow (m^3/S)	30	120	450	300	30

9.2.9 A rectangular detention basin is equipped with an outlet. The basin storage-elevation relationship and outflow-elevation relationship can be described by the following simple equations:

Storage-elevation relation: $S = 10 \times h$;
Outflow-elevation relation: $Q = 2 \times h^2$

in which S is the storage (in m^3/s-hr), Q is the outflow discharge (in m^3/s), and h is the water elevation (in m). With an initial water elevation being 0.25 m, route the following inflow hydrograph to determine:

(a) the percentage of reduction in peak discharge by the reservoir; and

(b) the peak water surface elevation.

Time	1:00 pm	2:00 pm	3:00 pm	4:00 pm	5:00 pm	6:00 pm
Inflow (m^3/s)	5	20	75	50	15	5

9.2.10 To investigate the effectiveness of a flood control reservoir, a 100-year design flood hydrograph is used as an input in the routing exercise. The reservoir has a surface area of 250 hectares and its only outlet is an uncontrolled spillway located 5 m above the datum. The design flood hydrograph is given in the following table and other physical characteristics of the reservoir are provided in the Figures P9.2.10a and b. Assuming that the initial reservoir level is 4 m above the datum, determine the effectiveness of the reservoir in terms of the reduction in inflow peak discharge.

Time (hr)	0	3	6	9	12	15	18	21
Inflow (m³/s)	50	200	400	600	300	200	100	50

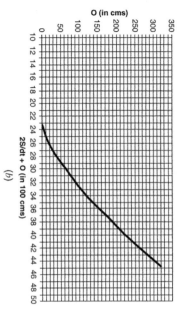
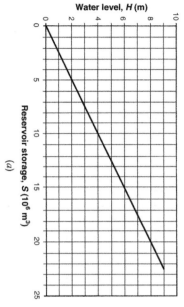
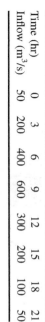

Figure P9.2.10 (*a*) Reservoir water level storage curve; (*b*) Reservoir routing curve.

9.2.11 To investigate the effectiveness of a flood control reservoir, a 100-year design flood hydrograph is used as input in the routing exercise. The reservoir has only one flow outlet located on the spillway crest with the elevation of 104 m. The reservoir has the elevation-storage-discharge relationship shown in Table 9.2.11(a). Given the design flood hydrograph as shown in Table 9.2.11(b) and assuming that the initial reservoir elevation level is at 103 m, determine the effectiveness of the reservoir in terms of the reduction in inflow peak discharge. (Note: 1 hectare = 0.01 km²)

Table 9.2.11(a) Reservoir Elevation-Storage-Outflow Relation

Elevation (m)	100	101	102	103	104	105	106	107
Storage (×10⁵ m³)	50	60	70	80	92	105	120	140
Outflow (m³/s)	0	0	0	0	8	17	27	40

Table 9.2.11(b) Inflow Hydrograph

Time (hr)	0	12	24	36	48	60	72	84
Inflow (m³/s)	10	20	30	40	30	25	15	10

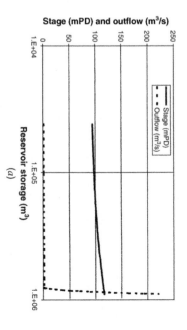

9.2.12 The Kowloon Bywash Reservoir (KBR) is located on the upstream of the Lai Chi Kok area. It is a small reservoir equipped with a tunnel with a maximum capacity of 2 m³/s delivering reservoir water to the downstream Tai Po Road water treatment plant. There is an uncontrolled spillway with the crest elevation at 115 m. The stage-volume-outflow relationships of KBR are shown in the attached table and Figure P9.2.12a. Further, the reservoir routing curve, i.e., $2S/\Delta t + O$ vs. O, for $\Delta t = 1$ hr is shown in the Figure P9.2.12b.

Elevation (mPD)	Outflow (m³/s)	Storage (m³)	$2S/\Delta t + O$ (m³/s)
109.73	2.00	531,000	297.0
112.78	2.00	679,182	379.3
115.06	2.00	801,442	447.2
115.22	4.02	809,759	453.9
115.37	12.11	818,107	466.6
115.52	22.21	826,488	481.4
115.67	36.36	834,901	500.2
115.83	51.50	843,347	520.0
115.98	69.72	851,824	543.0
116.13	89.93	860,333	567.9
116.28	114.20	868,874	596.9
116.44	136.43	877,447	623.9
116.59	162.72	886,051	655.0
116.74	193.03	894,688	690.1
116.89	228.40	903,355	730.3

Consider the inflow hydrograph given in the table below. Determine the peak outflow discharge from the KBR and the corresponding water surface elevation and the storage volume. Assume that the initial storage in the reservoir is 500,000 m³.

Time (hr)	1	2	3	4	5	6	7
Inflow (m³/s)	10	80	200	150	100	60	20

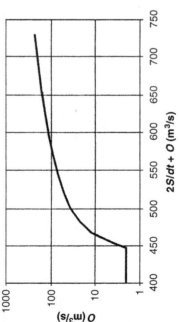

Figure P9.2.12 (a) Stage-Storage-Outflow relationship; (b) Reservoir routing curve

9.3.1 Rework example 9.3.4 using $\Delta t = 2$ hr.

9.3.2 Rework example 9.3.4 assuming $X = 0.3$.

9.3.3 Rework example 9.3.4 assuming $K = 1.4$ hr.

9.3.4 Calculate the Muskingum routing K and number of routing steps for a 1.25-mi long channel. The average cross-section dimensions for the channel are a base width of 25 ft and an average depth of 2.0 ft. Assume the channel is rectangular and has a Manning's n of 0.04 and a slope of 0.009 ft/ft.

9.3.5 Route the following upstream inflow hydrograph through a downstream flood control channel reach using the Muskingum method. The channel reach has a $K = 2.5$ hr and $X = 0.2$. Use a routing interval of 1 hr.

Time (h)	1	2	3	4	5	6	7
Inflow (cfs)	90	140	208	320	440	550	640

Time (h)	8	9	10	11	12	13	14
Inflow (cfs)	680	690	630	570	470	390	360

Time (h)	15	16	17	18	19	20
Inflow (cfs)	330	250	180	130	100	90

9.3.6 Use the U.S. Army Corps of Engineers HEC-HMS computer program to solve Problem 9.3.5.

9.3.7 A city engineer has called you wanting some information about how you would model a catchment called the Castro Valley catchment. In particular, he wants to know about how you would model the channel modifications. The 2.65-mi-long natural channel through subcatchment 1 has an approximate width of 25 ft and a Manning's n of 0.04. The slope is 0.0005 ft/ft. He thinks that a 25-ft wide trapezoidal concrete-lined channel (Manning's $n = 0.015$) would be sufficient to construct through subcatchment 1. So now you have to put together some information. Obviously you are going to have to tell him something about the channel routing procedure and the coefficients needed. He is particular and wants to know the procedure you will go through to determine these, including the number of routing steps. The natural and the concrete channels can be considered as wide-rectangular channels for your calculations so that the hydraulic radius can be approximated as the channel depth. At this time you don't know any peak discharges because you have not done any hydrologic calculations but yet you still need to approximate the X and K for the C_1, C_2, and C_3 and the number of steps for the routing method. You have decided that the approximate value of X for natural conditions is about 0.2 and for the modified conditions is about 0.3. What is the value of K and the number of routing steps needed?

9.3.8 Given the following flood hydrograph entering the upstream end of a river reach, apply the Muskingum routing procedure (with $K = 2.0$ and $X = 0.1$) to determine the peak discharge at the downstream end of the river reach and the time of its occurrence. Assume that the initial flow rate at the downstream end is 10 m³/s.

Time (hr)	0	3	6	9	12	15	18	21
Inflow (m³/s)	10	70	160	210	140	80	30	10

9.3.9 Given the following flood hydrograph entering the upstream end of a river reach, apply the Muskingum routing procedure (with $K = 6.0$ and $X = 0.2$) to determine the peak discharge at the downstream end of the river reach and the time of its occurrence. Assume that the initial flow rate at the downstream end is 20 m³/s.

Time (hr)	0	3	6	9	12	15	18	21
Inflow (m³/s)	20	260	380	580	320	180	80	20

9.3.10 Given the following flood hydrograph entering the upstream end of a river reach, apply the Muskingum routing procedure (with $K = 5.0$ hr and $X = 0.2$) to determine the peak discharge at the downstream end of the river reach and the time of its occurrence. Assume that the initial flow rate at the downstream end is 10 m³/s.

Time (hr)	0	2	4	6	8	10	12
Inflow (m³/s)	10	150	400	350	200	80	10

9.3.11 The table given below lists the inflow hydrograph.

Time (hr)	1	2	3	4	5	6
Instantaneous discharge (m³/s)	5	40	100	75	30	10

(a) Determine the percentage of attenuation in peak discharge as the hydrograph travels a distance of 10 km downstream using the Muskingum method with $X = 0.1$ and $K = 2.0$ hr. Assume that the initial outflow rate is 5 m³/s.

(b) Also, it is known that the channel bank-full capacity 10 km downstream is 50 m³/s, determine the overflow volume (in m³) of outflow hydrograph exceeding 50 m³/s.

9.3.12 From a storm event, the flood hydrographs at the upstream end and downstream end of a river reach were observed and are tabulated below.

Time (hr)	Inflow (m³/s)	Outflow (m³/s)
09:00	15	15
12:00	35	30
15:00	63	42
18:00	54	56
21:00	42	45
24:00	36	40

(a) Determine the Muskingum parameters K and X by an appropriate method of your choice.

(b) Determine the peak discharge for the following inflow hydrograph as it travels down the river.

Time (hr)	0	3	6	9	12	15	18	21	24	27
Inflow (m³/s)	10	40	80	100	60	50	40	30	20	10

9.3.13 The following table contains observed inflow and outflow hydrographs for a section of river.

Time (hr)	0	1	2	3	4	5	6
Inflow (m³/s)	200	400	700	550	400	300	200
Outflow (m³/s)	200	215	290	410	440	420	380

(a) Determine the parameters K and X in the Muskingum model by the least-squares method.

(b) Based on the K and X obtained in part (a), determine the outflow peak discharge for the following inflow hydrograph. What is the percentage of attenuation (reduction) in peak discharge?

Time (hr)	0	0.5	1.0	1.5	2.0	2.5	3.0
Inflow (m³/s)	100	400	300	200	100	100	100

9.3.14 From a storm event, the flood hydrographs at the upstream end and downstream end of a river reach are tabulated below.

Time (hr)	Inflow (m³/s)	Outflow (m³/s)
09:00	15	15
12:00	35	30
15:00	63	42
18:00	54	56
21:00	42	45

(a) Determine the Muskingum parameters K and X by the least-squares method of your choice.

(b) Based on the estimated values of K and X from part (a), determine the outflow peak discharge at the downstream end of the river reach for the following inflow hydrograph.

Time (hr)	0	2	4	6	8
Flow rate (m³/s)	20	70	50	40	30

9.3.15 Given the following flood hydrograph entering the upstream end of a river reach, apply the Muskingum routing procedure (with $K = 4.0$ hr and $X = 0.2$) to determine:

(a) the peak discharge at the downstream end of the river reach;

(b) the time of its occurrence; and

(c) the percentage of peak flow attenuation.

Assume that the initial flow rate at the downstream end is 10 m³/s.

Time (hr)	0	2	4	6	8	10	12
Inflow (m³/s)	10	250	570	320	180	70	10

9.3.16 Consider the following flood hydrograph entering the upstream end of a river reach. Apply the Muskingum routing procedure (with $K = 6.0$ and $X = 0.2$) to:

(a) determine the peak discharge at the downstream end of the river reach; and

(b) find the time to peak at the downstream section.

Assume that the initial flow rate at the downstream end is 50 m³/s.

Time (hr)	0	3	6	9	12	15	18	21
Inflow (m³/s)	50	150	300	500	300	150	100	50

9.3.17 The Castro Valley watershed has a total watershed area of 5.51 mi² and is divided into four subcatchments as shown in Figure P9.3.17. The following table provides existing characteristics of the subcatchments.

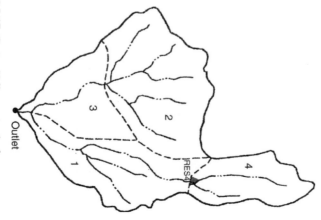

Figure P9.3.17 Castro Valley watershed

Subcatchment number	Area (mi²)	Watershed length (L) (mi)	Length to centroid (L_{CA}) (mi)	SCS curve number
1	1.52	2.65	1.40	70
2	2.17	1.85	0.68	84
3	0.96	1.13	0.60	80
4	0.86	1.49	0.79	70

Parameters for Snyder's synthetic unit hydrograph for existing conditions are $C_p = 0.25$ and $C_t = 0.38$. The Muskingum $K = 0.3$ hr for area 3 and $K = 0.6$ hr for area 1. The Muskingum X for each stream reach is 0.2. The rainfall to be used is the 100-year return period SCS type I storm pattern with a total rainfall of 10 in. Use the HEC-HMS model to determine the runoff hydrograph at the outlet of the watershed.

9.3.18 For the watershed described in problem 9.3.17, a residential development will be considered for area 4. This development will increase the impervious area so that the SCS curve number will be 85. The unit hydrograph parameters will change to $C_t = 0.19$ and $C_p = 0.5$. The natural channel through area 1 will be modified so that the Muskingum routing parameters will be $K = 0.4$ hr and $X = 0.3$. Use the HEC-HMS model to determine the change in the runoff hydrographs for area 4 and for the entire watershed.

9.3.19 Refer to problems 9.3.17 and 9.3.18. A detention basin is to be constructed at the outlet of area 4 with a low-level outlet and an overflow spillway (ogee type). The low-level outlet is a 5-ft-diameter pipe (orifice coefficient = 0.71) at a center line elevation of 391 ft above mean sea level (MSL). The overflow spillway has a length of 30 ft, crest elevation of 401.8 ft (above MSL), and a weir coefficient of 2.86. The characteristics of the detention basin are given in the following table.

Reservoir capacity (ac-ft)	Elevation (ft above MSL)
0	388.5
6	394.2
12	398.2
18	400.8
23	401.8
30	405.8

Use the HEC-HMS model to determine the runoff hydrograph at the watershed outlet for the developed conditions with the detention basin. Graphically show a comparison of the runoff hydrograph for the undeveloped, developed, and developed conditions with the detention basin.

9.3.20 Use the HEC-HMS model to solve problems 9.3.17, 9.3.18, and 9.3.19 considering the three as plans 1, 2, and 3 and solve through one simulation.

9.5.1 Determine the $\partial Q/\partial x$ on the time line $j + 1$ for the linear kinematic wave model. Consider a 100-ft-wide rectangular channel with a bed slope of 0.015 ft/ft and a Manning's $n = 0.035$. The distance between cross-sections is 3000 ft and the routing time interval is 10 min. $Q_i^{j+1} = 1000$ cfs, $Q_i^j = 800$ cfs, and $Q_{i+1}^j = 700$ cfs. Use the linear kinematic wave (conservation form) approach to compute $\partial Q/\partial x$ on time line $j + 1$.

9.5.2 Develop a flow chart of the linear kinematic wave (conservation form) method.

9.6.1 Determine the $\partial Q/\partial A$ using Q_i^{j+1} and Q_{i+1}^j for the Muskingum–Cunge model. Consider a 100-ft-wide rectangular channel with a bed slope of 0.015 ft/ft and a Manning's $n = 0.035$. The distance between cross-sections is 2000 ft and the routing time interval is 10 min. Given are $Q_i^{j+1} = 1000$ cfs, $Q_i^j = 800$ cfs, and $Q_{i+1}^j = 700$ cfs. Next compute K and x and then the routing coefficients.

9.6.2 Develop a flowchart of the Muskingum–Cunge method.

REFERENCES

Alley, W. M., and P. E. Smith, *Distributed Routing Rainfall-Runoff Model, Open File Report 82–344,* U.S. Geological Survey, Reston, VA, 1987.

Chow, V. T., *Open Channel Hydraulics,* McGraw-Hill, New York, 1959.

Chow, V. T., D. R. Maidment, and L. W. Mays, *Applied Hydrology,* McGraw-Hill, New York, 1988.

Cudworth, A. G., Jr., *Flood Hydrology Manual,* U. S. Department of the Interior, Bureau of Reclamation, Denver, CO, 1989.

Cunge, J. A., "On the Subject of a Flood Propagation Method (Muskingum Method)," *Journal of Hydraulics Research,* International Association of Hydraulic Research, vol. 7, no, 2, pp. 205–230, 1969.

Fread, D. L., *Numerical Properties of Implicit Form-Point Finite Difference Equation of Unsteady Flow,* NOAA Technical Memorandum NWS HYDRO 18, National Weather Service, NOAA, U.S. Dept. of Commerce, Silver Spring, MD, 1974.

Henderson, F. M., *Open Channel Flow,* Macmillan, New York, 1966.

Hewlett, J. D., *Principles of Forest Hydrology,* University of Georgia Press, Athens, GA, 1982.

Lighthill, M. J., and G. B. Whitham, "On Kinematic Waves, I: Flood Movement in Long Rivers," *Proc. Roy. Soc. London A,* vol. 229, no. 1178, pp. 281–316, 1955.

Masch, F. D., *Hydrology,* Hydraulic Engineering Circular No. 19, FHWA-10-84-15, Federal Highway Administration, U.S. Department of the Interior, McLean, VA, 1984.

Strelkoff, T., "The One-Dimensional Equations of Open-Channel Flow," *Journal of the Hydraulics Division,* American Society of Civil Engineers, vol. 95, no. Hy3, pp. 861–874, 1969.

U.S. Army Corps of Engineers, "Routing of Floods Through River Channels," *Engineer Manual,* 1110–2–1408, Washington, DC, 1960.

U.S. Army Corps of Engineers, Hydrologic Engineering Center, *HEC-1, Flood Hydrograph Package, User's Manual,* Davis, CA, 1990.

Woolhiser, D. A., R. E. Smith, and D. C. Goodrich, *KINEROS, A Kinematic Runoff and Erosion Model: Documentation and User Manual,* U.S. Department of Agricultural Research Service, ARS-77, Tucson, AZ, 1990.

Chapter 10

Probability, Risk, and Uncertainty Analysis for Hydrologic and Hydraulic Design

10.1 PROBABILITY CONCEPTS

This section very briefly covers probability concepts that are important in the probabilistic, risk, and uncertainty analysis for hydrologic and hydraulic design and analysis. Table 10.1.1 provides definitions of the various probability concepts needed for analysis. Many hydraulic and hydrologic variables must be treated as random variables because of the uncertainties involved in the respective hydraulic and hydrologic processes. As an example, the extremes that occur are random hydrologic events and can therefore be treated as such.

A *random variable* X is a variable described by a *probability distribution*. The distribution specifies the chance that an observation x of the variable will fall in a specified range of X. A set of observations $x_1, x_2, \ldots, x_n$ of the random variable X is called a *sample*. It is assumed that samples are drawn from a population (generally unknown) possessing constant statistical properties, while the properties of a sample may vary from one sample to another. The possible range of variation of all of the samples that could be drawn from the population is called the *sample space*, and an *event* is a subset of the sample space.

A *probability distribution* is a function representing the frequency of occurrence of the value of a random variable. By fitting a distribution to a set of data, a great deal of the probabilistic information in the sample can be compactly summarized in the function and its associated parameters. Fitting distributions can be accomplished by the method of moments or the method of maximum likelihood (see Chow et al., 1988). Between the two methods, the method of moments is more widely used, primarily for its computational simplicity. The method relates the parameters in a probability distribution model to the statistical moments to which the parameter-moment relationships for commonly used distributions in frequency analysis and reliability analysis are immediately available (see Table 10.1.2). In practice, the true mechanism that generates the observed random process is not entirely known. Therefore, to estimate the parameter values in a probability distribution model by the method of moments, sample moments are used.

Table 10.1.1   Various Parameters and Statistics Used to Describe Populations and Samples

Concept	Population value, Discrete case	Population value, Continuous case	Sample value
Cumulative distribution function (cdf)	Describes the probability that a random variable is less than or equal to a specified value x	Describes the probability that a random variable is less than or equal to a specified value x	Empirical distribution function (edf): describes the observed frequency of a random variable being less than or equal to a specified value x
Probability mass function (pmf) and probability density function (pdf)	pmf: the probability that X is equal to k	pdf: first derivative of the cumulative distribution function $$f(x) \equiv \frac{dF(x)}{dx}$$	Histogram: observed frequency with which random variable X falls into the assigned ranges
Mean, average, or expected value	$\sqrt{0.368}$	$$\mu \equiv \int_{-\infty}^{\infty} x f(x)\,dx$$	$$\overline{X} \equiv \sum_{i=1}^{n}\frac{X_i}{n}$$
Variance	$$\sigma^2 \equiv \sum_{i=1}^{\infty} P(X=x_i)(x_i-\mu)^2$$	$$\sigma^2 \equiv \int_{-\infty}^{\infty}(x-\mu)^2 f(x)\,dx$$	$$S^2 \equiv \sum_{i=1}^{n}\frac{(X_i-\overline{X})^2}{n-1}$$
kth central moment	$$M_k \equiv \sum_{i=1}^{\infty} P(X=x_i)(x_i-\mu)^k$$	$$M_k \equiv \int_{-\infty}^{\infty}(x-\mu)^k f(x)\,dx$$	$$\overline{M}_k \equiv \sum_{i=1}^{n}\frac{(X_i-\overline{X})^k}{n}$$
Standard deviation		$\sigma \equiv \sqrt{\sigma^2}$	$S \equiv \sqrt{S^2}$
Coefficient of variation or relative standard deviation (if $\mu \neq 0$)		$CV \equiv \dfrac{\sigma}{\mu}$	$CV \equiv \dfrac{S}{\overline{X}}$
Coefficient of skew (a measure of asymmetry)		$\gamma \equiv \dfrac{M_3}{\sigma^3}$	$G \equiv \dfrac{\widetilde{M}_3}{S^3}$
Quantiles		x_p is any value of X that has the properties that $$P[X<x_p] \le p$$ $$P[X>x_p] \le 1-p$$	$\widehat{x}_p$ is the pth quantile of edf
Median (useful for describing central tendency regardless of skewness)		$x_{0.5}$ Any value of X that has the property that $$P[X<x_p] \le 0.5$$ $$P[X>x_p] \le 0.5$$	$\widehat{X}_{0.5}$ The middle observation in a sorted sample, or the average of the two middle observations if the sample size is even.

Table 10.1.1 (*Continued*)

Concept	Population value, Discrete case	Population value, Continuous case	Sample value
Upper quartile, lower quartile, and hinges		Upper quartile = $x_{0.75}$ Lower quartile = $x_{0.25}$	Upper hinge = $\hat{X}_{0.75}$ This is an approximation to the sample upper quartile; it is defined as the median of all sample values of $X \le x_{0.50}$. The lower hinge, $\hat{X}_{0.25}$, is defined analogously.
Interquartile range (useful for describing spread of data regardless of symmetry)		$x_{0.75} - x_{0.25}$ Width of central region of population containing probability of 0.5	$\hat{X}_{0.75} - \hat{X}_{0.25}$ Width of central region of data set encompassing approximately half the data

Source: Hirsh, et al (1993).

Table 10.1.2 Probability Distributions Commonly Used

Distribution	Probability density function	Range	Parameter-Moment relations
Normal	$f(x) = \dfrac{1}{\sqrt{2\pi}\sigma} e^{-(x-\mu)^2/2\sigma^2}$	$-\infty < x < \infty$	
Log-normal	$f(x) = \dfrac{1}{\sqrt{2\pi}x\sigma_{\ln x}} e^{-(\ln x - \mu_{\ln x})^2/(2\sigma_{\ln x}^2)}$	$x > 0$	$\mu_{\ln x} = \dfrac{1}{2}\ln\left[\dfrac{\mu_x^2}{1+\Omega_x^2}\right]$ $\sigma_{\ln x}^2 = \ln(1+\Omega_x^2)$ $\Omega_x = \sigma_x/\mu_x$
Exponential	$f(x) = \lambda e^{-\lambda x}$	$x \ge 0$	$\lambda = \dfrac{1}{\mu_x}$
Gamma	$f(x) = \dfrac{\lambda^\beta x^{\beta-1} e^{-\lambda x}}{\Gamma(\beta)}$ where Γ = gamma function	$x \ge 0$	$\lambda = \dfrac{\mu_x}{\sigma_x^2}$, $\beta = \dfrac{\mu_x^2}{\sigma_x^2} = \dfrac{1}{C_v^2}$
Extreme Value Type I	$f(x) = \dfrac{1}{\alpha} e^{-(x-\beta)/\alpha - e^{-(x-\beta)/\alpha}}$	$-\infty < x < \infty$	$\alpha = \sqrt{6}\sigma_x/\pi$ $\beta = \mu_x - 0.5772\alpha$
Log Pearson Type III	$f(x) = \dfrac{\lambda^\beta (y-\epsilon)^{\beta-1} e^{-\lambda(y-\epsilon)}}{x\Gamma(\beta)}$ where $y = \log x$	$\log x \ge \epsilon$	$\lambda = \dfrac{s_y}{\sqrt{\beta}}$, $\beta = \left[\dfrac{2}{G_s(y)}\right]^2$ $\epsilon = \bar{y} - s_y\sqrt{\beta}$ (assuming $G_s(y)$ is positive)

10.2 COMMONLY USED PROBABILITY DISTRIBUTIONS

Of the distributions presented in Table 10.1.2, only the normal and log-normal distributions are discussed in this subsection. Section 10.4 discusses the Pearson Type III distribution. Section 10.6 discusses the exponential distribution.

10.2.1 Normal Distribution

The normal distribution is a well-known probability distribution, also called the *Gaussian distribution*. Two parameters are involved in a normal distribution: the mean and the variance. A normal random variable having a mean μ and a variance σ^2 is herein denoted as $X \sim N(\mu, \sigma^2)$ with a PDF (probability density function) of

$$f(x) = \frac{1}{\sqrt{2\pi}\sigma} \exp\left[-\frac{1}{2}\left(\frac{x-\mu}{\sigma}\right)^2\right] \quad \text{for } -\infty < x < \infty \qquad (10.2.1)$$

A normal distribution is bell-shaped and symmetric with respect to $x = \mu$. Therefore, the skew coefficient for a normal random variable is zero. A random variable Y that is a linear function of a normal random variable X is also normal. That is, if $X \sim N(\mu, \sigma^2)$ and $Y = aX + b$ then $Y \sim N(a\mu + b, a^2\sigma^2)$. An extension of this theorem is that the sum of normal random variables (independent or dependent) is also a normal random variable.

Probability computations for normal random variables are made by first transforming to the standardized variate as

$$Z = (X - \mu)/\sigma \qquad (10.2.2)$$

in which Z has a zero mean and unit variance. Since Z is a linear function of the random variable X, Z is also normally distributed. The PDF of Z, called the *standard normal distribution*, can be expressed as

$$\phi(z) = \frac{1}{\sqrt{2\pi}} \exp\left[-\frac{z^2}{2}\right] \quad \text{for } -\infty < z < \infty \qquad (10.2.3)$$

A table of the CDF of Z is given in Table 10.2.1. Computations of probability for $X \sim N(\mu, \sigma^2)$ can be performed using

$$P(X \le x) = P\left[\frac{X-\mu}{\sigma} \le \frac{x-\mu}{\sigma}\right] = P[Z \le z] = \Phi(z) \qquad (10.2.4)$$

where $\Phi(z)$ is the CDF of the standard normal random variable Z defined as

$$\Phi(z) = \int_{-\infty}^{z} \phi(z)\,dz \qquad (10.2.5)$$

10.2.2 Log-Normal Distribution

The log-normal distribution is a commonly used continuous distribution in hydrologic event analysis when random variables cannot be negative. A random variable X is said to be log-normally distributed if its logarithmic transform $Y = \ln(X)$ is normally distributed with mean $\mu_{\ln x}$ and variance $\sigma^2_{\ln x}$. The PDF of the log-normal random variable is

$$f(X) = \frac{1}{\sqrt{2\pi}X\sigma_{\ln X}} \exp\left[-\frac{1}{2}\left(\frac{\ln X - \mu_{\ln X}}{\sigma_{\ln X}}\right)^2\right] \quad \text{for } 0 < X < \infty \qquad (10.2.6)$$

Table 10.2.1 Cumulative Probability of the Standard Normal Distribution*

z	.00	.01	.02	.03	.04	.05	.06	.07	.08	.09
0	0.5000	0.5040	0.5080	0.5120	0.5160	0.5199	0.5239	0.5279	0.5319	0.5359
0.1	0.5398	0.5438	0.5478	0.5517	0.5557	0.5596	0.5636	0.5675	0.5714	0.5753
0.2	0.5793	0.5832	0.5871	0.5910	0.5948	0.5987	0.6026	0.6064	0.6103	0.6141
0.3	0.6179	0.6217	0.6255	0.6293	0.6331	0.6368	0.6406	0.6443	0.6480	0.6517
0.4	0.6554	0.6591	0.6628	0.6664	0.6700	0.6736	0.6772	0.6808	0.6844	0.6879
0.5	0.6915	0.6950	0.6985	0.7019	0.7054	0.7088	0.7123	0.7157	0.7190	0.7224
0.6	0.7257	0.7291	0.7324	0.7357	0.7389	0.7422	0.7454	0.7486	0.7517	0.7549
0.7	0.7580	0.7611	0.7642	0.7673	0.7704	0.7734	0.7764	0.7794	0.7823	0.7852
0.8	0.7881	0.7910	0.7939	0.7967	0.7995	0.8023	0.8051	0.8078	0.8106	0.8133
0.9	0.8159	0.8186	0.8212	0.8238	0.8264	0.8289	0.8315	0.8340	0.8365	0.8389
1.0	0.8413	0.8438	0.8461	0.8485	0.8508	0.8531	0.8554	0.8577	0.8599	0.8621
1.1	0.8643	0.8665	0.8686	0.8708	0.8729	0.8749	0.8770	0.8790	0.8810	0.8830
1.2	0.8849	0.8869	0.8888	0.8907	0.8925	0.8944	0.8962	0.8980	0.8997	0.9015
1.3	0.9032	0.9049	0.9066	0.9082	0.9099	0.9115	0.9131	0.9147	0.9162	0.9177
1.4	0.9192	0.9207	0.9222	0.9236	0.9251	0.9265	0.9279	0.9292	0.9306	0.9319
1.5	0.9332	0.9345	0.9357	0.9370	0.9382	0.9394	0.9406	0.9418	0.9429	0.9441
1.6	0.9452	0.9463	0.9474	0.9484	0.9495	0.9505	0.9515	0.9525	0.9535	0.9545
1.7	0.9554	0.9564	0.9573	0.9582	0.9591	0.9599	0.9608	0.9616	0.9625	0.9633
1.8	0.9641	0.9649	0.9656	0.9664	0.9671	0.9678	0.9686	0.9693	0.9699	0.9706
1.9	0.9713	0.9719	0.9726	0.9732	0.9738	0.9744	0.9750	0.9756	0.9761	0.9767
2.0	0.9772	0.9778	0.9783	0.9788	0.9793	0.9798	0.9803	0.9808	0.9812	0.9817
2.1	0.9821	0.9826	0.9830	0.9834	0.9838	0.9842	0.9846	0.9850	0.9854	0.9857
2.2	0.9861	0.9864	0.9868	0.9871	0.9875	0.9878	0.9881	0.9884	0.9887	0.9890
2.3	0.9893	0.9896	0.9898	0.9901	0.9904	0.9906	0.9909	0.9911	0.9913	0.9916
2.4	0.9918	0.9920	0.9922	0.9925	0.9927	0.9929	0.9931	0.9932	0.9934	0.9936
2.5	0.9938	0.9940	0.9941	0.9943	0.9945	0.9946	0.9948	0.9949	0.9951	0.9952
2.6	0.9953	0.9955	0.9956	0.9957	0.9959	0.9960	0.9961	0.9962	0.9963	0.9964
2.7	0.9965	0.9966	0.9967	0.9968	0.9969	0.9970	0.9971	0.9972	0.9973	0.9974
2.8	0.9974	0.9975	0.9976	0.9977	0.9977	0.9978	0.9979	0.9979	0.9980	0.9981
2.9	0.9981	0.9982	0.9982	0.9983	0.9984	0.9984	0.9985	0.9985	0.9986	0.9986
3.0	0.9987	0.9987	0.9987	0.9988	0.9988	0.9989	0.9989	0.9989	0.9990	0.9990
3.1	0.9990	0.9991	0.9991	0.9991	0.9992	0.9992	0.9992	0.9992	0.9993	0.9993
3.2	0.9993	0.9993	0.9994	0.9994	0.9994	0.9994	0.9994	0.9995	0.9995	0.9995
3.3	0.9995	0.9995	0.9995	0.9996	0.9996	0.9996	0.9996	0.9996	0.9996	0.9997
3.4	0.9997	0.9997	0.9997	0.9997	0.9997	0.9997	0.9997	0.9997	0.9997	0.9998

*To employ the table for z < 0, use $F_z(z) = 1 - F_z(|z|)$ where $F_z(|z|)$ is the tabulated value.

Source: Grant and Leavenworth (1972).

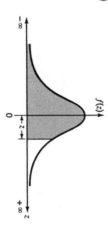

which can be derived from the normal PDF, that is, equation (10.2.1). Statistical properties of a log-normal random variable of the original scale can be computed from those of the log-transformed variable. To compute the statistical moments of X from those of ln X, the following formulas are useful:

$$\mu_x = \exp(\mu_{\ln x} + \sigma^2_{\ln x}/2)$$

(10.2.7)

$$\sigma_X^2 = \mu_X^2 [\exp(\sigma_{\ln X}^2) - 1] \tag{10.2.8}$$

$$\Omega_X^2 = \exp(\sigma_{\ln X}^2) - 1 \tag{10.2.9}$$

$$\lambda_X = \Omega_X^2 + 3\Omega_X \tag{10.2.10}$$

From equation (10.2.10), it is obvious that log-normal distributions are always positively skewed because $\Omega_X > 0$. Conversely, the statistical moments of $\ln X$ can be computed from those of X by

$$\mu_{\ln X} = \frac{1}{2} \ln \left[\frac{\mu_X^2}{1 + \Omega_X^2} \right] \tag{10.2.11}$$

$$\sigma_{\ln X}^2 = \ln(\Omega_X^2 + 1) \tag{10.2.12}$$

Since the sum of normal random variables is normally distributed, the multiplication of log-normal random variables is also log-normally distributed. Several properties of log-normal random variables are useful:

1. If X is a log-normal random variable and $Y = aX^b$, then Y has a log-normal distribution with mean $\mu_{\ln Y} = \ln a + b\mu_{\ln X}$ and variance $\sigma_{\ln Y}^2 = b^2 \sigma_{\ln X}^2$.
2. If X and Y are independently log-normally distributed, $W = XY$ has a log-normal distribution with mean $\mu_{\ln W} = \mu_{\ln X} + \mu_{\ln Y}$ and variance $\sigma_{\ln W}^2 = \sigma_{\ln X}^2 + \sigma_{\ln Y}^2$.
3. If X and Y are independent and log-normally distributed, then $R = X/Y$ is log-normal with $\mu_{\ln R} = \mu_{\ln X} - \mu_{\ln Y}$ and variance $\sigma_{\ln R}^2 = \sigma_{\ln X}^2 + \sigma_{\ln Y}^2$.

EXAMPLE 10.2.1

The annual maximum series of flood magnitudes in a river is assumed to follow a log-normal distribution with a mean of 6000 m³/s and a standard deviation of 4000 m³/s. (a) What is the probability in each year that a flood magnitude would exceed 7000 m³/s? (b) Determine the flood magnitude with a return period of 100 years.

SOLUTION

(a) Let Q be a random variable representing the annual maximum flood magnitude. Since Q is assumed to follow a log-normal distribution, $\ln(Q)$ is normally distributed with mean and variance that can be computed using $\mu_Q = 6000$ and $\Omega_Q = \sigma_Q/\mu_Q = 4000/6000 = 0.667$. By equations (10.2.11) and (10.2.12), respectively, we find

$$\mu_{\ln Q} = \frac{1}{2} \ln \left[\frac{\mu_Q^2}{1 + \Omega_Q^2} \right] = \frac{1}{2} \ln \left[\frac{6000^2}{1 + 0.667^2} \right] = 8.515$$

$$\sigma_{\ln Q}^2 = \ln(\Omega_Q^2 + 1) = \ln(0.667^2 + 1) = 0.368$$

The probability that the flood magnitude exceeds 7000 m³/s is

$$P(Q > 7000) = P(\ln Q > \ln 7000) = 1 - P(\ln Q \le \ln 7000)$$
$$= 1 - P\{[(\ln Q - \mu_{\ln Q})/\sigma_{\ln Q}] \le [(\ln 7000 - \mu_{\ln Q})/\sigma_{\ln Q}]\}$$
$$= 1 - P[Z \le (\ln 7000 - 8.515)/\sqrt{0.368}]$$
$$= 1 - P[Z \le 0.558]$$
$$= 1 - F(0.558) = 1 - 0.712 = 0.288$$

(b) A 100-year event in hydrology represents the event that occurs, on the average, once every 100 years. Therefore, the probability in every single year that a 100-year event is equaled or exceeded is 0.01, i.e., $P(Q \ge q_{100}) = 0.01$ in which q_{100} is the magnitude of the 100-year flood. This part of the problem is to determine q_{100}, which is the reverse of part (a).

$$P(Q \le q_{100}) = 1 - P(Q = q_{100}) = 1 - 0.01 = 0.99$$

Since $P(Q \le q_{100}) = P[\ln Q \le \ln q_{100}] = P[Z \le (\ln q_{100} - \mu_{\ln Q})/\sigma_{\ln Q}] = 0.99$,

$$0.99 = P[Z \le (\ln q_{100} - 8.515)/\sqrt{0.368}\,]$$

$$0.99 = \Phi\{(\ln q_{100} - 8.515)/\sqrt{0.368}\}$$

$$0.99 = \Phi(z)$$

From the standard normal probability table (Table 10.2.1), $z = 2.33$ for $\Phi(2.33) = 0.99$. Solving $z = (\ln q_{100} - 8.515)/\sqrt{0.368}$ for q_{100} first yields $\ln q_{100} = 9.928$, then $q_{100} = 20,500$ m^3/s.

10.2.3   Gumbel (Extreme Value Type I) Distribution

Gumbel (extreme value type I) distribution is another probability distribution function useful for hydrologic analysis, particularly rainfall analysis. The probability distribution function is presented in Table 10.1.2 and the CDF is

$$F(x) = \exp\{-\exp[-(x - \beta)/\alpha]\} \qquad -\infty < x < \infty \tag{10.2.13}$$

The parameters are estimated as

$$\alpha = 6^{0.5}\sigma_x/\pi \tag{10.2.14}$$

$$\beta = \mu_x - 0.5772\alpha \tag{10.2.15}$$

where β is the mode (point of maximum probability) of the distribution. The inverse of the CDF is

$$x = \beta - \alpha \ln[-\ln F] \tag{10.2.16}$$

EXAMPLE 10.2.2

Using an exceedance series of 30-min duration rainfall values, the mean is 1.09 in and the standard deviation is 0.343 in. Determine the 100-year 30-min duration rainfall value using the Gumbel (extreme value type I) distribution.

SOLUTION

Determine the parameters α and β.

$$\alpha = 6^{0.5}\sigma_x/\pi = 6^{0.5}(0.343)/3.14 = 0.268$$

$$\beta = \mu_x - 0.5772\alpha = 1.09 - 0.5772(0.268) = 0.935$$

Use the parameters α and β in the CDF with $F(x) = 1 - 0.01$ where 0.01 is the probability in any given year that the 100-year 30-min duration rainfall is equaled or exceeded, i.e., $P(x \ge x_{100})$ = 1/100 = 0.01.

$$F(x) = \exp\{-\exp[-(x - 0.935)/0.268]\} = 1 - 0.01 = 0.99$$

The inverse of the CDF $F(x)$ is $x = \beta - \alpha\ln[-\ln F]$ so that the 100-year 30-min duration rainfall is $x = 0.935 - 0.268\ln[-\ln 0.99] = 2.167$ in.

10.3   HYDROLOGIC DESIGN FOR WATER EXCESS MANAGEMENT

Hydrologic design is the process of assessing the impact of hydrologic events on a water resource system and choosing values for the key variables of the system so that it will perform adequately (Chow et al., 1988). This section focuses on water excess management; however, many of the concepts are applicable to water supply (use) management.

10.3.1   Hydrologic Design Scale

The *hydrologic design scale* is the range in magnitude of the design variable (such as the design discharge) within which a value must be selected to determine the inflow to the system (see Figure 10.3.1). The most important factors in selecting the design value are cost and safety. The optimal magnitude for design is one that balances the conflicting considerations of cost and safety. The practical upper limit of the hydrologic design scale is not infinite, since the global hydrologic cycle is a closed system; that is, the total quantity of water on earth is essentially constant. Although the true upper limit is unknown, for practical purposes an estimated upper limit may be determined. This *estimated limiting value* (ELV) is defined as the largest magnitude possible for a hydrologic event at a given location, based on the best available hydrologic information.

The concept of an estimated limiting value is implicit in the *probable maximum precipitation* (PMP) and the corresponding *probable maximum flood* (PMF). The probable maximum precipitation is defined by the World Meteorological Organization (1983) as a "quantity of precipitation that is close to the physical upper limit for a given duration over a particular basin." However, the return period varies geographically. Some arbitrarily assign a return period, say 10,000 years, to the PMP or PMF, but this has no physical basis.

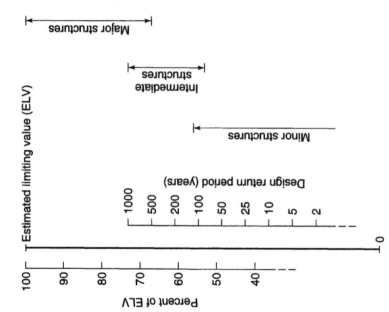

Figure 10.3.1 Hydrologic design scale. Approximate ranges of the design level for different types of structures are shown. Design may be based on a percentage of the ELV or on a design return period. The values for the two scales shown in the diagram are illustrative only and do not correspond directly with one another (from Chow et al. (1988)).

Table 10.3.1 Generalized Design Criteria for Water-Control Structures

Type of structure	Return period (Years)	ELV (%)
Highway culverts		
Low traffic	5–10	—
Intermediate traffic	10–25	—
High traffic	50–100	—
Highway bridges		
Secondary system	10–50	—
Primary system	50–100	—
Farm drainage		
Culverts	5–50	—
Ditches	5–50	—
Urban drainage		
Storm sewers in small cities	2–25	—
Storm sewers in large cities	25–50	—
Airfields		
Low traffic	5–10	—
Intermediate traffic	10–25	—
High traffic	50–100	—
Levees		
On farms	2–50	—
Around cities	50–200	—
Dams with no likelihood of loss of life (low hazard)		
Small dams	50–100	—
Intermediate dams	100+	—
Large dams	—	50–100
Dams with probable loss of life (significant hazard)		
Small dams	100+	50
Intermediate dams	—	50–100
Large dams	—	100
Dams with high likelihood of considerable loss of life (high hazard)		
Small dams	—	50–100
Intermediate dams	—	100
Large dams	—	100

Source: Chow et al. (1988).

Generalized design criteria for water-control structures have been developed, as summarized in Table 10.3.1. According to the potential consequence of failure, structures are classified as *major*, *intermediate*, and *minor*; the corresponding approximate ranges on the design scale are shown in Figure 10.3.1. The criteria for dams in Table 10.3.1 pertain to the design of spillway capacities, and are taken from the National Academy of Sciences (1983). The Academy defines a *small dam* as having 50–1000 acre-ft of storage or being 25–40 ft high, an *intermediate dam* as having 1000–50,000 acre-ft of storage or being 40–100 ft high, and a *large dam* as having more than 50,000 acre-ft of storage or being more than 100 ft high. In general, there would be considerable loss of life and extensive damage if a major structure failed. In the case of an intermediate structure, a small loss of life would be possible and the damage would be within the financial capability of the owner. For minor structures, there generally would be no loss of life, and the damage would be of the same magnitude as the cost of replacing or repairing the structure.

10.3.2 Hydrologic Design Level (Return Period)

A *hydrologic design level* on the design scale is the magnitude of the hydrologic event to be considered for the design of a structure or project. As it is not always economical to design structures and projects for the estimated limiting values, the ELV is often modified for specific design purposes. The final design value may be further modified according to engineering judgment and the experience of the designer or planner. Table 10.3.1 presents generalized criteria for water-control structures. A large number of the structures are designed using return periods.

An extreme hydrologic event is defined to have occurred if the magnitude of the event X is greater than or equal to some level x_T, i.e., $X \geq x_T$. The *return period* T of the event $X = x_T$ is the expected value of the *recurrence interval* (time between occurrences). The expected value $E(\ )$ is the average value measured over a very large number of occurrences. Consequently, the return period of a hydrologic event of a given magnitude is defined as the *average recurrence interval* between events that equal or exceed a specified magnitude.

10.3.3 Hydrologic Risk

The *probability of occurrence* $P(X \geq x_T)$ of the hydrologic event $(X \geq x_T)$ for any observation is the inverse of the return period, i.e.,

$$P(X \geq x_T) = \frac{1}{T} \qquad (10.3.1)$$

For a 100-year peak discharge, the probability of occurrence in any given year is $P(X \geq x_{100}) = 1/100 = 0.01$.

The probability of nonexceedance is

$$P(X < x_T) = 1 - \frac{1}{T} \qquad (10.3.2)$$

Because each hydrologic event is considered independent, the *probability of nonexceedance* for n years is

$$P(X < x_T \text{ each year for } n \text{ years}) = \left(1 - \frac{1}{T}\right)^n$$

The complement, the *probability of exceedance* at least once in n years, is

$$P(X \geq x_T \text{ at least once in } n \text{ years}) = 1 - \left(1 - \frac{1}{T}\right)^n$$

which is the probability that a T-year return period event will occur at least once in n years. This is also referred to as the *natural, inherent,* or *hydrologic risk of failure* $\bar{R}$:

$$\bar{R} = 1 - \left(1 - \frac{1}{T}\right)^n = 1 - [1 - P(X \geq x_T)]^n \qquad (10.3.3)$$

where n is referred to as the expected life of the structure. The hydrologic risk relationship is plotted in Figure 10.3.2.

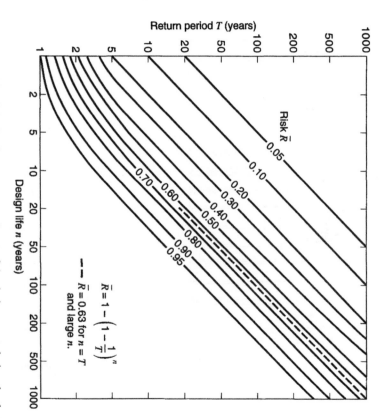

Figure 10.3.2 Risk of at least one exceedance of the design event during the design life (from Chow et al. (1988)).

EXAMPLE 10.3.1

Determine the hydrologic risk of a 100-year flood occurring during the 30-year service life of a project.

SOLUTION

Use equation (10.3.3) to determine the risk:

$$\bar{R} = 1 - \left(1 - \frac{1}{T}\right)^n = 1 - (1 - 1/100)^{30} = 0.26$$

10.3.4 Hydrologic Data Series

Figure 10.3.3a shows all the data available (that have been collected) for a hydrologic event. This represents a *complete-duration series*. A *partial-duration series* includes data that are selected so that their values are greater than some base value. An *annual-exceedance series* has a base value so that the number of values in the series is equal to the number of years of record. Figure 10.3.3b illustrates the annual exceedance series. An *extreme-value series* consists of the largest or smallest values occurring in each of the equally long time intervals of the record. If the time interval length is one year, the series is an *annual series*. An *annual maximum series* over the largest values in each respective year (Figure 10.3.3c) consists of the largest annual values in each of the respective years. An *annual minimum series* consists of the smallest annual values in each of the respective years. Figure 10.3.4 illustrates the annual-exceedance series and the annual maximum series of the hypothetical data in Figure 10.3.3.

The return periods for annual exceedance series T_E are related to the corresponding annual maximum series return period T by (Chow, 1964)

$$T_E = \left[\ln\left(\frac{T}{T-1}\right)\right]^{-1} \tag{10.3.4}$$

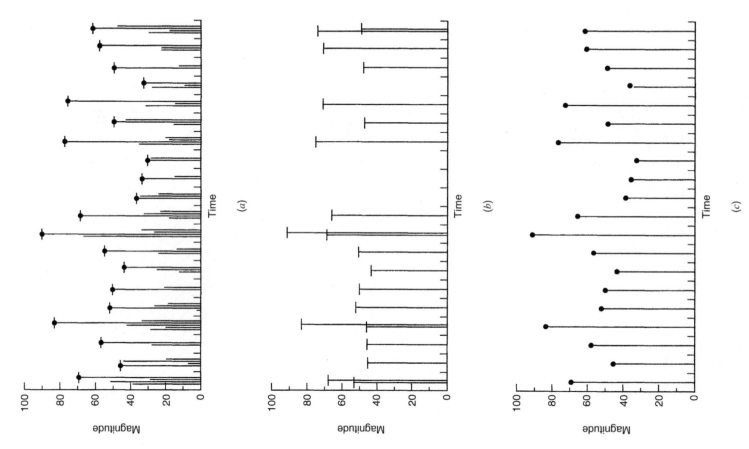

Figure 10.3.3 Hydrologic data arranged by time of occurrence. (*a*) Original data: $N = 20$ years; (*b*) Annual exceedances; (*c*) Annual maxima (from Chow (1964)).

10.4 HYDROLOGIC FREQUENCY ANALYSIS

10.4.1 Frequency Factor Equation

One of the primary objectives of the frequency analysis of hydrologic data is to determine the recurrence interval of a hydrologic event of a given magnitude. The *recurrence interval*, which is the same as the return period, may also be defined as the average interval of time within which the magnitude of a hydrologic event will be equaled or exceeded once, on the average. The term "frequency" is often used interchangeably with "recurrence interval"; however, it should not be construed to indicate a regular or stated interval of occurrence or recurrence. *Hydrologic frequency analysis* is the approach of using probability and statistical analysis to estimate future frequencies (probabilities of hydrologic events occurring) based upon information contained in hydrologic records. Through the use of statistical methods, observed data is analyzed so as to provide not only a

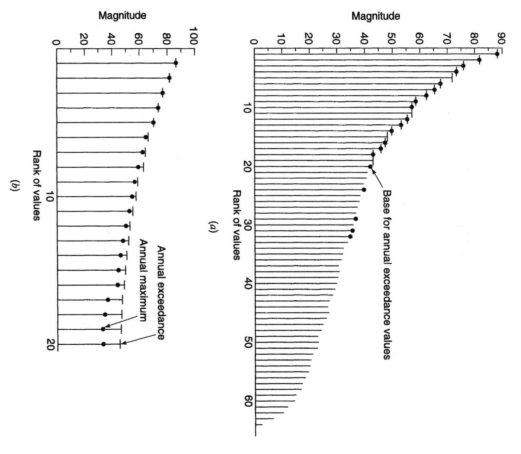

Figure 10.3.4 Hydrologic data arranged in the order of magnitude. (a) Original data; (b) Annual exceedance and maximum values (from Chow (1964)).

more accurate estimate of future frequencies than is indicated by the observed data, but also criteria for determining the reliability of frequency estimates. The emphasis of hydrologic frequency analysis in this section is for the determination of flood frequency curves for streams and rivers.

The results of flood flow frequency analysis can be used for many engineering purposes: (1) for the design of dams, bridges, culverts, water supply systems, and flood control structures; (2) to determine the economic value of flood control projects; (3) to determine the effect of encroachments in the floodplain; (4) to determine a reservoir stage for real estate acquisition and reservoir-use purposes; (5) for the selection of runoff magnitudes for interior drainage, pumping plant, and local protection project design; and (6) for flood-plain zoning, etc.

In the application of statistical methods to hydrologic frequency analysis, theoretical probability distributions are utilized. The hydrologic events that have occurred are assumed to constitute a random sample (observed set of events) and then are used to make inferences about the true population (all possible events) for the theoretical distribution considered. These inferences are subject to considerable uncertainty because a set of observed hydrologic events represents only a sample or small subset of the many sets of physical conditions that could represent the population described by the theoretical probability distribution.

The existing methods of frequency analysis are numerous, with many diverse and confusing viewpoints and theories. Several types of probability distributions have been used in the past for hydrologic frequency determination. The most popular of these for flood flow frequency determination have been the log-normal, Gumbel (extreme value type I), and log-Pearson Type III distribution (see Table 10.2.2). Because of the range of uncertainty and diversity of methods in determining flood flow estimates and the varying results that can be obtained using the various methods, the U.S. Water Resources Council (1981) attempted to promote a uniform or consistent approach to flood-flow frequency studies.

Computation of the magnitudes of extreme events, such as flood flows, requires that the probability distribution function be invertible; that is, given a value for T or $[f(x_T) = T/(T-1)]$, the corresponding value of x_T can be determined. Some probability distribution functions are not readily invertible, including the normal and Pearson Type III distributions, and an alternative method of calculating the magnitudes of extreme events is required for these distributions.

The magnitude x_T of a hydrologic event may be represented as the mean $\bar{x}$ plus a departure of the variate from the mean. This departure is equal to the product of the standard deviation S_x and a frequency factor K_T. The departure Δx_T and the frequency factor K_T are functions of the return period and the type of probability distribution to be used in the analysis. Chow (1951) proposed the following frequency factor equation:

$$x_T = \bar{x} + K_T S_x \tag{10.4.1}$$

When the variable analyzed is $y = \log x$, then the same method is applied to the statistics for the logarithms of the data, using

$$y_T = \bar{y} + K_T S_y \tag{10.4.2}$$

and the required value of x_T is found by taking the antilog of y_T. For a given distribution, a K–T relationship can be determined between the frequency factor and the corresponding return period.

10.4.2 Application of Log-Pearson III Distribution

For the log-Pearson Type III distribution, the first step is to compute the logarithms of the hydrologic data, $y = \log x$. Usually, logarithms to base 10 are used. The mean $\bar{y}$, standard deviation S_y, and coefficient of skewness G_s are calculated for the logarithms of the data. Note that while γ is generally used for skew coefficient, G_s is used for a sample space (see Table 10.1.1). The frequency factor depends on the return period T and the coefficient of skewness G_s. When $G_s = 0$, the frequency factor is equal to the standard normal variable z. When $G_s \neq 0$, K_T is approximated by Kite (1977) as

$$K(T, G_s) = z + (z^2 - 1)k + \frac{1}{3}(z^3 - 6z)k^2 - (z^2 - 1)k^3 + zk^4 + \frac{1}{3}k^5 \qquad (10.4.3)$$

where $k = G_s/6$. Table 10.4.1 lists values of the frequency factor for the Pearson Type III (and log-Pearson Type III) distribution for various values of the return period and coefficient of skewness.

The U.S. Water Resources Council (WRC) recommended that the log-Pearson Type III be used as a base method for flood flow frequency studies (U.S. Water Resources Council, 1981). This was an attempt to promote a consistent, uniform approach to flood flow frequency determination for use in all federal planning involving water and related land resources. This choice of the log-Pearson Type III is, however, subjective to some extent, in that no rigorous statistical criteria exist on which a comparison of distributions can be made.

The frequency factor equation for the log-Pearson Type III distribution is written in terms of discharge as

$$\log Q_T = \bar{y} + K(T, G_s) \cdot S_y \qquad (10.4.4)$$

where Q_T is the discharge for the T-year return period.

The steps in the procedure to compute the discharge Q_T of return period T are as follows:

Step 1 Transform all discharge values to log Q_1, log $Q_2, \ldots$, log Q_n.

Step 2 Determine the mean ($\bar{y}$), standard deviation (S_y) and skew (G_s) of the log-transformed values.

Step 3 Use Table 10.4.1 to determine the frequency factors for the return periods of interest.

Step 4 Apply the frequency factor equation (10.4.4) and compute Q_T = antilog (log Q_T).

Table 10.4.1 K_T Values for Pearson Type III Distribution

Skew coeff.	Recurrence interval (Yr)										
	1.0101	1.0526	1.1111	1.2500	2	5	10	25	50	100	200
	Exceedance probability										
	.99	.95	.90	.80	.50	.20	.10	.04	.02	.01	.005
3.0	−0.667	−0.665	−0.660	−0.636	−0.396	0.420	1.180	2.278	3.152	4.051	4.970
2.9	−0.690	−0.688	−0.681	−0.651	−0.390	0.440	1.195	2.277	3.134	4.013	4.909
2.8	−0.714	−0.711	−0.702	−0.666	−0.384	0.460	1.210	2.275	3.114	3.973	4.847
2.7	−0.740	−0.736	−0.724	−0.681	−0.376	0.479	1.224	2.272	3.093	3.932	4.783
2.6	−0.769	−0.762	−0.747	−0.696	−0.368	0.499	1.238	2.267	3.071	3.889	4.718
2.5	−0.799	−0.790	−0.771	−0.711	−0.360	0.518	1.250	2.262	3.048	3.845	4.652
2.4	−0.832	−0.819	−0.795	−0.725	−0.351	0.537	1.262	2.256	3.023	3.800	4.584
2.3	−0.867	−0.850	−0.819	−0.739	−0.341	0.555	1.274	2.248	2.997	3.753	4.515
2.2	−0.905	−0.882	−0.844	−0.752	−0.330	0.574	1.284	2.240	2.970	3.705	4.444
2.1	−0.946	−0.914	−0.869	−0.765	−0.319	0.592	1.294	2.230	2.942	3.656	4.372
2.0	−0.990	−0.949	−0.895	−0.777	−0.307	0.609	1.302	2.219	2.912	3.605	4.298
1.9	−1.037	−0.984	−0.920	−0.788	−0.294	0.627	1.310	2.207	2.881	3.553	4.223
1.8	−1.087	−1.020	−0.945	−0.799	−0.282	0.643	1.318	2.193	2.848	3.499	4.147
1.7	−1.140	−1.056	−0.970	−0.808	−0.268	0.660	1.324	2.179	2.815	3.444	4.069
1.6	−1.197	−1.093	−0.994	−0.817	−0.254	0.675	1.329	2.163	2.780	3.388	3.990
1.5	−1.256	−1.131	−1.018	−0.825	−0.240	0.690	1.333	2.146	2.743	3.330	3.910
1.4	−1.318	−1.168	−1.041	−0.832	−0.225	0.705	1.337	2.128	2.706	3.271	3.838
1.3	−1.383	−1.206	−1.064	−0.838	−0.210	0.719	1.339	2.108	2.666	3.211	3.745
1.2	−1.449	−1.243	−1.086	−0.844	−0.195	0.732	1.340	2.087	2.626	3.149	3.661

(Continued)

Table 10.4.1 (*Continued*)

Skew coeff.	Recurrence interval (yr)										
	1.0101	1.0526	1.1111	1.2500	2	5	10	25	50	100	200
	Exceedance probability										
	.99	.95	.90	.80	.50	.20	.10	.04	.02	.01	.005
1.1	−1.518	−1.280	−1.107	−0.848	−0.180	0.745	1.341	2.066	2.585	3.087	3.575
1.0	−1.588	−1.317	−1.128	−0.852	−0.164	0.758	1.340	2.043	2.542	3.022	3.489
0.9	−1.660	−1.353	−1.147	−0.854	−0.148	0.769	1.339	2.018	2.498	2.957	3.401
0.8	−1.733	−1.388	−1.166	−0.856	−0.132	0.780	1.336	1.993	2.453	2.891	3.312
0.7	−1.806	−1.423	−1.183	−0.857	−0.116	0.790	1.333	1.967	2.407	2.824	3.223
0.6	−1.880	−1.458	−1.200	−0.857	−0.099	0.800	1.328	1.939	2.359	2.755	3.132
0.5	−1.955	−1.491	−1.216	−0.856	−0.083	0.808	1.323	1.910	2.311	2.686	3.041
0.4	−2.029	−1.524	−1.231	−0.855	−0.066	0.816	1.317	1.880	2.261	2.615	2.949
0.3	−2.104	−1.555	−1.245	−0.853	−0.050	0.824	1.309	1.849	2.211	2.544	2.856
0.2	−2.178	−1.586	−1.258	−0.850	−0.033	0.830	1.301	1.818	2.159	2.472	2.763
0.1	−2.252	−1.616	−1.270	−0.846	−0.017	0.836	1.292	1.785	2.107	2.400	2.670
0.0	−2.326	−1.645	−1.282	−0.842	0	0.842	1.282	1.751	2.054	2.326	2.576
−0.1	−2.400	−1.673	−1.292	−0.836	0.017	0.846	1.270	1.716	2.000	2.252	2.482
−0.2	−2.472	−1.700	−1.301	−0.830	0.033	0.850	1.258	1.680	1.945	2.178	2.388
−0.3	−2.544	−1.726	−1.309	−0.824	0.050	0.853	1.245	1.643	1.890	2.104	2.294
−0.4	−2.615	−1.750	−1.317	−0.816	0.066	0.855	1.231	1.606	1.834	2.029	2.201
−0.5	−2.686	−1.774	−1.323	−0.808	0.083	0.856	1.216	1.567	1.777	1.955	2.108
−0.6	−2.755	−1.797	−1.328	−0.800	0.099	0.857	1.200	1.528	1.720	1.880	2.016
−0.7	−2.824	−1.819	−1.333	−0.790	0.116	0.857	1.183	1.488	1.663	1.806	1.929
−0.8	−2.891	−1.839	−1.336	−0.780	0.132	0.856	1.166	1.448	1.606	1.733	1.837
−0.9	−2.957	−1.858	−1.339	−0.769	0.148	0.854	1.147	1.407	1.549	1.660	1.749
−1.0	−3.022	−1.877	−1.340	−0.758	0.164	0.852	1.128	1.366	1.492	1.588	1.664
−1.1	−3.087	−1.894	−1.341	−0.745	0.180	0.848	1.107	1.324	1.435	1.518	1.581
−1.2	−3.149	−1.910	−1.340	−0.732	0.195	0.844	1.086	1.282	1.379	1.449	1.501
−1.3	−3.211	−1.925	−1.339	−0.719	0.210	0.838	1.064	1.240	1.324	1.383	1.424
−1.4	−3.271	−1.938	−1.337	−0.705	0.225	0.832	1.041	1.198	1.270	1.318	1.351
−1.5	−3.330	−1.951	−1.333	−0.690	0.240	0.825	1.018	1.157	1.217	1.256	1.282
−1.6	−3.388	−1.962	−1.329	−0.675	0.254	0.817	0.994	1.116	1.166	1.197	1.216
−1.7	−3.444	−1.972	−1.324	−0.660	0.268	0.808	0.970	1.075	1.116	1.140	1.155
−1.8	−3.499	−1.981	−1.318	−0.643	0.282	0.799	0.945	1.035	1.069	1.087	1.097
−1.9	−3.553	−1.989	−1.310	−0.627	0.294	0.788	0.920	0.996	1.023	1.037	1.044
−2.0	−3.605	−1.996	−1.302	−0.609	0.307	0.777	0.895	0.959	0.980	0.990	0.995
−2.1	−3.656	−2.001	−1.294	−0.592	0.319	0.765	0.869	0.923	0.939	0.946	0.949
−2.2	−3.705	−2.006	−1.284	−0.574	0.330	0.752	0.844	0.888	0.900	0.905	0.907
−2.3	−3.753	−2.009	−1.274	−0.555	0.341	0.739	0.819	0.855	0.864	0.867	0.869
−2.4	−3.800	−2.011	−1.262	−0.537	0.351	0.725	0.795	0.823	0.830	0.832	0.833
−2.5	−3.845	−2.012	−1.250	−0.518	0.360	0.711	0.771	0.793	0.798	0.799	0.800
−2.6	−3.889	−2.013	−1.238	−0.499	0.368	0.696	0.747	0.764	0.768	0.769	0.769
−2.7	−3.932	−2.012	−1.224	−0.479	0.376	0.681	0.724	0.738	0.740	0.740	0.741
−2.8	−3.973	−2.010	−1.210	−0.460	0.384	0.666	0.702	0.712	0.714	0.714	0.714
−2.9	−4.013	−2.007	−1.195	−0.440	0.390	0.651	0.681	0.683	0.689	0.690	0.690
−3.0	−4.051	−2.003	−1.180	−0.420	0.396	0.636	0.660	0.666	0.666	0.667	0.667

EXAMPLE 10.4.1

The mean, standard deviation, and skew of the log transformed discharges for the Medina River, Texas, are 3.639, 0.394, and 0.200, respectively, where the discharges are in ft³/s. Compute the 10-year and 100-year peak discharges.

SOLUTION

Assume that the peak discharges follow a log-Pearson Type III distribution and the procedures recommended by WRC will be used. First, we need to determine the frequency factors for 10-year and 100-year storm events from Table 10.4.1. To determine the frequency factors, we need to know the exceedance probability and the skew coefficient. Since the exceedance probability for a T-year storm event is $1/T$, the exceedance probabilities for 10-year and 100-year storm events are 0.1 and 0.01, respectively. Since the skew coefficient is given as $G_s = 0.200$, the frequency factors for the 10-year and 100-year storm events can be found from Table 10.4.1 as $K_{10} = 1.301$ and $K_{100} = 2.472$. Next, compute the 10-year and 100-year peak discharges:

$$\log Q_T = \bar{y} + K(T, G_s) S_y$$
$$\log Q_{10} = 3.639 + 1.301(0.394)$$

$Q_{10} = 14,180$ cfs

$$\log Q_{100} = 3.639 + 2.472(0.394)$$

$Q_{100} = 41,020$ cfs

EXAMPLE 10.4.2

The annual maximum series for the U.S.G.S. gauge on the Wichita River near Cheyenne, Oklahoma, is listed in Table 10.4.2. The objective is to compute the 25-year and 100-year peak discharges using the log-Pearson Type III distribution. Also compute the plotting position using the Weibull plotting position formula, $P(X > x_T) = 1/T = m/(n+1)$, where m is the rank of descending values and n is the number of peaks in the annual maximum series. (Adapted from Cudworth, 1989.)

SOLUTION

Step 1 Transform data using logarithms, computed as given in Table 10.4.2.

Step 2 Determine the mean, standard deviation, and skew of the log-transformed values:

$$\bar{y} = \frac{\sum y}{n} = \frac{156.876}{47} = 3.338$$

$$S_y = \sqrt{\frac{\sum y^2 - \frac{(\sum y)^2}{n}}{n-1}} = \sqrt{\frac{543.222 - \frac{(156.876)^2}{47}}{47-1}} = 0.653$$

$$G_s = \frac{n^2 \left(\sum y^3\right) - 3n \left(\sum y\right)\left(\sum y^2\right) + 2\left(\sum y\right)^3}{n(n-1)(n-2)\left(S_y^3\right)}$$

$$= \frac{(47)^2(1940.423) - (3)(47)(156.876)(543.222) + (2)(156.876)^3}{(47)(47-1)(47-2)(0.653)^3}$$

$$= -0.294$$

$$G_s \approx -0.3$$

Step 3 Determine frequency factors for the 25-year and the 100-year events using Table 10.4.1:

$K(25, -0.3) = 1.643$

$K(100, -0.3) = 2.104$

Table 10.4.2 Annual Peak Discharges for Each Year of Record for the Drainage Area Above Foss Dam

(1) Year	(2) Annual Peak Discharge, ft³/s	(3) Ranked Annual Peak Discharge, ft³/s	(4) Weibull Plotting Position	(5) Logarithm of Discharge y	(6) y^2	(7) y^3
1938	14,600	69,800	0.021	4.84386	23.46298	113.65139
1939	3070	40,000	0.042	4.60206	21.17896	97.46683
1940	1080	14,600	0.063	4.16435	17.34181	72.21737
1941	40,000	14,000	0.083	4.14613	17.19039	71.27359
1942	14,000	11,900	0.104	4.07555	16.61011	67.69533
1943	2190	9900	0.125	3.99563	15.96506	63.79047
1944	1240	8900	0.146	3.94939	15.59768	61.60133
1945	9900	8900	0.167	3.94939	15.59768	61.60133
1946	8900	8450	0.189	3.92686	15.42023	60.55308
1947	7100	7310	0.208	3.86392	14.92988	57.68785
1948	8900	7100	0.229	3.85126	14.83220	57.12267
1949	11,900	6420	0.250	3.80754	14.49736	55.19928
1950	8450	5830	0.271	3.76567	14.18027	53.39822
1951	5040	5040	0.292	3.70243	13.70799	50.75287
1952	465	4710	0.313	3.67302	13.49108	49.55299
1953	3550	4650	0.333	3.66839	13.45709	49.36584
1954	69,800	4470	0.354	3.65031	13.32476	48.63952
1955	5830	4210	0.375	3.62428	13.13541	47.60639
1956	3890	3890	0.396	3.58995	12.88774	46.26635
1957	4210	3550	0.417	3.55023	12.60413	44.74757
1958	1750	3070	0.438	3.48714	12.16015	42.40413
1959	6420	2990	0.458	3.47567	12.08028	41.98707
1960	1510	2930	0.479	3.46687	12.01919	41.66896
1961	7310	2280	0.500	3.35793	11.27569	37.86299
1962	2930	2190	0.521	3.34044	11.15854	37.27443
1963	574	1960	0.542	3.29226	10.83898	35.68473
1964	159	1800	0.563	3.25527	10.59678	34.49539
1965	1400	1750	0.583	3.24304	10.51731	34.10805
1966	1800	1510	0.604	3.17898	10.10591	32.12650
1967	2990	1420	0.625	3.15229	9.93693	31.32409
1968	4470	1400	0.646	3.14613	9.89813	31.14082
1969	2280	1360	0.667	3.13354	9.81907	30.76846
1970	734	1240	0.688	3.09342	9.56925	29.60170
1971	4710	1080	0.708	3.03342	9.20164	27.91243
1972	1360	1050	0.729	3.02119	9.12759	27.57618
1973	265	734	0.750	2.86570	8.21224	23.53381
1974	592	592	0.771	2.77232	7.68576	21.30738
1975	1050	574	0.792	2.75891	7.61158	20.99968
1976	1960	560	0.813	2.74819	7.55255	20.75584
1977	4660	465	0.833	2.66745	7.11529	18.97968
1978	297	427	0.854	2.63043	6.91916	18.20037
1979	400	400	0.875	2.60206	6.77072	17.61781
1980	560	297	0.896	2.47276	6.11454	15.11980
1981	38	265	0.917	2.42325	5.87214	14.22967
1982	1420	159	0.938	2.20140	4.84616	10.66834
1983	427	119	0.958	2.07555	4.30791	8.94128
1984	119	38	0.979	1.57978	2.49570	3.94266
			Totals	156.87558	543.2220	1940.4225

Source: Cudworth (1989).

Step 4 Apply frequency factor equations to determine Q_{25} and Q_{100}

$$\log Q_{25} = \bar{y} + K(25, -0.3)S_y$$

$$= 3.338 + 1.643(0.653)$$

$$= 4.411$$

$$Q_{25} = \text{antilog}\ (4.411) = 25{,}765\ \text{cfs}$$

$$\log Q_{100} = \bar{y} + K(100, -0.3)S_y$$

$$= 3.338 + 2.104(0.653)$$

$$= 4.712$$

$$Q_{100} = \text{antilog}\ (4.712) = 51{,}525\ \text{cfs}$$

The annual peak discharges are ranked in descending order in column (3) of Table 10.4.2 and the Weibull plotting positions are listed in column (4) for the corresponding discharge in column (3). As an example, the largest discharge has a plotting position of $m = 1$, so $1/T = 1/(47 + 1) = 0.021$. The second largest, $m = 2$, has a Weibull plotting position of $2/(47 + 1) = 0.042$.

10.4.3 Extreme Value Distribution

For the extreme value type I (Gumbel) distribution, the frequency factor can be derived by substituting the frequency equation (10.4.1), $x_T = \bar{x} + K_T S_x$ into the cumulative distribution function, equation (10.2.13), $F(x_T) = \exp\{-\exp[-(x_T - \beta)/\alpha]\}$ so that

$$F(x) = \exp\{-\exp[-(\bar{x} + K_T S_x - \beta)/\alpha]\} = 1 - 1/T \qquad (10.4.5)$$

Substitute the parameters α and β (equations (10.2.14) and (10.2.15)) into equation (10.4.5) and solve for K_T in terms of the return period T,

$$K_T = -(6^{0.5}/\pi)\{0.5772 + \ln[\ln(T/(T-1))]\} \qquad (10.4.6)$$

The return period T can also be expressed in terms of K_T as

$$T = \{1 - \exp[-\exp(-0.5772 - \pi K_T/6^{0.5})]\}^{-1} \qquad (10.4.7)$$

EXAMPLE 10.4.3

Following example 10.2.2, the exceedance series of 30-min duration rainfall values has a mean of 1.09 in and standard deviation of 0.343 in. Determine the 100-year 30-min duration rainfall value using the Gumbel (extreme value type I) distribution with the frequency factor method.

SOLUTION

First solve for the frequency factor using equation (10.4.6), $K_T = -(6^{0.5}/\pi)\{0.5772 + \ln[\ln(T/(T-1))]\}$ with $T = 100$, giving $K_T = 3.138$. Using the frequency factor equation (10.4.1),

$$x_T = \bar{x} + K_T S_x = 1.09 + 3.138(0.343) = 2.166\ \text{in}$$

The 100-year 30-min duration rainfall is 2.166 in.

10.5 U.S. WATER RESOURCES COUNCIL GUIDELINES FOR FLOOD FLOW FREQUENCY ANALYSIS

Flood flow frequency analysis is another method of discharge determination using statistical methods when gauged data are available to develop annual maximum series. Section 10.4 introduced hydrologic frequency analysis for flood flow. This section extends the concepts to the so-called *U.S. Water Resources Council (WRC) method*.

10.5.1 Procedure

The skew coefficient is very sensitive to the size of the sample; thus, it is difficult to obtain an accurate estimate from small samples. Because of this, the U.S. Water Resources Council (1981) recommended using a generalized estimate of the skew coefficient when estimating the skew for short records. As the length of record increases, the skew is usually more reliable. The guidelines recommend the use of a *weighted skew* G_w, based upon the equation

$$G_w = WG_s + (1 - W)G_m \qquad (10.5.1)$$

where W is a weight, G_s is the skew coefficient computed using the sample data, and G_m is a map skew, values of which for the United States are found in Figure 10.5.1. The generalized skew is derived as a weighted average between skew coefficients computed from sample data (sample skew) and regional or map skew coefficients (referred to as a generalized skew in U.S. Water Resources Council (1981).

A weighting procedure was derived that is a function of the variance of the sample skew and the variance of the map skew. Such a procedure considers the uncertainty of deriving skew coefficients from both sample data and regional or map values to obtain a generalized skew that minimizes uncertainty based upon information known.

The estimates of the sample skew coefficient and the map skew coefficient in equation (10.5.1) are assumed to be independent with the same mean and respective variances. Assuming independence of G_s and G_m, the variance (mean square error) of the weighted skew, $V(G_w)$, can be expressed as

$$V(G_w) = W^2 \cdot V(G_s) + (1 - W)^2 \cdot V(G_m) \qquad (10.5.2)$$

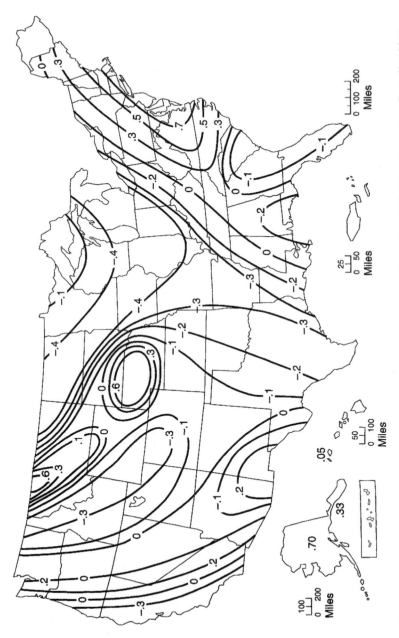

Figure 10.5.1 Generalized skew coefficients of annual maximum streamflow (from U.S. Water Resources Council (1981)).

where $V(G_s)$ is the variance of the sample skew and $V(G_m)$ is the variance of the map skew. The skew weight that minimizes the variance of the weighted skew can be determined by differentiating equation (10.5.2) with respect to W and solving $d[V(G_w)]/dW = 0$ for W to obtain

$$W = \frac{V(G_m)}{V(G_s) + V(G_m)} \qquad (10.5.3)$$

Equation (10.5.3) is a convex function and the second derivative of equation (10.5.2) is greater than 0, proving that equation (10.5.3) gives the value of W that minimizes the weighted skew.

Determination of W using equation (10.5.3) requires the values of $V(G_s)$ and $V(G_m)$. $V(G_m)$ can be estimated from the map of the skew coefficients as the squared value of the standard deviation of station values of skew coefficients about the isolines of the skew map. The value of $V(G_m)$, estimated for the skew map in U.S. Water Resources Council (1981), is 0.3025. Alternatively, $V(G_m)$ could be derived from a regression study relating the skew to physiographical and meteorological characteristics of the basins and determining $V(G_m)$ as the square of the standard error of the regression equation (Tung and Mays, 1981).

The weighted skew G_w can be determined by substituting equation (10.5.3) into equation (10.5.1), resulting in

$$G_w = \frac{V(G_m) \cdot G_s + V(G_s) \cdot G_m}{V(G_m) + V(G_s)} \qquad (10.5.4)$$

The variance (mean square error) of the station skew for log-Pearson Type III random variables can be obtained from the results of Monte Carlo experiments by Wallis et al. (1974). Their results showed that $V(G_s)$ of the logarithmic station skew is a function of record length and population skew. For use in calculating G_w, this function $V(G_s)$ can be approximated with sufficient accuracy using

$$V(G_s) = 10^{[A - B \log_{10}(n/10)]} \qquad (10.5.5)$$

where

$$A = -0.33 + 0.08 \, |G_s| \quad \text{if} \quad |G_s| \le 0.90 \qquad (10.5.6a)$$

$$A = -0.52 + 0.30 \, |G_s| \quad \text{if} \quad |G_s| > 0.90 \qquad (10.5.6b)$$

$$B = 0.94 - 0.26 \, |G_s| \quad \text{if} \quad |G_s| \le 1.50 \qquad (10.5.7a)$$

$$B = 0.55 \quad \text{if} \quad |G_s| > 1.50 \qquad (10.5.7b)$$

in which $|G_s|$ is the absolute value of the station skew (used as an estimate of population skew) and n is the record length in years. The same steps used in section 10.4 can be used to compute the discharge Q_T of return period T.

10.5.2 Testing for Outliers

The WRC method recommends that outliers be identified and adjusted according to their recommended methods. *Outliers* are data points that depart significantly from the trend of the remaining data. The retention or deletion of these outliers can significantly affect the magnitude of statistical parameters computed from the data, especially for small samples. Procedures for treating outliers require judgment involving both mathematical and hydrologic considerations. According to the U.S. Water Resources Council (1981), if the station skew is greater than +0.4, tests for high outliers are considered first. If the station skew is less than −0.4, tests for low outliers are considered first. Where the station skew is between ±0.4, tests for both high and low outliers should be applied before eliminating any outliers from the data set.

Table 10.5.1 Outlier Test K Values: 10 Percent Significance Level K Values

Sample size	K value	Sample size	K value	Sample size	K value	Sample size	K value
10	2.036	45	2.727	80	2.940	115	3.064
11	2.088	46	3.736	81	2.945	116	3.067
12	2.134	47	2.744	82	2.949	117	3.070
13	2.175	48	2.753	83	2.953	118	3.073
14	2.213	49	2.760	84	2.957	119	3.075
15	2.247	50	2.768	85	2.961	120	3.078
16	2.279	51	2.775	86	2.966	121	3.081
17	2.309	52	2.783	87	2.970	122	3.083
18	2.335	53	2.790	88	2.973	123	3.086
19	2.361	54	2.798	89	2.977	124	3.089
20	2.385	55	2.804	90	2.981	125	3.092
21	2.408	56	2.811	91	2.984	126	3.095
22	2.429	57	2.818	92	2.989	127	3.097
23	2.448	58	2.824	93	2.993	128	3.100
24	2.467	59	2.831	94	2.996	129	3.102
25	2.486	60	2.837	95	3.000	130	3.104
26	2.502	61	2.842	96	3.003	131	3.107
27	2.519	62	2.849	97	3.006	132	3.109
28	2.534	63	2.854	98	3.011	133	3.112
29	2.549	64	2.860	99	3.014	134	3.114
30	2.563	65	2.866	100	3.017	135	3.116
31	2.577	66	2.871	101	3.021	136	3.119
32	2.591	67	2.877	102	3.024	137	3.122
33	2.604	68	2.883	103	3.027	138	3.124
34	2.616	69	2.888	104	3.030	139	3.126
35	2.628	70	2.893	105	3.033	140	3.129
36	2.639	71	2.897	106	3.037	141	3.131
37	2.650	72	2.903	107	3.040	142	3.133
38	2.661	73	2.908	108	3.043	143	3.135
39	2.671	74	2.912	109	3.046	144	3.138
40	2.682	75	2.917	110	3.049	145	3.140
41	2.692	76	2.922	111	3.052	146	3.142
42	2.700	77	2.927	112	3.055	147	3.144
43	2.710	78	2.931	113	3.058	148	3.146
44	2.719	79	2.935	114	3.061	149	3.148

This table contains one-sided 10 percent significance level K values for a normal distribution.

Source: U.S. Water Resources Council (1981).

The following frequency equation can be used to detect high outliers:

$$y_H = \bar{y} + K_n S_y \qquad (10.5.8)$$

where y_H is the high outlier threshold in log units and K_N is the K value from Table 10.5.1 for sample size N. If the logarithms of peaks in the sample are greater than y_H in the above equation, then they are considered high outliers. Flood peaks considered high outliers should be compared with historic flood data and flood information at nearby sites. According to the U.S. Water Resources Council (1981), if information is available indicating that a high outlier(s) is the maximum in an extended period of time, the outlier(s) is treated as historic flood data. If useful historic information is not available to adjust for high outliers, then the outliers should be retained as part of the systematic record.

The following frequency equation can be used to detect low outliers:

$$y_L = \bar{y} - K_n S_y \qquad (10.5.9)$$

where y_L is the low outlier threshold in log units. Flood peaks considered low outliers are deleted from the record and a conditional probability adjustment described in the U.S. Water Resources Council (1981) is applied. Use of the K values in Table 10.5.1 is equivalent to a one-sided test that detects outliers at the 10 percent level of significance. The K values are based on a normal distribution for detection of single outliers.

EXAMPLE 10.5.1

Using the data for example 10.4.2 (annual maximum series for the gauge on the Wichita River near Cheyenne, Oklahoma), compute the 25-year and the 100-year peak discharges using the U.S. Water Resource Council (1981) guidelines. The map skew for this location is −0.015.

SOLUTION

Steps 1–2 Transform data using logarithms and compute statistics as in example 10.4.2, where $\bar{y} = 3.338$, $S_y = 0.653$, and $G_s = -0.294 \approx -0.3$.

Step 3 Compute the weighted skew coefficient, G_w.

Step 3a Compute A and B using equations (10.5.6a) and (10.5.7a):

$$A = -0.33 + 0.08|-0.294| = -0.306$$

$$B = 0.94 - 0.26|-0.294| = 0.864$$

Step 3b Compute $V(G_s)$.

$$V(G_s) = 10^{A - B\log(n/10)} = 10^{-0.306 - 0.864\log(47/10)} = 0.130$$

Step 3c Use equation (10.5.4) to compute the weighted skew coefficient using $V(G_m) = 0.302$ (as estimated in U.S. Water Resources Council (1981)):

$$G_w = \frac{0.302(-0.294) + 0.130(-0.015)}{0.302 + 0.130} = -0.210$$

Step 4 Use Table 10.4.1 to obtain the frequency factors using the weighted skew:

$$K(25, -0.210) = 1.676$$

$$K(100, -0.210) = 2.171$$

Step 5 Apply the frequency factor equation to determine Q_{25} and Q_{100}:

$$\log Q_{25} = \bar{y} + K(25, -0.251)S_y$$

$$= 3.338 + 1.676(0.653) = 4.432$$

$$Q_{25} = 27,070 \text{ cfs}$$

$$\log Q_{100} = 3.338 + 2.171(0.653) = 4.756$$

$$Q_{100} = 56,975 \text{ cfs}$$

Note that in example 10.4.2 we computed $Q_{25} = 25,765$ cfs and $Q_{100} = 51,525$ cfs when the sample skew was used.

EXAMPLE 10.5.2

Using the data in examples 10.4.2 and 10.5.1, determine whether any outliers exist in the annual maximum series (Table 10.4.2).

SOLUTION

Determine whether the data include any high or low outliers. First, consider high outliers. From Table 10.5.1, $K_n = 2.744$ for $n = 47$, and using equation (10.5.8) to determine the threshold value y_H, we find

$$y_H = \bar{y} + K_n S_y = 3.338 + 2.744(0.653) = 5.130$$

so

$$Q_H = 10^{y_H} = (10)^{5.130} = 134,900 \text{ cfs}$$

The largest recorded value of 69,800 cfs for the year 1954 in Table 10.4.2 does not exceed the threshold value of 134,900 cfs.

Next consider low outliers and use equation (10.5.9) to determine the threshold:

$$y_L = \bar{y} + K_N S_y = 3.338 - 2.744(0.653) = 1.546$$

so

$$Q_L = 10^{y_L} = (10)^{1.546} = 35 \text{ cfs}$$

The smallest recorded peak, 38 cfs for 1963, is larger than 35 cfs, so there are no low outliers by this methodology.

10.6   ANALYSIS OF UNCERTAINTIES

In the design and analysis of hydrosystems, many quantities of interest are functionally related to a number of variables, some of which are subject to uncertainty. For example, hydraulic engineers frequently apply weir flow equations such as $Q = CLH^{1.5}$ to estimate spillway capacity in which the coefficient C and head H are subject to uncertainty. As a result, discharge over the spillway is not certain. A rather straightforward and useful technique for the approximation of such uncertainties is the *first-order analysis of uncertainties*, sometimes called the *delta method*.

The use of the first-order analysis of uncertainties is quite popular in many fields of engineering because of its relative ease in application to a wide array of problems. First-order analysis is used to estimate the uncertainty in a deterministic model formulation involving parameters that are uncertain (not known with certainty). More specifically, first-order analysis enables one to estimate the mean and variance of a random variable that is functionally related to several other variables, some of which are random. By using first-order analysis, the combined effect of uncertainty in a model formulation, as well as the use of uncertain parameters, can be assessed.

Consider a random variable y that is a function of k random variables (multivariate case):

$$y = g(x_1, x_2, \ldots, x_k) \tag{10.6.1}$$

This can be a deterministic equation such as the weir equation mentioned above, or the rational formula or Manning's equation; or this function can be a complex model that must be solved on a computer. The objective is to treat a deterministic model that has uncertain inputs in order to determine the effect of the uncertain parameters $x_1, \ldots, x_k$ on the model output y.

Equation (10.6.1) can be expressed as $y = g(\mathbf{x})$ where $\mathbf{x} = x_1, x_2, \ldots, x_k$. Through a Taylor series expansion about k random variables, ignoring the second- and higher-order terms, we get

$$y \approx g(\bar{\mathbf{x}}) + \sum_{i=1}^{k} \left[\frac{\partial g}{\partial x_i}\right]_{\bar{\mathbf{x}}} (X_i - \bar{x}_i) \tag{10.6.2}$$

The derivation $[\partial g / \partial x_i]_{\bar{\mathbf{x}}}$ are the *sensitivity coefficients* that represent the rate of change of the function value $g(\mathbf{x})$ at $\mathbf{x} = \bar{\mathbf{x}}$.

Assuming that the k random variables are independent, then the variance of y is approximated as

$$\sigma_y^2 = \text{Var}[y] = \sum a_i^2 \sigma_{x_i}^2 \tag{10.6.3}$$

and the coefficient of variation is Ω_y:

$$\Omega_y = \left[\sum_{\tau=1}^{k} a_i^2 \left(\frac{x_i}{\mu_y} \right)^2 \Omega_{x_i}^2 \right]^{1/2} \qquad (10.6.4)$$

where $a_i = (\partial g/\partial x)_{\bar{x}}$. Refer to Mays and Tung (1992) for a detailed derivation of equations (10.6.3) and (10.6.4).

EXAMPLE 10.6.1

Apply the first-order analysis to the rational equation $Q = CiA$, in which C is the runoff coefficient, i is the rainfall intensity in in/m, and A is the drainage area in acres, to determine formulas for σ_Q and Ω_Q. Consider C, i, and A to be uncertain.

SOLUTION

The first-order approximation of Q is determined using equation (10.6.2), so that

$$Q \approx \bar{Q} + \left[\frac{\partial Q}{\partial C} \right]_{\bar{C}, \bar{i}, \bar{A}} (C - \bar{C}) + \left[\frac{\partial Q}{\partial i} \right]_{\bar{C}, \bar{i}, \bar{A}} (i - \bar{i}) + \left[\frac{\partial Q}{\partial A} \right]_{\bar{C}, \bar{i}, \bar{A}} (A - \bar{A})$$

For $C = \bar{C}$, $i = \bar{i}$, and $A = \bar{A}$, $\bar{Q} = \bar{C}\bar{i}\bar{A}$. The variance is computed using equation (10.6.3):

$$\sigma_Q^2 = \left[\frac{\partial Q}{\partial C} \right]^2_{\bar{C}, \bar{i}, \bar{A}} \sigma_C^2 + \left[\frac{\partial Q}{\partial i} \right]^2_{\bar{C}, \bar{i}, \bar{A}} \sigma_i^2 + \left[\frac{\partial Q}{\partial A} \right]^2_{\bar{C}, \bar{i}, \bar{A}} \sigma_A^2$$

$$\sigma_Q = \left\{ \left[\frac{\partial Q}{\partial C} \right]^2_{\bar{C}, \bar{i}, \bar{A}} \sigma_C^2 + \left[\frac{\partial Q}{\partial i} \right]^2_{\bar{C}, \bar{i}, \bar{A}} \sigma_i^2 + \left[\frac{\partial Q}{\partial A} \right]^2_{\bar{C}, \bar{i}, \bar{A}} \sigma_A^2 \right\}^{1/2}$$

The coefficient of variation is computed using equation (10.6.4):

$$\Omega_Q^2 = \left[\frac{\partial Q}{\partial C} \right]^2 \left(\frac{\bar{C}}{\bar{Q}} \right)^2 \Omega_C^2 + \left[\frac{\partial Q}{\partial i} \right]^2 \left(\frac{\bar{i}}{\bar{Q}} \right)^2 \Omega_i^2 + \left[\frac{\partial Q}{\partial A} \right]^2 \left(\frac{\bar{A}}{\bar{Q}} \right)^2 \Omega_A^2$$

$$= (\bar{i}\bar{A})^2 \left(\frac{\bar{C}}{\bar{Q}} \right)^2 \Omega_C^2 + (\bar{C}\bar{A})^2 \left(\frac{\bar{i}}{\bar{Q}} \right)^2 \Omega_i^2 + (\bar{C}\bar{i})^2 \left(\frac{\bar{A}}{\bar{Q}} \right)^2 \Omega_A^2$$

$$= \Omega_C^2 + \Omega_i^2 + \Omega_A^2$$

$$\Omega_Q = \left[\Omega_C^2 + \Omega_i^2 + \Omega_A^2 \right]^{1/2}$$

EXAMPLE 10.6.2

Determine the mean, coefficient of variation, and standard deviation of the runoff using the rational equation with the following parameter values:

Parameter	Mean	Coefficient of variation
C	0.8	0.09
i	100 mm/h	0.5
A	0.1 km²	0.005

SOLUTION

$Q = KCiA$, where $K = 0.28$ for SI units for i in mm/hr and A in km².

Using $\bar{Q} = 0.28 \bar{C}\bar{i}\bar{A}$ from example (10.6.1), we find

$\bar{Q} = 0.28(0.8)(100)(0.1) = 2.24 \, \text{m}^3/\text{s}$

Using $\Omega_Q = [\Omega_C^2 + \Omega_i^2 + \Omega_A^2]^{1/2}$ from example 10.6.1, we find

$$\Omega_Q = [0.09^2 + 0.5^2 + 0.005^2]^{1/2} = 0.508$$

The standard deviation of Q can be determined using

$$\sigma_Q = \bar{Q}\,\Omega_Q = 2.24(0.508) = 1.138 \ \text{m}^3/\text{s}$$

Apply the first-order analysis to Manning's equation for full pipe flow, given in U.S. customary units as

$$Q = \frac{0.463}{n} S^{1/2} D^{8/3}$$

to determine equations for computing σ_Q and Ω_Q. Consider the diameter D to be deterministic without any uncertainty, and consider n and S to be uncertain.

SOLUTION

Since n and S are uncertain, Manning's equation can be rewritten as

$$Q = Kn^{-1}S^{1/2}$$

where $K = 0.463D^{8/3}$. The first-order approximation of Q is determined using equation (10.6.2), so that

$$Q \approx \bar{Q} + \left[\frac{\partial Q}{\partial n}\right]_{\bar{n},\bar{S}} (n - \bar{n}) + \left[\frac{\partial Q}{\partial S}\right]_{\bar{n},\bar{S}} (S - \bar{S})$$

$$= \bar{Q} + \left[-K\bar{n}^{-2}\bar{S}^{1/2}\right](n - \bar{n}) + \left[0.5K\bar{n}^{-1}\bar{S}^{-1/2}\right](S - \bar{S})$$

where $\bar{Q} = K\bar{n}^{-1}\bar{S}^{-1/2}$.

Next compute the variance of the pipe capacity using equation (10.6.3):

$$\sigma_Q^2 = \left[\frac{\partial Q}{\partial n}\right]_{(\bar{n},\bar{S})}^2 \sigma_n^2 + \left[\frac{\partial Q}{\partial S}\right]_{(\bar{n},\bar{S})}^2 \sigma_S^2$$

$$\sigma_Q = \left\{\left[\frac{\partial Q}{\partial n}\right]_{(\bar{n},\bar{S})}^2 \sigma_n^2 + \left[\frac{\partial Q}{\partial S}\right]_{(\bar{n},\bar{S})}^2 \sigma_S^2\right\}^{1/2}$$

Determine the coefficient of variation of Q using equation (10.6.4):

$$\Omega_Q^2 = \sum_{i=1}^{2} \left[\frac{\partial Q}{\partial x_i}\right]^2 \left[\frac{\bar{x}_i}{\bar{Q}}\right]^2 \Omega_{x_i}^2$$

$$= \left[\frac{\partial Q}{\partial n}\right]^2 \left[\frac{\bar{n}}{\bar{Q}}\right]^2 \Omega_n^2 + \left[\frac{\partial Q}{\partial S}\right]^2 \left[\frac{\bar{S}}{\bar{Q}}\right]^2 \Omega_S^2$$

$$= \left[\frac{-K\bar{S}^{1/2}}{\bar{n}^2}\right]^2 \left[\frac{\bar{n}}{\bar{Q}}\right]^2 \Omega_n^2 + \left[\frac{0.5K}{\bar{n}\bar{S}^{1/2}}\right]^2 \left[\frac{\bar{S}}{\bar{Q}}\right]^2 \Omega_S^2$$

$$= \left[\frac{-K\bar{S}^{1/2}}{\bar{Q}}\right]^2 \left[\frac{1}{\bar{n}^2}\right] \Omega_n^2 + \left[0.5\right]^2 \left[\frac{K}{\bar{n}\bar{S}^{1/2}}\right]^2 \left[\frac{\bar{S}}{\bar{Q}}\right]^2 \Omega_S^2$$

$$= \left[\frac{\bar{n}}{\bar{n}^2}\right] \left[\frac{1}{\bar{n}^2}\right] \Omega_n^2 + 0.25 \left[\frac{1}{\bar{S}}\right]^2 [\bar{S}]^2 \Omega_S^2$$

$$= \Omega_n^2 + 0.25\Omega_S^2$$

$$\Omega_Q = [\Omega_n^2 + 0.25\Omega_S^2]^{1/2}$$

Determine the mean capacity of a storm sewer pipe, the coefficient of variation of the pipe capacity, and the standard deviation of the pipe capacity using Manning's equation for full pipe flow. Refer to example 10.6.3. The following parameter values are to be considered:

Parameter	Mean	Coefficient of variation
n	0.015	0.01
D	1.5 m	0
S	0.001	0.05

SOLUTION

Manning's equation in SI units for full pipe flow is

$$Q = \frac{0.311}{n} S^{1/2} D^{8/3}$$

so for first-order analysis we have

$$\bar{Q} = \frac{0.311}{\bar{n}} S^{1/2} D^{8/3} = \frac{0.311}{0.015} (0.001)^{1/2} (1.5)^{8/3}$$

$$= 1.93 \, \text{m}^3/\text{s}$$

Using example 10.6.3, we find

$$\Omega_Q = \left[(0.01)^2 + 0.25(0.05)^2 \right]^{1/2} = 0.027$$

$$\sigma_Q = \bar{Q}\Omega_Q = 1.93(0.027)$$

$$= 0.052 \, \text{m}^3/\text{s}$$

10.7   RISK ANALYSIS: COMPOSITE HYDROLOGIC AND HYDRAULIC RISK

The *resistance* or strength of a component is defined as the ability of the component to fulfill its required purpose satisfactorily without a failure when subjected to an external stress. *Stress* is the loading of the component, which may be a mechanical load, an environmental exposure, a flow rate, temperature fluctuation, etc. The stress or loading tends to cause failure of the component. When the strength of the component is less than the stress imposed on it, failure occurs. This type of analysis can be applied to the reliability analysis of components of various hydraulic systems. The *reliability of a hydraulic system* is defined as the probability of the resistance to exceed the loading, i.e., the probability of survival. The terms "stress" and "strength" are more meaningful to structural engineers, whereas the terms "loading" and "resistance" are more descriptive to water resources engineers. The *risk of a hydraulic component*, subsystem, or system is defined as the probability of the loading exceeding the resistance, i.e., the *probability of failure*. The mathematical representation of the reliability R can be expressed as

$$R = P(r > \ell) = P(r - \ell > 0) \tag{10.7.1}$$

where $P()$ refers to probability, r is the resistance, and ℓ is the loading. The relationship between reliability R and risk $\bar{R}$ is

$$R = 1 - \bar{R} \tag{10.7.2}$$

The resistance of a hydraulic system is essentially the flow-carrying capacity of the system, and the loading is essentially the magnitude of flows through or pressure imposed on the system by demands. Since the loading and resistance are random variables due to the various hydraulic and demand uncertainties, a knowledge of the probability distributions of r and ℓ is required to develop

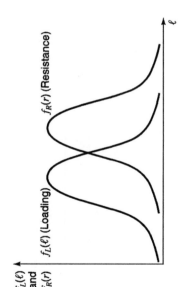

Figure 10.7.1 Load-resistance interference reliability analysis.

reliability models. The computation of risk and reliability can be referred to as *loading-resistance interference*. Probability distributions for loading and resistance are illustrated in Figure 10.7.1. The *reliability* is the probability that the resistance is greater than the loading for all possible values of the loading.

The word "static," from the reliability computation point of view, represents the worst single stress, or load, applied. Actually, the loading applied to many hydraulic systems is a random variable. Also, the number of times a loading is imposed is random.

10.7.1 Reliability Computation by Direct Integration

Following the reliability definition given in equation (10.7.1), the reliability can be expressed as

$$R = \int_0^\infty f_R(r) \left[\int_0^r f_L(\ell)d\ell \right] dr = \int_0^\infty f_R(r) F_L(r) dr \qquad (10.7.3)$$

in which $f_R()$ and $f_L()$ represent the probability density functions of resistance and loading, respectively. The reliability computations require the knowledge of the probability distributions of loading and resistance. A schematic diagram of the reliability computation by equation (10.7.3) is shown in Figure 10.7.2.

To illustrate the computation procedure involved, we consider that the loading ℓ and the resistance r are distributed exponentially, i.e.,

$$f_L(\ell) = \lambda_\ell e^{-\lambda_\ell \ell}, \quad \ell \geq 0 \qquad (10.7.4)$$

$$f_R(r) = \lambda_r e^{\lambda_r r}, \quad r \geq 0 \qquad (10.7.5)$$

Then the static reliability can be derived by applying equation (10.7.3) in a straightforward manner as

$$R = \int_0^\infty \lambda_r e^{-\lambda_r r} \left[\int_0^r \lambda_\ell e^{-\lambda_\ell \ell} d\ell \right] dr = \int_0^\infty \lambda_r e^{-\lambda_r r} \left[1 - e^{-\lambda_\ell r} \right] dr = \frac{\lambda_\ell}{\lambda_r + \lambda_\ell} \qquad (10.7.6)$$

For some special combinations of load and resistance distributions, the static reliability can be derived analytically in the closed form. In cases in which both the loading ℓ and resistance r are log-normally distributed, the reliability can be computed as (Kapur and Lamberson, 1977)

$$R = \int_{-\infty}^\infty \phi(z) dz = \Phi(z) \qquad (10.7.7)$$

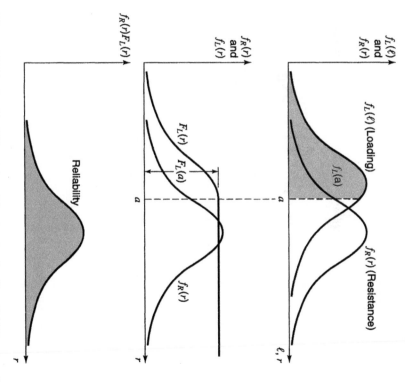

Figure 10.7.2 Graphical illustration of the steps involved in reliability computation, by equation (10.7.3).

where $\phi(z)$ and $\Phi(z)$ are the probability density function and the cumulative distribution function, respectively, for the standard normal variate z given as

$$z = \frac{\mu_{\ln r'} - \mu_{\ln \ell}}{\sqrt{\sigma_{\ln r'}^2 + \sigma_{\ln \ell}^2}} \qquad (10.7.8)$$

The values of the cumulative distribution function $\Phi(z)$ for the standard normal variate are given in Table 10.2.1.

In cases in which the loading ℓ is exponentially distributed and the resistance is normally distributed, the reliability can be expressed as (Kapur and Lamberson, 1977)

$$R = 1 - \Phi\left(\frac{\mu_r}{\sigma_r}\right) - \exp\left[-\frac{1}{2}\left(2\mu_r\lambda_\ell - \lambda_\ell^2\sigma_r^2\right)\right] \times \left[1 - \Phi\left(-\frac{\mu_r - \lambda_\ell\sigma_r^2}{\sigma_r}\right)\right] \qquad (10.7.9)$$

EXAMPLE 10.7.1

Consider a water distribution system (see Figure 10.7.3) consisting of a storage tank serving as the source and a 2-ft diameter cast-iron pipe 1 mi long, leading to a user. The head elevation at the source is maintained at a constant height of 100 ft above the elevation at the user end. It is also known that, at the user end, the required pressure head is fixed at 20 psi with variable demand on flow rate. Assume that the demand in flow rate is random, having a log-normal distribution with mean 1 cfs and standard deviation 0.3 cfs. Because of the uncertainty in pipe roughness, the supply to the user is not certain. We know that the

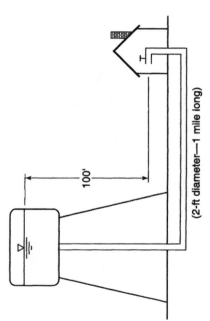

(2-ft diameter—1 mile long)

Figure 10.7.3 Example system.

pipe has been installed for about three years. Therefore, our estimation of the pipe roughness in the Hazen–Williams equation is about 130, with some errors of ±20. Again, we further assume that the Hazen–Williams C coefficient has a log-normal distribution with a mean of 130 and a standard deviation of 20. It is required to estimate the reliability with which the water demand by the user will be satisfied.

SOLUTION

In this example, the resistance of the system is the water supply from the source, while the loading is the water demand by the user. Both supply and demand are random variables. By the Hazen–Williams equation, the supply is calculated as

$$r = Q_s = \frac{C}{149.2} \left(\frac{\Delta h}{L}\right)^{0.54} D^{2.63}$$

where Δh is the head difference (in feet) between the source and the user, D is the pipe diameter in feet, and L is the pipe length in feet. Because the roughness coefficient C is a random variable, so is the supply. Due to the multiplicative form of the Hazen–Williams equation, the logarithmic transformation leads to a linear relation among variables, i.e.,

$$\ln Y = \ln C - \ln(149.2) + 0.54 \ln\left[\frac{100 - \dfrac{(20)(144)}{62.4}}{5280}\right] + 2.63 \ln(2)$$

$$= \ln C - 5.659$$

Assume that the roughness coefficient C is log-normally distributed; then $\ln C$ is normally distributed, as is the log-transformed water supply (resistance). From the moment relations given in Table 10.1.2 for log-normal distribution, the mean and the standard deviation of $\ln C$ are determined as follows. From Table 10.1.2,

$$\mu_{\ln r} = \mu_{\ln C} = \ln C - \sigma_{\ln r}^2/2$$

where $\sigma_{\ln r}^2 = \sqrt{\ln(1 + \Omega_C^2)} = \sqrt{\ln[1 + (20/130)]} = 0.153$. Thus, $\mu_{\ln r} = \ln 130 - (0.153^2/2) = 4.856$. From these results, the mean and the standard deviation of $\ln Y$ are -0.803 and 0.153, respectively.

Because the water demand (loading) has a log-normal distribution, the mean and standard deviation of its log-transformed scale can be calculated in the same manner as for roughness coefficient C. That is,

$$\sigma_{\ln \ell} = \sqrt{\ln\left[1 + (0.3/1)^2\right]} = 0.294 \text{ and } \mu_{\ln \ell} = 1 - (0.294^2/2) = -0.043.$$

Knowing the distributions and statistical properties of the load (water demand) and resistance (water supply), both log-normal in this example, we can calculate the reliability of the system by equation (10.7.7) as

$$R = \Phi\left[z = \frac{-0.803 - (-0.043)}{\sqrt{0.153^2 + 0.294^2}}\right] = \Phi(z = -2.293) = 1 - \Phi(2.293) = 0.013$$

This means that the water demanded by the user will be met with a probability of 1.3 percent.

10.7.2   Reliability Computation Using Safety Margin/Safety Factor

Safety Margin

The *safety margin* (SM) is defined as the difference between the project capacity (resistance) and the value calculated for the design loading, $SM = r - \ell$. The reliability is equal to the probability that $r > \ell$, or equivalently,

$$R = p(r - \ell > 0) = P(SM > 0) \tag{10.7.10}$$

If r and ℓ are independent random variables, then the mean value of SM is given by

$$\mu_{SM} = \mu_r - \mu_\ell \tag{10.7.11}$$

and its variance by

$$\sigma_{SM}^2 = \sigma_r^2 + \sigma_\ell^2 \tag{10.7.12}$$

If the safety margin is normally distributed, then

$$z = (SM - \mu_{SM})/\sigma_{SM} \tag{10.7.13}$$

is a standard normal variate z. By subtracting μ_{SM} from both sides of the inequality in equation (10.7.10) and dividing both sides by σ_{SM}, it can be seen that

$$R = P\left(z > \frac{\mu_{SM}}{\sigma_{SM}}\right) = \Phi\left(\frac{\mu_{SM}}{\sigma_{SM}}\right) \tag{10.7.14}$$

The key assumption of this analysis is that it considers that the safety margin is normally distributed but does not specify the distributions of loading and capacity. Ang (1973) indicated that, provided $R > 0.001$, R is not greatly influenced by the choice of distribution for r and ℓ and the assumption of a normal distribution for SM is satisfactory. For lower risk than this (e.g., $R = 0.00001$), the shape of the tails of the distributions for r and ℓ becomes critical, in which case accurate assessment of the distribution of SM of direct integration procedure should be used to evaluate the risk or probability of failure.

EXAMPLE 10.7.2

Apply the safety margin approach to evaluate the reliability of the simple water distribution system described in Example 10.7.1.

SOLUTION

Calculate the mean and standard deviation of the resistance (i.e., water supply) as

$$\mu_r = \exp\left(\mu_{\ln \ell} + \frac{1}{2}\sigma_{\ln \ell}^2\right) = \exp\left[-0.803 + \frac{1}{2}(0.153)^2\right] = 0.453 \text{ cfs}$$

and

$$\sigma_r = \sqrt{\mu_r^2\left[\exp(\sigma_{\ln \ell}^2) - 1\right]} = \sqrt{0.453^2\left[\exp(0.153)^2 - 1\right]} = 0.070 \text{ cfs}$$

From the problem statement, the mean and standard deviation of the load (water demand) are $\mu_\ell = 1$ cfs and $\sigma_\ell = 0.3$ cfs, respectively. Therefore, the mean and variance of the safety margin can be calculated as

$$\mu_{SM} = \mu_r - \mu_\ell = 0.453 - 1.0 = -0.547 \text{ cfs}$$

$$\sigma_{SM}^2 = \sigma_r^2 + \sigma_\ell^2 = (0.070)^2 + (0.3)^2 = 0.095 (\text{cfs})^2$$

Now, the reliability of the system can be assessed, by the safety margin approach, as

$$R = \Phi\left[\frac{-0.453}{\sqrt{0.095}}\right] = \Phi[-1.470] = 1 - \Phi[1.470] = 1 - 0.929 = 0.071$$

The reliability computed by the safety margin method is not identical to that of direct integration although the difference is practically negligible. It should, however, be pointed out that the distribution of the safety margin in this example is not exactly normal, as assumed. Thus, the reliability obtained should be regarded as an approximation to the true reliability.

EXAMPLE 10.7.3

Determine the risk (probability) that the surface runoff (loading) exceeds the capacity of the storm sewer pipe for the problems in examples 10.6.2 and 10.6.4. Use the safety margin approach.

SOLUTION

From example 10.6.2, $\bar{Q}_\ell = 2.24$ m³/s and $\sigma_\ell = 1.138$ m³/s and $\sigma_\ell = 1.138$ m³/s. From example 10.6.4, $\bar{Q}_r = 1.93$ m³/s and $\sigma_r = 0.052$ m³/s. Compute μ_{SM} using equation (10.7.11):

$$\mu_{SM} = \mu_r - \mu_\ell = \bar{Q}_r - \bar{Q}_\ell = 1.93 - 2.24 = -0.31 \text{ m}^3/\text{s}$$

Compute σ_{SM} using equation (10.7.12):

$$\sigma_{SM} = \left[\sigma_r^2 + \sigma_\ell^2\right]^{1/2} = \left[(0.052)^2 + (1.138)^2\right]^{1/2} = 1.139 \text{ m}^3/\text{s}$$

Compute the reliability using equation (10.7.14):

$$R = \Phi\left(\frac{\mu_{SM}}{\sigma_{SM}}\right) = \Phi\left(\frac{-0.310}{1.139}\right) = \Phi(-0.272) = 0.607$$

The risk (the probability of $Q_\ell > Q_r$) is thus

$$\bar{R} = 1 - R = 1 - 0.607$$
$$= 0.393 \text{ or } 39.3\%$$

Safety Factor

The *safety factor* (SF) is given by the ratio r/ℓ and the reliability can be specified by $P(SF > 1)$. Several safety factor measures and their usefulness in hydraulic engineering are discussed by Mays and Tung (1992) and Yen (1978). By taking logarithms of both sides of this inequality, we find

$$R = P(SF > 1) = P[\ln(SF) > 0] = P[\ln(r/\ell) > 0]$$

$$= P\left(z \le \frac{\mu_{\ln SF}}{\sigma_{\ln SF}}\right) = \Phi\left(\frac{\mu_{\ln SF}}{\sigma_{\ln SF}}\right)$$

(10.7.15)

If the resistance and loading are independent and log-normally distributed, then the risk can be expressed as

$$\bar{R} = \Phi\left\{\frac{\ln\left[\frac{\mu_r}{\mu_\ell}\sqrt{\frac{1+\Omega_\ell^2}{1+\Omega_r^2}}\right]}{\ln\left[(1+\Omega_\ell^2)(1+\Omega_r^2)\right]^{1/2}}\right\} \tag{10.7.16}$$

EXAMPLE 10.7.4   Determine the risk (probability) that the surface runoff (loading) exceeds the capacity of the storm sewer pipe for the problems in examples 10.6.2 and 10.6.4. Use the safety factor approach.

SOLUTION $\bar{Q}_\ell = 2.24\ \text{m}^3/\text{s}, \Omega_\ell = 0.508, \bar{Q}_r = 1.93\ \text{m}^3/\text{s}, \Omega_r = 0.027$. Use equation (10.7.16) to compute the risk:

$$\bar{R} = \Phi\left\{\frac{\ln\left[\frac{\bar{Q}_r}{\bar{Q}_\ell}\sqrt{\frac{1+\Omega_\ell^2}{1+\Omega_r^2}}\right]}{\ln\sqrt{(1+\Omega_\ell^2)(1+\Omega_r^2)}}\right\}$$

$$= \Phi\left\{\frac{\ln\left[\frac{1.93}{2.24}\sqrt{\frac{1+0.508^2}{1+0.027^2}}\right]}{\ln\sqrt{(1+0.508^2)(1+0.027^2)}}\right\}$$

$$= \Phi(-0.300)$$

$$= 0.382 = 38.2\%$$

Note that the risk values (magnitudes) calculated for the same problem by the safety margin approach in example 10.7.3 and by the safety factor approach in this example are very close.

10.8   COMPUTER MODELS FOR FLOODFLOW FREQUENCY ANALYSIS

Table 10.8.1 describes the features of the HEC-FFA (U.S. Army Corps of Engineers, 1992) and PEAKFQ (U.S. Geological Survey) models that are used for flood flow frequency analysis, based upon fitting the log-Pearson Type III distribution to observed annual maximum flood series. The

Table 10.8.1   HEC-FFA and PEAKFQ Features

Feature	Analysis Procedure
Parameter estimation	Estimate parameters with method of moments; this assumes sample mean, standard deviation, skew coefficient = parent population mean, standard devition, and skew coefficient. To account for variability in skew computed from small samples, use weighted sum of station skew and regional skew.

(Continued)

Table 10.8.1 (*Continued*)

Feature	Analysis Procedure
Outliers	These are observations that "... depart significantly from the trend of the remaining data." Models identify high and low outliers. If information available indicates that a high outlier is maximum in the extended time period, it is treated as historical flow. Otherwise, they are treated as part of a systematic sample. Low outliers are deleted from the sample, and conditional probability adjustment is supplied.
Zero flows	If the annual maximum flow is zero (or below a specified threshold), the observations are deleted from the sample. The model parameters are estimated with the remainder of the sample. The resulting probability estimates are adjusted to account for the conditional probability of exceeding a specified discharge, given that a nonzero flow occurs.
Historical flood information	If information is available indicating that an observation represents the greatest flow in a period longer than that represented by the sample, model parameters are computed with "historically" weighted moments.
Broken record	If observations are missed due to "... conditions not related to flood magnitude," different sample segments are analyzed as a single sample with the size equal to the sum of the sample sizes.
Expected probability adjustment	This adjustment is made to the model results "... to incorporate the effects of uncertainty in application of the [frequency] curve."

Source: Ford and Hamilton (1996).

World Wide Web (www) site for the U.S. Army Corps of Engineers is www.usace.army.mil and for the U.S. Geological Survey is www.usgs.gov.

PROBLEMS

10.2.1 Solve example 10.2.1 to determine the flood magnitude having a return period of 50 years.

10.2.2 The annual maximum series of flood magnitude in a river has a log-normal distribution with a mean of 8000 m^3/s and a standard deviation of 3000 m^3/s. (a) What is the probability in each year that a flood magnitude would exceed 12,000 m^3/s? (b) Determine the flood magnitude for return periods of 25 and 100 years.

10.2.3 Solve problem 10.2.2 for a mean of 5000 m^3/s.

10.2.4 Determine the rainfall depths for the 2-, 5-, 10-, and 25-year storms for a 60-min duration storm for a mean depth of 1.5 inches and a standard deviation of 0.5 in. Use the Gumbel (extreme value type I) distribution.

10.2.5 In the Yuen-Long area of Hong Kong, there is an observatory rain gauge from which the annual maximum 3-hr rainfall depth from 1971–1975 is extracted in the table below. Conduct a frequency analysis to determine the average 3-hr rainfall intensity (mm/h) associated with an annual exceedance probability of 1% by (a) using the Gumbel (extreme value type I) distribution; and (b) using the log-normal distribution.

Year	3-hr rainfall (mm)
1971	108.2
1972	93.0
1973	104.8
1974	61.3
1975	70.3

10.2.6 The following table contains annual maximum 15-min rainfall data from a rain gauge. Assume that the data follow the Gumbel (extreme type I) distribution.

(a) Determine the magnitude of 100-year rainfall intensity (mm/hr).
(b) Determine the return period corresponding to a rainfall depth of 60 mm.

Year	15-min annual max. depth (mm)
1991	25
1992	37
1993	30
1994	33
1995	45